DFG

Occupational Toxicants

DFG Deutsche Forschungsgemeinschaft

Occupational Toxicants

Critical Data Evaluation for MAK Values and Classification of Carcinogens

Volume 18

Edited by Helmut Greim

Commission for the Investigation of Health Hazards of Chemical Compounds in the Work Area

WILEY-VCH

Prof. Dr. Helmut Greim
Senatskommission
zur Prüfung gesundheitsschädlicher Arbeitsstoffe
der Deutschen Forschungsgemeinschaft
Technische Universität München
Hohenbachernstraße 15-17
85354 Freising

Translation and layout: Dr. A. E. Wild and J. Handwerker-Sharman

Library of Congress Card No. applied for.

A catalogue record for this book is available from the British Library.

Deutsche Bibliothek Cataloguing-in-Publication Data:
Occupational toxicants : critical data evaluation for MAK
values and classification of carcinogens / DFG, Deutsche
Forschungsgemeinschaft ; Commission for the Investigation of
Health Hazards of Chemical Compounds in the Work Area.
- Weinheim ; New York ; Chichester ; Brisbane ; Singapore ; Toronto : Wiley-VCH
Einheitssacht : Gesundheitsschädliche Arbeitsstoffe <engl.>
Erscheint unregelmäßig. - Aufnahme nach Vol. 1 (1991)
Vol. 1 (1991) -

Printing: betz druck GmbH, Darmstadt
Bookbinding: J. Schäffer GmbH & Co. KG, Grünstadt
Printed in the Federal Republic of Germany.

Contents of Volume 18

Allyl chloride

MAK value	–
Peak limitation	–
Absorption through the skin (1998)	**H**
Sensitization	–
Carcinogenicity (1978)	**Category 3B**
Prenatal toxicity	–
Germ cell mutagenicity	–
BAT value	–
Synonyms	3-chloropropylene chlorallylene
Chemical name (CAS)	3-chloro-1-propene
CAS number	107-05-1
Structural formula	$H_2C{=}CH{-}CH_2Cl$
Molecular formula	C_3H_5Cl
Molecular weight	76.5
Melting point	–134.5°C
Boiling point	45°C
Density at 20°C	0.938 g/cm^3
Vapour pressure at 20°C	392 hPa
log P_{ow}*	2.1
1 ml/m^3 (ppm) ≙ 3.18 mg/m^3	**1 mg/m^3 ≙ 0.314 ml/m^3 (ppm)**

The present document is based on a review of the toxicological data for allyl chloride by the Beratergremium für umweltrelevante Altstoffe (Advisory Committee on Existing Chemicals of the German Chemical Society; BUA 1995). The end points relevant for classification of the substance are described in detail.

* *n*-octanol/water distribution coefficient

1 Toxic Effects and Mode of Action

At room temperature allyl chloride is a colourless to pale yellow liquid with a garlic-like odour. The values given for the odour threshold lie between 1.4 and 75 mg/m^3. The substance serves as intermediate in the production of epichlorohydrin and glycerin, and as starting material for the synthesis of pesticides. Exposure is therefore usually to a mixture of substances.

Allyl chloride is absorbed via inhalation, ingestion and through the skin, and is distributed in the whole organism with only a moderate tendency towards accumulation in adipose tissue. Via a saturable route, the substance can be degraded to form CO_2 and exhaled. Alternatively the substance can be conjugated with glutathione and excreted in the urine as the mercapturic acid derivative.

Allyl chloride is of only moderate acute toxicity. The 4-hour LC_{50} values for various species are in the range of 1000 to 2600 ml/m^3. Intoxication leads to narcosis, pulmonary oedema and congestion in the liver and kidneys. Oral LD_{50} values are between 300 and 1000 mg/kg body weight, dermal values about 2000 mg/kg body weight. The substance has only slight irritative effects on the skin and mucous membranes.

In man, irritative effects are described after short-term exposure to the substance; after repeated exposure to relatively high allyl chloride concentrations functional disturbances of the nervous system, liver and kidneys, and toxic effects on reproduction are seen. It is unclear, however, to what extent the substance is responsible for the effects. After repeated inhalation or oral exposure, the main effects seen in experimental animals were disturbances in kidney and liver function and in some cases also toxic effects on the peripheral nervous system. For rats exposed by inhalation the NOEC (no observed effect concentration) is in the range of 50 ml/m^3, for mice the NOEC is above 250 ml/m^3.

Allyl chloride forms DNA adducts *in vitro* as a result of direct alkylation and is weakly mutagenic in prokaryotes. It does not induce chromosomal aberrations or unscheduled DNA synthesis in eukaryotic cell lines. The genotoxicity studies carried out *in vivo* do not permit a conclusive evaluation, as the doses used were too low.

Studies of the carcinogenic potential of allyl chloride showed the substance to have weak local effects: in mice, but not in rats, long-term intragastric administration of allyl chloride led to a dose-dependent increase in the incidence of squamous cell papillomas and carcinomas in the forestomach. On mouse skin and with phorbol myristate acetate for promotion, allyl chloride was shown to have initiating effects.

In high doses allyl chloride is spermatotoxic in mice and therefore possibly impairs fertility. In rats and rabbits the substance was not found to have embryotoxic, foetotoxic or teratogenic effects. Only after maternally toxic doses is the number of resorptions increased, and subcutaneous oedema, changed body lengths and delayed bone growth are observed.

2 Mechanism of Action

The mutagenic effects of allyl chloride in prokaryotes are probably the result of the spontaneous or enzyme-catalysed alkylating effects of the substance. On the basis of its directly mutagenic effects in the *Salmonella* mutagenicity test and its electrophilia, which correlates with the mutagenicity, the alkylating potential of allyl chloride was regarded as weak compared to that of other alkylating substances. It was further reduced by the addition of S9 mix (Eder *et al.* 1982, Neudecker *et al.* 1980). Incubation of isolated salmon roe DNA with allyl chloride yielded 50 % alkylation of the DNA after 360 hours; the half-time for aqueous hydrolysis was 124 hours, which confirms that the alkylating potential of the substance is relatively weak. N^6-Allyl adenine, 3-allyl adenine, O^6-allyl guanine, N^2-allyl guanine and 7-allyl guanine were identified as products of the reaction of allyl chloride with DNA bases (Eder *et al.* 1987). The concentration-dependent formation of these adducts in the DNA of hepatocytes was also demonstrated after perfusion of isolated rat livers with allyl chloride. The formation of O^6-allyl guanine was regarded as of particular importance as the O^6-guanine adducts in general are thought to have a high cancer-inducing potential (Eder and Zugelder 1990). It has also been suggested that the hypothetical allyl chloride metabolite epichlorohydrin participates in the local carcinogenic effects (De Rooij *et al.* 1996, Van Duuren *et al.* 1979). To what extent also epigenetic mechanisms, in particular irritative and thus proliferation-enhancing effects of the substance, are involved in the local carcinogenic effects is unclear. The findings in the forestomach of mice given intragastric doses of allyl chloride (NCI 1978; see also Section 5.6) suggest that such a mechanism is possible.

The mechanism of the toxic effects on the peripheral nervous system of repeated exposure to allyl chloride is still unclear. After intraperitoneal administration of high doses to rats, a reduction in the size of microtubules and neurofilaments in nerve fibres was observed; the myelin sheaths were not affected (Sun *et al.* 1986a). At the same time a reduction in the number of presynaptic vesicles, and mitochondrial degeneration were found (Sun *et al.* 1986b), and reduced ATPase and cholinesterase activity in the neurons (Sun *et al.* 1987). These findings were attributed to the prevention of axonal transport as a result of the denaturation of neurofilaments (Sun *et al.* 1989a). In addition, a temporal relationship was observed between the exposure to allyl chloride and reversible muscle atrophy and reduced cholinesterase activity (Sun *et al.* 1989b). After incubation with neurofilament-rich protein fractions *in vitro,* not only allyl chloride but also the hypothetical metabolites acrolein and epichlorohydrin led to chemical changes in the proteins of the axonal cytoskeleton (He *et al.* 1995).

The mechanisms of neurotoxicity of allyl chloride and 2,5-hexanedione were compared by investigating groups of 24 male 7-week-old Donryu rats given subcutaneous doses of allyl chloride (2 mmol/kg body weight) or 2,5-hexanedione (2.5 mmol/kg body weight) on 5 days a week for 3 months. The body weight gains of the animals in the 2,5-hexanedione group were decreased, and in both groups clinical symptoms of neurotoxicity were observed, which was also detectable electrophysiologically. Unlike in the animals treated with 2,5-hexanedione, in those treated with allyl chloride no significant accumulation of neurofilamentary proteins in the spinal cord and no inhibition of neurofilament degradation by the Ca^{2+}-activated neutral protease were detected. A pathomechanism like that for 2,5-hexanedione is therefore unlikely (Nagano *et al.* 1993).

3 Toxicokinetics and Metabolism

3.1 Absorption and distribution

Exposure of male F344 rats in a head-only chamber to allyl chloride concentrations of 31.3, 313, 3130 or 6260 mg/m^3 (10, 100, 1000 and 2000 ml/m^3) yielded concentration-dependent rates of absorption from the air of 0.89, 9.83, 63.9 and 67.5 µg/minute. During the 6-hour exposure period the steady state in blood was not quite reached (CMA 1982c). Kinetic studies of dermal absorption are not available. Absorption can, however, be regarded as very good, at least if evaporation is prevented, as the relationship between oral and dermal LD_{50} values is known for some species and is above 0.2. The deep pain often described beneath the site of skin contact with allyl chloride has also been considered to be evidence that the substance readily penetrates the skin, as was the fact that workers exposed to allyl chloride excrete the toxicant for several days not only via the lungs, but also via the skin (BUA 1995, Häusler and Lenich 1968). Also in workers exposed to allyl chloride, evidence was found that simultaneous dermal exposure can increase the amount of the allyl chloride metabolite *S*-allyl mercapturic acid excreted with the urine (no other details) (De Rooij *et al.* 1997).

The *in vitro* distribution coefficients for allyl chloride between air and various rat tissues at equilibrium were determined as 17.3 ± 0.6 for blood/air, 101 ± 2 for adipose tissue/air, 38.9 ± 4.5 for liver tissue/air and 11.0 ± 0.2 for muscle tissue/air (Gargas *et al.* 1989). Correspondingly, even distribution of the substance in the organism and, compared to that found with other strongly lipophilic substances, an only moderate tendency to accumulation in adipose tissue are to be expected. In F344 rats given oral doses of ^{14}C-labelled allyl chloride of 1 or 100 mg/kg body weight, 1.6 % and 0.9 % of the radioactivity was still detectable in blood 48 hours after administration of the substance. In the blood-drained animal carcasses 1.6 % and 3.7 % of the radioactivity was still found (CMA 1982c).

3.2 Metabolism and excretion

To date, the only available allyl chloride metabolism studies are studies with rats in which the urinary metabolites were analysed after dosing by gavage (76 mg/kg body weight; Sanduja *et al.* 1989), subcutaneous injection (400 mg/kg body weight in arachis oil; Kaye *et al.* 1972) or intraperitoneal injection (5–45 mg/kg body weight; De Rooij *et al.* 1996). According to these studies, allyl chloride or its metabolites are mainly conjugated with GSH and the conjugates degraded to form the corresponding mercapturic acid derivatives. In F344 rats exposed to 10, 100, 1000 or 2000 ml/m^3, from concentrations of 100 ml/m^3 a marked decrease was observed in the non-protein sulfhydryl groups in the kidneys, lungs and liver, from 1000 ml/m^3 also in blood (CMA 1982c); this is possibly the result of the involvement of glutathione in the biotransformation of the substance. On the basis of the metabolites detected to date, the metabolic pathway shown in Figure 1 was deduced. In addition, further metabolic pathways such as degradation to form CO_2 (see below; CMA 1982c) must be assumed to exist but have not yet been investigated.

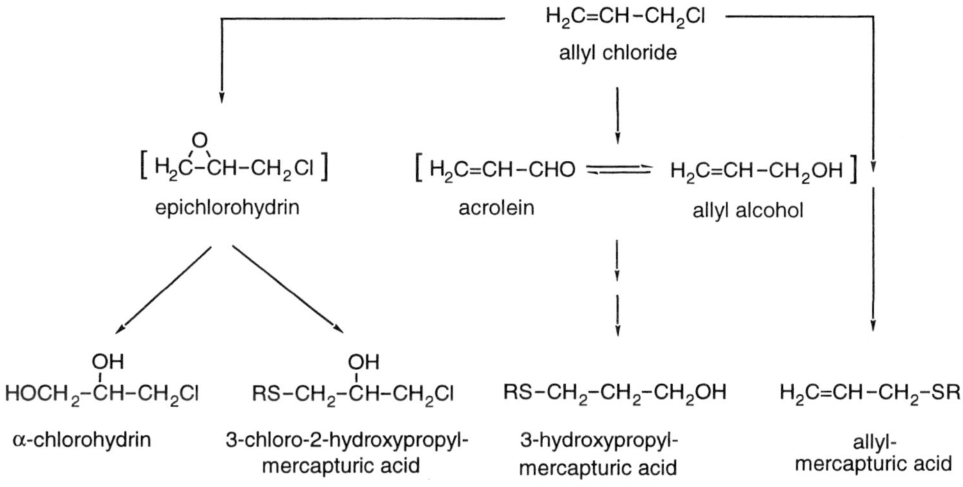

Figure 1. Metabolism of allyl chloride in the rat (from De Rooij *et al.* 1996)

Allyl alcohol and acrolein have been suggested as intermediates in the formation of 3-hydroxypropyl mercapturic acid (Neudecker and Henschler 1986, Sanduja *et al.* 1989), in particular as 3-hydroxypropyl mercapturic acid can also be formed from these allyl compounds. *S*-Allyl mercapturic acid was also detected and shown not to form 3-hydroxypropyl mercapturic acid (Kaye *et al.* 1972). The formation of *S*-allyl mercapturic acid is possibly the result of a GSH-*S*-alkene transferase-transmitted reaction with GSH (Neudecker and Henschler 1986). As a chloride ion can readily be cleaved from allyl chloride to form an allyl cation, also the non-enzymatic reaction with GSH according to an SN_1 mechanism seems possible (Eder *et al.* 1980). This also seems probable in view of the physico-chemical properties of allyl chloride (Boerth *et al.* 1991).

The data in the various studies mentioned above for the amounts of 3-hydroxypropyl mercapturic acid and *S*-allyl mercapturic acid excreted relative to the dose administered are contradictory and remain unexplained (De Rooij *et al.* 1996). In workers exposed to allyl chloride, *S*-allyl mercapturic acid was identified in urine as the main metabolite, and in smaller amounts also 3-hydroxypropyl mercapturic acid. A linear relationship between the 8-hour mean allyl chloride concentration and the concentration of *S*-allyl mercapturic acid in post-shift urine was found. Garlic consumption increased the amount of *S*-allyl mercapturic acid excreted and had therefore to be taken into consideration as a confounder (De Rooij *et al.* 1997).

The epoxidation of allyl chloride to form epichlorohydrin was not detectable *in vitro* even with the use of a metabolic activation system (Neudecker and Henschler 1986). In view of the different metabolite profiles in the urine of rats given oral doses of allyl chloride or epichlorohydrin of 100 mg/kg body weight, it was concluded that epichlorohydrin is not formed, or is formed only in small amounts as a metabolite of allyl chloride (CMA 1982c). The two early metabolism studies (Kaye *et al.* 1972, Sanduja *et al.* 1989) did not reveal metabolites which could have been formed from epichlorohydrin, but in a recently published study α-chlorohydrin (0.13 % of the administered dose) and 3-chloro-2-hydroxypropyl mercapturic acid (0.21 %) were detected (De Rooij *et al.* 1996). Thus, the epoxidation of the olefinic double bond in allyl chloride, previously regarded as improbable, has been shown to be a possible metabolic pathway, at least for the rat. In the workers, the corresponding metabolites in urine were not analysed because simultaneous exposure to allyl chloride and epichlorohydrin was assumed, and thus the source of any α-chlorohydrin and 3-chloro-2-hydroxypropyl mercapturic acid would have been unclear (De Rooij *et al.* 1997).

Allyl chloride and its metabolites are excreted mainly via the kidneys and the lungs, and only in small amounts via the intestines. After oral administration of ^{14}C-labelled allyl chloride doses of 1 and 100 mg/kg body weight to F344 rats, within 48 hours in both dose groups regardless of the sex of the animals around 36 % of the administered radioactivity was excreted with the urine. In the low dose group, 34 % was exhaled as CO_2 and 1.5 % unchanged, after doses of 100 mg/kg body weight 5.5 % as CO_2 and 18.1 % as allyl chloride. At the most 5 % of the administered radioactivity was excreted with the faeces; biliary excretion of allyl chloride and its metabolites with partial reabsorption from the intestine is therefore assumed. The elimination half-times of allyl chloride from the plasma of F344 rats after oral and intravenous administration of 100 mg/kg body weight were found to be 2.58 hours and 23.5 minutes, respectively. The elimination half-time from the blood after exposure by inhalation was concentration-dependent. After exposure of rats for 6 hours to allyl chloride concentrations of 31.3 and 313 mg/m^3, similar values of about 30 minutes were determined (CMA 1982c).

4 Effects in Man

The published values for the odour threshold of allyl chloride are in the range from 1.4 to 75 mg/m^3; the upper value is at the same time the irritation threshold. Allyl chloride vapour in concentrations of 150 to 300 mg/m^3 caused irritation of the eyes, higher concentrations caused eye pain and photophobia. In addition, after short-term exposures (level of exposure not specified) the described damage included toxic pulmonary oedema.

Depending on the duration of exposure, the damage caused by direct skin contact with liquid allyl chloride ranged from erythema to burn-like damage and was followed a few hours later by deep pains in the bones. Direct eye contact caused damage to the cornea and "deep eye pain" (no other details) (BUA 1995).

There are various studies available of the effects of repeated exposure to allyl chloride in workers from factories which produced or processed allyl chloride. After exposure to relatively high allyl chloride concentrations, functional disturbances of the nervous system, liver and kidneys, and toxic effects on reproduction (incidence of spontaneous abortion and perinatal mortality increased, birth weights decreased) were described. Generally, these studies are limited by the fact that there was no differentiation between smokers and non-smokers. Furthermore, the exposure is usually inadequately characterized, and exposure to a mixture of substances must be assumed. Therefore it cannot be decided to what extent allyl chloride is responsible for the observed effects. An extensive review and evaluation of the studies can be found in the BUA report (BUA 1995).

A cohort study of 863 employees from two epichlorohydrin plants revealed increased mortality from cardio-vascular diseases. The subcohort with the highest incidence of deaths due to cardio-vascular diseases (evaluation of death certificates) was the group exposed simultaneously to the highest concentrations of epichlorohydrin and allyl chloride for the longest period of time. For this reason also a relationship between exposure to allyl chloride and increased mortality due to cardio-vascular diseases was suspected (Enterline *et al.* 1990). In the follow-up investigation of this collective (Tsai *et al.* 1996) and in another cohort study with 1064 male employees exposed between 1957 and 1986 for at least one month to epichlorohydrin and allyl chloride (Olsen *et al.* 1994) this could not be confirmed.

5 Animal Experiments and *in vitro* Studies

5.1 Acute toxicity

To determine the acute toxicity of allyl chloride numerous studies with various species have been carried out; these studies are only summarized below, but are reviewed in detail in the BUA report (BUA 1995).

5.1.1 Inhalation

For the rat 4-hour LC_{50} values of 3506 mg/m^3 (about 1100 ml/m^3) and 8200 mg/m^3 (about 2600 ml/m^3) were reported, for the mouse 3130 mg/m^3 (about 1000 ml/m^3). The animals were narcotized and histopathological examination revealed toxic pulmonary oedema and congestion of the liver and kidneys.

In male CF1 mice an RD_{50} value of 7410 mg/m^3 (2330 ml/m^3) was determined after exposure for 10 minutes to allyl chloride.

5.1.2 Ingestion, dermal absorption and intraperitoneal administration

For the rat, mouse and rabbit, oral LD_{50} values in the range from 300 to 700 mg/kg body weight were given. The clinical symptoms observed were hypoactivity, lethargy and symptoms of peripheral neuropathy (paralysis of the hind limbs, tremor, in some animals spasms). Death was attributed to respiratory arrest, and in addition to irritation of the mucous membranes of the gastrointestinal tract, histological examination revealed damage especially in the liver and kidneys.

With occlusive epicutaneous application, dermal LD_{50} values of about 2000 mg/kg body weight were determined for rats and rabbits. The intraperitoneal LD_{50} values were 248 mg/kg body weight for the rat and 370 mg/kg body weight for the mouse.

5.2 Subacute, subchronic and chronic toxicity

5.2.1 Inhalation

Groups of 10 male and 10 female $B6C3F_1$ mice were exposed to allyl chloride concentrations of 782 mg/m^3, 6 hours/day, for 4 days and examined extensively on day 5. The body weights of the male mice and the relative liver and kidney weights of the female rats were significantly increased. Histopathological examination revealed minimal changes in the epithelial cells of the renal cortex in the rats (CMA 1982b). Exposure of 20 rats to nominal concentrations of allyl chloride of 400 mg/m^3 (125 ml/m^3), 4 hours daily, 5 days a week for 4 weeks, led to a significant increase in the amount of chloride excreted in the urine, reduced body weights and central nervous depression. More exact details of the methods were not given (Shamilov and Abasov 1973).

After exposure of 5 male and 5 female rats, 4 male guinea pigs and one female rabbit to allyl chloride concentrations of 7.8 ml/m^3, 7 hours daily, 5 days a week for 5 weeks, in the livers of all animals dilation of the sinusoids, cloudy swelling and focal necrosis were found, and in the kidneys glomerular changes, necrosis of the tubule epithelium and proliferation of the intestinal tissue (Torkelson *et al.* 1959). On the other hand, in a 90-day study with groups of 10 F344 rats and 10 $B6C3F_1$ mice of each sex exposed to allyl chloride concentrations of 1, 3, 10 or 20 ml/m^3 (exposure for 6 hours a day on 5 days a week) no substance-related toxic effects were found (CMA 1982a), and so the experiment was repeated with higher concentrations (50, 100 and 250 ml/m^3) and a larger number of animals (25 animals of each sex per group). To summarize the results, changes in kidney tissue, interpreted as physiological adaptation reactions, were found only in rats in the middle concentration group; in the high concentration group slight nephrotoxic effects were found (CMA 1982b). According to these studies, repeated exposure to allyl chloride led above all to disturbances in kidney and liver functions. The histopathological changes in the kidney and liver tissue of rats exposed to relatively low concentrations which were described in the earlier study could not be confirmed for this concentration range in more recent studies. Therefore, with medium-term exposure to allyl chloride, NOECs of 50 ml/m^3 and of more than 250 ml/m^3 were deduced for F344 rats and B6C3F$_1$ mice, respectively.

In another medium-term inhalation study with groups of 6 male rabbits and 2 female cats exposed dynamically, 6 hours daily, 6 days weekly for 3 months to air or allyl chloride concentrations of about 65 ml/m^3, evidence of neurotoxic effects was found. During the second month of exposure, denervation potentials in the electromyogram and the corresponding clinical symptoms (unsteady gait, paralysis and weight loss) were observed in rabbits. In the cats, the symptoms were less pronounced, and only one of the animals had an unsteady gait at the end of the exposure period. The liver and kidney parameters (alanine aminotransferase, level of creatinine and thiol groups in serum, urine status) were not significantly changed relative to the control values. Histopathological examination revealed in the rabbits, in addition to signs of secondary degeneration of the ischias nerve, also dilation of the sinusoids and vacuolar degeneration of the liver; hyperaemia, cloudy swelling and fatty degeneration of the epithelium of the renal tubules, and thickening of the alveolar septa were also found (Lu *et al.* 1982).

Studies of long-term exposure have been carried out only with relatively low concentrations or only to investigate selected end points, so that no conclusions can be drawn regarding a concentration–effect relationship or the target organs.

After inhalation exposure of 24 rats, 3 rabbits, 9 guinea pigs and 1 male and 1 female dog to allyl chloride concentrations of 3 ml/m^3 for 6 months, only in the female rats was slight, reversible, centrilobular liver cell degeneration observed (Torkelson *et al.* 1959). In another study, 5 male and 1 female rabbit and 10 male rats were exposed to allyl chloride concentrations of 17.5 ± 0.6 mg/m^3 (about 5.5 ml/m^3), 6 hours daily, 6 days weekly for 5 months. Apart from occasional slight changes in the electromyogram in 3 rabbits, no differences relative to the corresponding control values were found for any of the parameters investigated (behaviour, body weights, relative organ weights, necropsy and histopathology) in any of the animals during or at the end of the exposure period (Lu *et al.* 1982). Neurotoxic effects were specifically investigated in groups of 5 Donryu rats exposed to allyl chloride concentrations of 31, 157 or 313 mg/m^3 (10, 50, 100 ml/m^3) for 8 hours daily on 5 days a week for 34 weeks; other parameters were, however, not sufficiently taken into consideration. Only the animals of the high concentration group were found from week 28 to have a significantly ($p < 0.01$) reduced motor and sensory nerve conduction velocity in the proximal and distal part of the tail nerve, and reduced nerve action potentials. These changes occurred simultaneously with clinical symptoms of neuropathy, e.g. weakness of the hind limbs. Body weight development was slightly reduced in all groups after week 20 (Nagano *et al.* 1991).

An extensive review of these studies, and of other studies that cannot be evaluated because of various methodological shortcomings, can be found in the BUA report (BUA 1995).

5.2.2 Ingestion

In animals given repeated oral doses of allyl chloride, above all damage to the kidneys and liver were found, and in some cases also toxic effects on the peripheral nervous system. The studies are not suitable for determining a dose without effects. After short-term and medium-term exposure, first effects were described at the lowest doses tested, 45 and 300 mg/kg body weight and day. An extensive review can be found in the BUA report (BUA 1995).

5.3 Local effects on skin and mucous membranes

The irritative effects of allyl chloride on rabbit skin were found to be comparable to those caused by acetone, that is weak; the experiment was, however, insufficiently documented. When the tails of mice were dipped into undiluted allyl chloride for 3 to 5 hours, marked local skin damage or even necrosis resulted.

In the rabbit eye, the irritative effects of 0.5 ml undiluted allyl chloride were evaluated as weak at the 24-hour reading (BUA 1995).

There are no data available for the allergenic effects of allyl chloride.

5.4 Reproductive and developmental toxicity

5.4.1 Fertility

For allyl chloride there is evidence of spermatotoxic and gonadotoxic effects which may impair the fertility of the animals. The studies, however, either have methodological shortcomings or they revealed effects only in the range of the LD_{50}. An extensive review can be found in the BUA report (BUA 1995). Since the BUA report was written, another study has appeared in which groups of 50 male ICR mice were given single subcutaneous doses of allyl chloride of 126 mg/kg body weight or the vehicle DMSO, and the testes of groups of 10 animals were subjected to histopathological examination 4, 9, 14, 22 and 39 days after injection of the substance. In the treated animals 4, 9 and 14 days after injection of the substance, various of the cell types produced during spermatogenesis were no longer detectable and the seminiferous tubules were atrophied. The authors concluded from the time-dependent pattern of damage that allyl chloride causes specific damage to pachytene spermatocytes. Even on day 39 this damage could still be detected in some of the animals. The dose was about a fifth of the LD_{50} determined in an earlier study (Zhao *et al.* 1998).

5.4.2 Developmental toxicity

An inhalation study with Sprague-Dawley rats, mated by the breeder and delivered 3 days after mating, revealed no increase in the number of malformations. Groups of 25 animals were exposed from days 6 to 15 after mating to allyl chloride concentrations of 30 or 300 ml/m^3 for 7 hours a day; the control group consisted of 39 animals. Autopsy was carried out on day 21 of gestation. Actually pregnant were 34/39 animals in the control group, 24/25 in the low concentration group and 22/25 in the high concentration group. In the dams of the high concentration group the absolute liver and kidney weights were significantly increased, in the low concentration group only the absolute liver weights. The number of viable pups was unaffected and foetotoxicity was not observed. The significantly increased incidence of delayed bone growth (non-ossified sternebrae and delayed ossification of the body of the vertebrae) in the high concentration group was not regarded as evidence of teratogenic effects because of the slight maternal toxicity of this dose. In the same study also groups of 20 to 25 pregnant New Zealand White rabbits were exposed to allyl chloride concentrations of 30 or 300 ml/m^3 for 7 hours daily from days 6 to 18 after mating. In the high concentration group, in which slight maternal toxicity (increased liver weights, reduced body weight gains) was observed, there was also a not significant increase in the incidence of resorptions and in a few cases there were severe external malformations which, however, had been observed also in historical controls and therefore were not regarded as substance-related (CMA 1980, John *et al.* 1983).

A study with groups of 10 pregnant rats exposed during the whole gestation period (days 1–20 after mating) for 4 hours daily to allyl chloride concentrations of 0.1 and 1 ml/m^3, low compared to those used in the experiments described above, revealed at the high concentration, which was also maternally toxic, an increase in post-implantation losses, resorptions, reduced average foetal weights and umbilical hernias, organ anomalies, subcutaneous oedema, haemorrhage and in some cases unilateral hydronephrosis. The findings at the low concentration, which was not maternally toxic, were not abnormal (Alizadeh *et al.* 1982). The validity of this study cannot be evaluated as there is no information available about the analytical procedures or the purity of the allyl chloride. There is still no explanation for the maternally toxic effects at the relatively low concentration of 1 ml/m^3.

Pregnant Sprague-Dawley rats were given intraperitoneal doses of allyl chloride in corn oil of 80 mg/kg body weight on days 1 to 15 after mating; this dose had been determined to be the maximum tolerated dose (MTD) in a pilot study with females that were not pregnant. On day 21 of gestation the dams were found to have significantly increased heart, liver, spleen and kidney weights, but there were no histopathological changes. The number of resorptions in the litters of the treated animals was significantly increased. Visceral or skeletal malformations were not found. Foetuses with subcutaneous oedema and with short muzzles and lolling tongues were significantly more frequent than in the control group (Hardin *et al.* 1981).

In general, in rats and rabbits, allyl chloride doses in the maternally non-toxic range were not found to have embryotoxic, foetotoxic or teratogenic potential. Only at maternally toxic concentrations or doses did embryotoxic effects occur (BUA 1995).

5.5 Genotoxicity

5.5.1 *In vitro*

In genotoxicity tests with prokaryotes, allyl chloride yielded inconsistent results. This could be because the volatility of the substance was not always sufficiently taken into account, or because of different preincubation times in the *Salmonella* mutagenicity tests. A weak mutagenic potential was found for allyl chloride in those studies in which suitable methods were used.

In the standard plate test with the strain TA1535 and allyl chloride concentrations up to 10 µl/plate (about 9.4 mg/plate), no mutagenic effects were observed either in the presence or absence of a metabolic activation system (S9 mix) (Mc Coy *et al.* 1978). On the other hand, allyl chloride in concentrations up to 40 µl/plate (about 38.6 mg/plate) was found to be mutagenic in the strains TA1535 and TA100, and the addition of S9 mix had a markedly enhancing effect on the formation of his^+ revertants (Bignami *et al.* 1980). The results of the two experiments, however, were not sufficiently reproducible and therefore cannot be evaluated. In another standard plate test with the strains TA1535, TA1537, TA1538, TA98 and TA100, allyl chloride in concentrations up to 2 mg/plate (purity after gas chromatographic analysis 98 %) with and without the addition of S9 mix as activation system was not found to be mutagenic (Dean *et al.* 1985).

To prevent the rapid evaporation of the substance when the 45°C warm soft agar was added, 0.1, 1 or 10 µl aliquots of allyl chloride were applied to filter paper and the filters were not placed on the plates until they were cool; the plates were then incubated in plastic bags. With this technique, marginal effects were found with TA100 and TA1535. With the preincubation technique, which also reduces evaporation losses, and concentrations up to 10 µl/plate, a weak mutagenic effect was observed with strain TA1535; the experiment was not carried out with TA100 (Mc Coy *et al.* 1978). In modified preincubation tests (90 minutes at 37°C) with tightly sealed tubes, allyl chloride without S9 activation was found to produce concentration-dependent mutagenicity in TA100. Allyl chloride concentrations up to 15 µmol/plate (1.1 mg/plate) were tested. The mutagenic potency was given as 9 revertants/µmol or 0.12 revertants/µg. In experiments with activation by S9 mix, the mutagenic activity was reduced to 1 revertant/µmol (Eder *et al.* 1980, 1982, Neudecker *et al.* 1980). With an extension of the preincubation time to two and three hours, the same research group was able to detect both direct and S9-dependent mutagenic effects. It was concluded that metabolic activation is slow, but also the cell division that takes place during the long preincubation period may be at least partially responsible for this effect (Neudecker and Henschler 1985).

With *Streptomyces coelicolor*, allyl chloride induced reversion even at the lowest concentration tested of 9.4 mg/plate at which survival was 78 %. With higher concentrations the effect was more pronounced, with reduction of survival to 20 %. The experiments were carried out without metabolic activation and according to the standard plate technique, so that considerable substance losses as a result of evaporation can be assumed (Bignami *et al.* 1980). In a test for differential killing with the *Escherichia coli* strains WP_2 and WP_2uvrA, allyl chloride yielded positive results which were not concentration-dependent (Dean *et al.* 1985). With DNA-polymerase proficient and deficient *E. coli* strains, allyl chloride applied on paper discs tended to inhibit the growth of the deficient strain, which indicates a DNA-modifying effect (Mc Coy *et al.* 1978). The experiment is insufficiently documented.

In the same study, the induction of gene conversion in *Saccharomyces cerevisiae* by allyl chloride concentrations of 4.7 to 23.4 µg/ml culture solution is also reported. The tests were carried out without the addition of a metabolic activation system. No details of toxicity were given (Mc Coy *et al.* 1978). Mitotic gene conversion in *Saccharomyces cerevisiae* as a result of exposure to allyl chloride was observed in another study with and without S9 mix; the exact test concentrations, however, are not given (Dean *et al.* 1985). In haploid cells of *Aspergillus nidulans*, allyl chloride was not found to be toxic or mutagenic in the concentration range of 9.4 to 37.5 mg/ml (Bignami *et al.* 1980). When *Aspergillus nidulans* conidia on agar were exposed to allyl chloride concentrations of 29490 mg/m^3 for 24 hours, marked toxicity (survival concentration-dependent, 7 % to 42 %) and an increase in the incidence of mitotic crossing-over and non-disjunction was observed (Crebelli *et al.* 1984). The meaningfulness of this short-term test for the induction of genome mutation is questionable.

With the rat liver cell lines RL_1 and RL_4 incubation with allyl chloride concentrations of 25 µg/ml for 24 hours did not lead to chromosomal changes (Dean *et al.* 1985). A test for unscheduled DNA repair (UDS) in a human embryonal cell line (Flow 11000) with autoradiographical evaluation did not reveal increased repair activity either with or without metabolic activation and non-toxic allyl chloride concentrations (NIOSH 1981). With He-La cells, UDS was slightly increased at concentrations of 76.5 to 382.5 µg/ml. Higher concentrations were cytotoxic. Unlike in the experiment described above, the evaluation was carried out using a liquid scintillation counter (Schiffmann *et al.* 1983).

5.5.2 *In vivo*

To date there is only one publication available on the *in vivo* genotoxicity of allyl chloride. All the tests described yielded negative results. It has been suggested, however, that higher concentrations should have been used (NIOSH 1981). The tests carried out were a chromosomal aberration test with bone marrow cells from CD rats exposed once or on 5 consecutive days for 7 hours to allyl chloride concentrations of 1 or 25 ml/m^3, a dominant lethal test in which the male animals were exposed to the above concentrations for 7 hours daily on 5 days, and a sperm anomaly test with $B6C3F_1$ mice, also treated as described above. Also in the test for X-chromosomal recessive lethal mutations (SLRL test) in *Drosophila melanogaster*, no lethal mutations were found in the F_2 generation. An extensive review with details of the methods can be found in the BUA report (BUA 1995).

5.6 Carcinogenicity

Although all the available studies of the carcinogenic potential of allyl chloride have methodological shortcomings, they do indicate that the substance has local effects. A conclusive evaluation of the carcinogenic potential on the basis of these studies is, however, not possible.

Groups of 50 male and 50 female Osborne-Mendel rats and $B6C3F_1$ mice were given gavage doses of technical grade allyl chloride in corn oil on 5 days a week for 78 weeks. The initial high dose was 110 mg/kg body weight and day for rats, and for female and male mice, respectively, 300 and 400 mg/kg body weight and day. The doses were increased during the first weeks as a result of a lack of toxicity and later reduced because of high mortality. The doses averaged over time for the male rats were 57 and 77 mg/kg body weight and day, for female rats 55 and 73 mg/kg body weight and day. For male mice the doses averaged over time were 129 and 258 mg/kg body weight and day, and for the females 172 and 199 mg/kg body weight and day. One untreated group and one group given the vehicle, each with 20 animals of each sex, were used as controls. The recovery period for the rats was 30 to 33 weeks, for the mice 14 weeks, with the exception of the high dose group which was not observed after the end of exposure.

In the low dose group within 77 weeks and in the high dose group within 14 weeks 50 % of the male rats died. The incidence of mortality of the female rats reached 50 % in the low dose group after 99 weeks, in the high dose group after 38 weeks. Mortality was clearly dose-dependent, but not tumour-associated, and in some of the control groups was also very high. A reason for the increased mortality was not given; it may have been the result of the chronic pneumonia which was frequent in all groups. There were no statistically significant differences in the frequency and distribution of tumours in the various groups.

The mice were in a very poor state during the last 6 months. The mortality of the male animals of the high dose group had already reached 48 % by week 27, while 50 % of the male animals of the low dose group survived for 86 weeks. Of the female mice, 68 % of the high dose group and 80 % of the low dose group survived the last exposure. Histological examination revealed exposure-related damage only in the gastrointestinal tract (acanthosis and hyperkeratosis of the mucous membranes of the stomach). There were no statistically significant differences between the various groups in the frequency and distribution of tumours. The incidence of squamous cell carcinomas and papillomas of the forestomach was, however, increased in the treated female mice (controls, vehicle controls, low dose, high dose: 0/20, 0/19, 3/47, 3/45). The incidence of acanthosis and hyperkeratosis of the forestomach was increased in a dose-dependent manner in the male animals (0/18, 0/20, 9/46, 19/50) and female animals (0/20, 0/19, 17/48, 25/45). The above-mentioned tumours had occurred spontaneously only rarely in various studies carried out for the NCI (in 1/180 male and 1/180 female animals). The increase in forestomach tumours in this study is therefore probably not coincidental. It is not clear to what extent this is the result of the irritative effects of allyl chloride. The incidence of alveolar-bronchial adenomas and malignant lymphomas in female mice also seems to be increased, but the increase is not statistically significant. This may be due to the fact that the control groups were too small. The pulmonary adenomas and malignant lymphomas are not discussed by the authors because the increases in incidence were not significant (NCI 1978).

After application of single doses of allyl chloride of 94 mg in 0.2 ml acetone to the skin of 30 Ha:ICR-Swiss mice and subsequent promotion with 0.005 mg phorbol myristyl acetate in 0.2 ml acetone 3 times a week for about 500 days, the incidence of papillomas was significantly increased relative to that in controls given only the promoter. In all, 10 tumours were found in 7/30 treated mice, but only 7 tumours in 6/90 control animals. Repeated application of allyl chloride (31 or 94 mg in 0.2 ml acetone, 3 times weekly, for 500 days) without a promoter had no effect (Van Duuren *et al.* 1979).

Groups of 10 Strain A mice per dose and sex were given intraperitoneal doses of allyl chloride dissolved in tricaprylin (50, 122 or 245 mg/kg body weight) 3 times a week for 8 weeks. Immediately after a recovery period of 24 weeks, which all animals survived, autopsy was carried out; the incidence of pulmonary adenomas was increased relative to that in the control group, but the increase was neither significant nor dose-dependent (0.6 adenomas/mouse compared with 0.19/mouse in the vehicle control group). The authors speak of borderline findings (Theiss *et al.* 1979).

6 Manifesto (MAK value/classification)

According to the available animal studies, allyl chloride has a weak and only locally effective carcinogenic potential. The mechanism is not yet fully understood. The initiating effects on mouse skin may be the result of the mutagenic effects of the substance shown in various *in vitro* studies. The locally irritative and thus proliferation-enhancing effects probably play a decisive role in the formation of forestomach tumours in mice.

Because it is suspected of having carcinogenic effects, allyl chloride is classified in Carcinogen category 3B in Section III of the *List of MAK and BAT Values*. As the involvement of genotoxic mechanisms in the carcinogenic effects cannot be excluded, the previous MAK value for allyl chloride of 1 ml/m^3 has been withdrawn. The data available for man and animals, however, indicates that this value is sufficiently low to protect against irritative and reversible systemic-toxic effects.

There is evidence that allyl chloride penetrates the skin readily (see Section 3). The substance is therefore designated with an "H" although no confirmatory experimental data are available.

The potential of allyl chloride to cause germ cell mutations has not been sufficiently investigated and cannot, therefore, be evaluated. There is also no information available on the sensitizing effects of allyl chloride; for this reason no decision can be made on the designation of the substance with an "S".

7 References

Alizadeh SM, Guseinov FG, Denisko LI (1982) Experimental study of embryotrophic influence of chlorous allyl (Russian). *Azerb Med Zh 59*: 66–70

Bignami M, Conti G, Conti L, Crebelli R, Misuraca F, Puglia AM, Randazzo R, Sciandrello G, Carere A (1980) Mutagenicity of halogenated aliphatic hydrocarbons in *Salmonella typhimurium, Streptomyces coelicolor* and *Apergillus nidulans. Chem Biol Interact 30*: 9–23

Boerth DW, Eder E, Rasul G, Morais J (1991) Theoretical structure-activity study of mutagenic allyl chlorides. *Chem Res Toxicol 4*: 368–372

BUA (GDCh-Advisory Committee on Existing Chemicals of Environmental Relevance) (1995) Allyl chloride. *BUA Report 186*, S. Hirzel Wissenschaftliche Verlagsgesellschaft, Stuttgart

CMA (Chemical Manufacturer's Association) (1980) *The effects of inhaled allyl chloride on embryonal and fetal development in rats and rabbits.* NTIS/OTS 0539393, EPA/OTS Doc ID 88-920001939, NTIS, Springfield, VA, USA

CMA (Chemical Manufacturer's Association) (1982a) *Allyl chloride – subchronic studies. Ia. 90-day subchronic study in laboratory rodents.* NTIS/OTS 0000199-0, EPA/OTS Doc ID FYI-AX-0782-0199, NTIS, Springfield, VA, USA

CMA (Chemical Manufacturer's Association) (1982b) *Allyl chloride – subchronic studies. Ib. Results of an inhalation 4-day probe and 90-day subchronic study in laboratory rodents.* NTIS/OTS 0000199-0, EPA/OTS Doc ID FYI-AX-0782-0199, NTIS, Springfield, VA, USA

CMA (Chemical Manufacturer's Association) (1982c) *Allyl chloride: pharmacokinetics and metabolism following administration to CDF-Fischer 344 rats by three routes.* NTIS/OTS 0000199-0, EPA/OTS Doc ID FYI-AX-0782-0199, NTIS, Springfield, VA, USA

Crebelli R, Conti G, Conti L, Carere A (1984) Induction of somatic segregation by halogenated aliphatic hydrocarbons in *Aspergillus nidulans. Mutat Res 138*: 33–38

Dean BJ, Brooks TM, Hodson-Walker G, Hutson DH (1985) Genetic toxicology testing of 41 industrial chemicals. *Mutat Res 153*: 57–77

De Rooij BM, Boogard PJ, Commandeur JNM, van Sittert NJ, Vermeulen NPE (1997) Allylmercapturic acid as urinary biomarker of human exposure to allyl chloride. *Occup Environ Med 54*: 653–661

De Rooij BM, Commandeur JNM, Groot EJ, Boogard PJ, Vermeulen NPE (1996) Biotransformation of allyl chloride in the rat. Influence of inducers on the urinary metabolic profile. *Drug Metab Dispos 24*: 765–772

Eder E, Zugelder JP (1990) DNA-binding studies with allyl chloride and allyl bromide using the isolated rat liver technique. *Toxicol In Vitro 4*: 661–665

Eder E, Lutz D, Jörns M (1987) Allylic compounds bind directly to DNA: investigations of the binding mechanisms *in vitro. Chem Biol Interact 61*: 97–108

Eder E, Neudecker T, Lutz D, Henschler D (1980) Mutagenic potential of allyl and allylic compounds. *Biochem Pharmacol 29*: 993–998

Eder E, Neudecker T, Lutz D, Henschler D (1982) Correlation of alkylating and mutagenic activities of allyl and allylic compounds: standard alkylation test vs. kinetic investigations. *Chem Biol Interact 38*: 303–315

Enterline PE, Henderson V, Marsh G (1990) Mortality of workers potentially exposed to epichlorohydrin. *Br J Ind Med 47*: 269–276

Gargas ML, Burgess RJ, Voisard DE, Cason GH, Andersen ME (1989) Partition coefficients of low-molecular weight volatile chemicals in various liquids and tissues. *Toxicol Appl Pharmacol 98*: 87–99

Hardin BD, Band GP, Sikov MR, Andrew FD, Beliles RP, Niemeier RW (1981) Testing of selected workplace chemicals for teratogenic potential. *Scand J Work Environ Health 7*: 66–75

Häusler M, Lenich R (1968) Zur Wirkung von Allylchlorid bei chronischer gewerblicher Exposition (On the effects of long-term occupational exposure to allyl chloride) (German). *Arch Toxikol 23*: 209–214

He Y, Nagano M, Yamamoto H, Miyamoto E, Futatsuka M (1995) Modifications of neurofilament proteins by possible metabolites of allyl chloride *in vitro*. *Drug Chem Toxicol 18*: 315–331

John JA, Gushow TS, Ayres JA, Hanley Jr TR, Quast JF, Rao KS (1983) Teratologic evaluation of inhaled epichlorohydrin and allyl chloride in rats and rabbits. *Fundam Appl Toxicol 3*: 437–442

Kaye CM, Clapp JJ, Young L (1972) The metabolic formation of mercapturic acids from allyl halides. *Xenobiotica 2*: 129–139

Lu B, Dong S, Yu A, Xian Y, Geng T, Chui T (1982) Studies on the toxicity of allyl chloride. *Ecotoxicol Environ Saf 6*: 19–27

Mc Coy EC, Burrows L, Rosenkranz HS (1978) Genetic activity of allyl chloride. *Mutat Res 57*: 11–15

Nagano M, Hitoshi T, Futatsuka M (1991) Neurotoxicity of allyl chloride in rats. *Jpn J Ind Health 33*: 73–80

Nagano M, Yamamoto H, Harada K, Miyamoto E, Futatsuka M (1993) Comparative study of modification and degradation of neurofilament proteins in rats subchronically treated with allyl chloride, acrylamide or 2,5-hexanedione. *Environ Res 63*: 229–240

NCI (National Cancer Institute) (1978) *Bioassay of allyl chloride for possible carcinogenicity, CAS No. 107-05-1*. NCI-CG-TR-73, DHEW Publication No. (NIH) 78–1323, US Department of Health, Education, and Welfare, Public Health Service, NIH, Bethesda, MD, USA

Neudecker T, Henschler D (1985) Mutagenicity of chloro-olefines in the *Salmonella*/mammalian microsome test. *Mutat Res 157*: 145–148

Neudecker T, Henschler D (1986) Mutagenicity of chloro-olefins in the *Salmonella*/mammalian microsome test. Part III. Metabolic activation of the allylic chloropropenes allyl chloride, 1,3-dichloropropene, 2,3-dichloro-1-propene, 1,2,3-trichloropropene by S9 mix via two different metabolic pathways. *Mutat Res 170*: 1–9

Neudecker T, Lutz D, Eder E, Henschler D (1980) Structure-activity relationship in halogen and alkyl substituted allyl and allylic compounds: correlation of alkylating and mutagenic properties. *Biochem Pharmacol 29*: 2611–2617

NIOSH (National Institute for Occupational Safety and Health) (1981) *Tier II mutagenic screening of 13 NIOSH priority compounds. Individual compound report 3-chloropropene.* NTIS/OTS 86-239845, NTIS, Springfield, VA, USA

Olsen GW, Lacy SE, Chamberlin SR, Albert DL, Arceneaux TG, Bullard LF, Stafford BA, Boswell JM (1994) Retrospective cohort mortality study of workers with potential exposure to epichlorohydrin and allyl chloride. *Am J Ind Med 25*: 205–218

Sanduja R, Ansari GAS, Boor PJ (1989) 3-Hydroxypropylmercapturic acid – a biological marker of exposure to allylic and related compounds. *J Appl Toxicol 9*: 235–238

Schiffmann D, Eder E, Neudecker T, Henschler D (1983) Induction of unscheduled DNA synthesis in HeLa cells by allylic compounds. *Cancer Lett 20*: 263–269

Shamilov TA, Abasov DM (1973) Effects of allyl chloride on the organism of experimental animals (Russian). *Trudy Azerb Nauchno-Issledovat Inst Gig Truda Prof Zabol 8*: 12–15

Sun K, Yu L, Yang D, Zhang X, Gao T, Li B (1986a) Effect of allyl chloride on the ultrastructure of rat peripheral nerve fibers (Chinese). *Shandong Yike Daxue Xuebao 24*: 2–8; cited from: *Chem Abstr 105*: 37015w (1986)

Sun K, Yu L, Yang D, Zhang X, Gao T, Li B (1986b) Effect of allyl chloride on the ultrastructure of motor endplates in rat (Chinese). *Shandong Yike Daxue Xuebao 24*: 19–22; cited from: *Chem Abstr 106*: 1416j (1987)

Sun K, Yang D, Li M, Zhang X, Yu L, Zhu Z, Chen H (1987) Histochemical observation of the activity changes of ATPase and ChE in nerve fibers caused by allyl chloride (Chinese). *Shandong Yike Daxue Xuebao 25*: 42–45; cited from: *Chem Abstr 108*: 89095d (1988)

Sun K, Yang D, Li M, Zhang X, Yu L, Chen H (1989a) Cholinesterase activity of neurocyte in the spinal cord of normal rats exposed to allyl chloride, a histochemical study (Chinese). *Shandong Yike Daxue Xuebao 27*: 13–15; cited from: *Chem Abstr 110*: 226793n (1989)

Sun K, Yang D, Li M (1989b) Histochemical changes of muscle tissue and cholinesterase (ChE) activity in motor endplate caused by allyl chloride (Chinese). *Shandong Yike Daxue Xuebao 27*: 34–37; cited from: *Chem Abstr 112*: 173761h (1990)

Theiss JC, Shimkin MB, Poirier LA (1979) Induction of pulmonary adenomas in strain A mice by substituted organohalides. *Cancer Res 39*: 391–395

Torkelson TR, Wolf MA, Oyen F, Rowe VK (1959) Vapor toxicity of allyl chloride as determined on laboratory animals. *J Am Ind Hyg Assoc 20*: 217–223

Tsai SP, Gilstrap EL, Ross CE (1996) Mortality study of employees with potential exposure to epichlorohydrin: a 10 year update. *Occup Environ Med 53*: 299–304

Van Duuren BL, Goldschmidt BM, Loewengart G, Smith AC, Melchionne S, Seidman I, Roth D (1979) Carcinogenicity of halogenated olefinic and aliphatic hydrocarbons in mice. *J Nat Cancer Inst 63*: 1433–1439

Zhao M, Omura M, Tokunaga S, Romero Y, Inoue N (1998) Histopathological changes within the testis caused by allyl chloride exposure in mice. *Bull Environ Contam Toxicol 60*: 494–501

completed 10.07.1997

Amitrole

MAK value (1983)	**0.2 mg/m³ I (inhalable aerosol fraction)**
Peak limitation (1998)	**Category III**
Absorption through the skin	**–**
Sensitization	**–**
Carcinogenicity (1998)	**Category 3B**
Prenatal toxicity (1999)	**Pregnancy risk group C**
Germ cell mutagenicity	**–**
BAT value	**–**
Synonyms	aminotriazole
Chemical name (CAS)	1*H*-1,2,4-triazol-3-amine
CAS number	61-82-5

There is extensive documentation available on the toxicological properties of amitrole (IARC 1986, WHO 1994, see also the 1983 MAK documentation in Volume 4 of the present series). This supplement describes and evaluates new findings for the mechanism of action, genotoxicity, carcinogenicity, reproductive toxicity and sensitizing effects of the substance.

1 Toxic Effects and Mode of Action

Amitrole is absorbed rapidly from the gastrointestinal tract. In rats and mice it is found above all in the liver. The substance is excreted rapidly, mainly in unchanged form with the urine. Amitrole is of moderate acute toxicity and causes slight irritation of the skin and mucous membranes.

The target organ of the effects of amitrole is the thyroid gland, in particular in the rat. Amitrole inhibits thyroid peroxidase and thus reduces the formation of the thyroid hormone thyroxine (T_4). After long-term exposure of rats the incidence of thyroid tumours is increased. In mice, an increase is observed in the incidence of liver tumours.

Amitrole has been tested in numerous genotoxicity tests. The results are mainly negative, but the test systems used were in most cases probably unsuitable for detecting changes which could lead to the formation of reactive metabolites *in vivo*, for example, as a result of peroxidase-dependent activation of amitrole. Amitrole was found to be mutagenic in mammalian cells expressing peroxidase.

2 Mechanism of Action

The thyroid gland is the main target organ, particularly in rats, but also in mice and dogs. Investigations with lactoperoxidase, which is like thyroid peroxidase, and H_2O_2 indicate that reactive metabolites can be formed during the metabolism of amitrole by peroxidases (Doerge and Niemczura 1989). The nature of the reactive intermediates is unknown. Peroxidases metabolize aryl amines, however, mostly via *N*-oxidation (Eling *et al.* 1988). The reactive metabolites can bind covalently and thus cause irreversible inactivation of thyroid peroxidase (Andrae 1996, Doerge and Niemczura 1989). *In vitro* it was also demonstrated that thyroid peroxidase, lactoperoxidase and prostaglandin-*H*-synthase transform amitrole into reactive metabolites which bind to proteins and RNA (Krauss and Eling 1987). Per lactoperoxidase equivalent, 7 amitrole equivalents are bound to histidine and tyrosine residues of the enzyme (Doerge and Niemczura 1989). As a result of the inhibition of thyroid peroxidase the formation of the thyroid hormone T_4 is reduced. Because rats do not have specific T_4-binding globulins (Capen and Martin 1989, Curran and DeGroot 1991), which are responsible for the storage and transport of thyroid hormones, fluctuations in the hormone level cannot be buffered. For this reason rats react very sensitively to changes in hormone concentrations. A reduced level of T_4 leads, via a feedback mechanism, to the increased formation of thyrotropin (TSH) in the pituitary gland, and this stimulates growth of the thyroid gland. This mechanism may be responsible for the thyroid tumours observed in rats after long-term exposure to amitrole.

3 Effects in Man

There are no data available for prenatal toxicity or adverse effects on reproduction caused by amitrole in man.

3.1 Repeated exposures

In accordance with the previously described occupational-medical and epidemiological studies of the effects of amitrole on the thyroid gland in man, determination of the serum concentrations of triiodothyronine (T_3) and T_4, and the thyroid gland scintigrams of five workers employed for between 3 and 16 years in the production and filling of amitrole yielded no evidence of functional alterations in the thyroid gland (BAYER 1983).

3.2 Allergenic effects

A case was described of a weed control operator with contact dermatitis who produced reactions in patch tests with 0.1 % and 1 % amitrole in petrolatum (English *et al.* 1986). Other cases are not known to have been reported during the decades in which the substance has been in use in industry and agriculture.

4 Animal Experiments and *in vitro* Studies

4.1 Subacute, subchronic and chronic toxicity

Groups of beagle dogs (4 animals per sex and dose) were given amitrole (purity 97 %) in doses of 0, 10, 500 and 1500 mg/kg diet over a period of one year. The highest concentration without effects was found in this experiment to be 10 mg/kg diet (0.3 mg/kg body weight and day). In dogs fed concentrations of 500 and 1500 mg/kg (13 and 32 mg/kg body weight and day), slight hypertrophy of the pituitary gland, enlargement and weight increase of the thyroid gland, ectopic thyroid tissue and follicular hyperplasia were observed. The concentrations of T_4 and T_3 were reduced, the concentration of TSH was unchanged. At the highest administered concentration of 1500 mg/kg diet (32 mg/kg body weight and day), amitrole had no effects on survival or body weight changes but food consumption was reduced and lethargy was detected in both sexes. In addition, clinical neurological examination revealed a moderate decrease in muscle mass and the haematological profile demonstrated a mild, nonregenerative normochromatic anaemia (BAYER 1994).

In a study with rabbits treated dermally with amitrole doses of 0, 25 and 100 mg/kg body weight and day for 3 weeks, no increase in thyroid weights was detected. Histopathological examination did not yield evidence of changes in the thyroid gland (BAYER 1984a).

4.2 Allergenic effects

In an open epicutaneous test amitrole was not found to be sensitizing in any of 24 guinea pigs treated dermally 20 times during 4 weeks for induction (3 %, 10 % or 30 %,) and 4 times for challenge (1 %, 3 %, 10 % and 30 % in physiological saline) (BAYER 1985). In the maximization test with guinea pigs, however, sensitizing effects were observed (intradermal induction 2.5 %, dermal induction 25 %, dermal challenge 12 %, each in distilled water) (BAYER 1984b).

4.3 Reproductive and developmental toxicity

4.3.1 Fertility

In a preliminary one-generation reproduction study, groups of 10 male and 10 female Sherman rats were fed amitrole at dietary levels of 0, 500 and 1000 ppm (about 0, 36, 75 mg/kg body weight and day; values calculated from the published data). The offspring were weaned on day 21. The parent rats were killed after a total of 107 and 108 days for the males and 110 days for the females. Both amitrole concentrations were toxic for the parents and pups. In the parents and pups growth was delayed and the thyroid glands were hyperplastic, in the pups in addition the thymus and spleen were atrophic and mortality after weaning was considerably increased. No effects on fertility were observed (Gaines *et al.* 1973).

In the subsequent 2-generation reproduction study, groups of 10 male and 10 female Sherman rats were fed amitrole at dietary levels of 0, 25 and 100 ppm (about 0, 2, 7 mg/kg body weight and day; values calculated from the published data). The F_0 animals were mated after 61 and 173 days pretreatment, the F_{1b} generation (groups of 12 males and 12 females per dose) after 108 days. In the animals of the 25 ppm group, the incidence of thyroid hyperplasia in the F_0 generation females (4/10 animals) and in the F_1 generation animals of both sexes (♀ 4/10 animals; ♂ 6/10 animals) was increased (about 50 %). All animals treated with 100 ppm were found to have hyperplasia of the thyroid gland. In a few cases reduced relative kidney weights were found. No effects on fertility or the development of the offspring were observed (Gaines *et al.* 1973).

4.3.2 Developmental toxicity

Rats

Groups of 8 pregnant Sherman rats were given amitrole by stomach tube at dosage levels of 0, 20 and 100 mg/kg body weight and day from days 7 to 15 of gestation. The pups were weaned at the age of 21 days. The treatment had no effect on fertility, litter size, body weight gains of the parent animals or on body weight at birth or weaning, survival or gross abnormalities of the pups (Gaines *et al.* 1973).

Groups of 20 Long-Evans rats were given doses of amitrole of 0, 100, 300 and 1000 mg/kg body weight and day from days 6 to 15 of gestation by oral intubation. On day 20 of gestation the dams underwent caesarean section. Adverse effects on the dams with regard to appearance, behaviour, body weight development or mortality were not observed. Also the development of the foetuses was not impaired. Malformations observed in one foetus of the low and middle exposure group, respectively, and in 3 foetuses of 2 dams of the high exposure group were evaluated by the authors as not exposure-related (Bayer AG 1977b).

In an earlier study, doses of amitrole of 500 mg/kg body weight and day (one animal) and 1000 mg/kg body weight and day (two animals) were given to pregnant rats on days 8 to 13 of gestation. The body weight gains of the three dams were slightly reduced (< 10 %). In one dam (1000 mg/kg body weight and day) the number of foetuses was reduced. Malformations were not observed. After administration of gavage doses of amitrole of 100 mg/kg body weight and day to 5 female rats from 7 days before conception until day 15 of gestation, body weight gains were reduced. In 3 animals implantation losses or resorptions were observed, but no malformations. The treated pups—in particular the males—developed more slowly than the control animals. Mating experiments with the offspring did not reveal any adverse effects on fertility (Hapke 1967). Because of the low number of animals, and the fact that the findings could not be reproduced in subsequent studies, the effects described are not regarded as relevant.

Mice

Groups of NMRI mice were treated from days 6 to 18 of gestation with amitrole in concentrations of 0, 500, 1000, 2500 and 5000 ppm in the drinking water (estimated doses about 70, 140, 360 and 710 mg/kg body weight and day). On day 18 of gestation the dams underwent caesarean section. At concentrations of 500 ppm there was no evidence of damage to the dams or foetuses. From 1000 ppm the body weight gains of the dams were reduced. The weights of the foetuses were also reduced and skeletal development delayed. At 5000 ppm the incidence of resorptions was increased (Tjälve 1974).

In a Chernoff-Kavlock teratogenicity screening test carried out as part of a validation study, gavage doses of amitrole of 0 and 6500 mg/kg body weight and day were given to ICR/SIM mice from days 8 to 12 of gestation. The body weight gains of the treated dams were slightly reduced. The offspring were reared up to day 3. No effects on survival or the body weight gains of the offspring were observed (Seidenberg *et al.* 1986).

4.4 Genotoxicity

A detailed description of the studies of the genotoxicity of amitrole may be found in the WHO review (WHO 1994).

4.4.1 *In vitro*

The genotoxic potential of amitrole has been investigated in numerous test systems using various test protocols. Negative results were obtained in most of the studies (*Salmonella* mutagenicity test with various strains; diverse tests with *Escherichia coli*, *Saccharomyces cerevisiae* and *Aspergillus nidulans*; mutations at the thymidine kinase locus in L5178Y mouse lymphoma cells; chromosomal aberrations in human fibroblasts) (WHO 1994).

Some studies which yielded positive results cannot, however, be included in the evaluation as a result of methodological short-comings or inadequate documentation: rec assay with *Bacillus subtilis* (positive results only after the addition of S9 from fish or clam, but not after activation with rat S9; Kada 1981); one positive result in a *Salmonella* mutagenicity test with TA98 and TA100 (Venitt and Crofton-Sleigh 1981) compared to 8 similar tests with negative results; RK test with *E. coli* CHY832 (previously unused method; substance markedly cytotoxic; positive also with DMSO; Hayes *et al.* 1984); DNA repair and mitotic gene conversion in *S. cerevisiae* (studies not validated; positive results also with sugar and vitamin C; Sharp and Parry 1981a, 1981b); mitotic nondisjunction in *Aspergillus nidulans* (results not described; Morpurgo *et al.* 1979); UDS test with HeLa cells (no experimental data; Martin and McDermid 1981); SCE test with CHO cells (not correlated with the dose, no control data; Perry and Thomson 1981)[1]. It must be borne in mind, however, that the *in vitro* methods used were in most cases probably unsuitable for detecting changes involving the formation of reactive metabolites *in vivo*, e.g. as a result of peroxidase-dependent activation.

With embryonal cells of the Syrian hamster (SHE cells), which express prostaglandin-*H*-synthase, mutations were induced very efficiently at the HPRT and Na^+/K^+-ATPase loci (Tsutsui *et al.* 1984). In these cells, amitrole also caused morphological transformation (see Table 1).

4.4.2 *In vivo*

Host mediated assays (Swiss-Webster mice) with *S. typhimurium* TA1530, TA1535, TA1538 and *S. cerevisiae* D3 yielded positive results which, however, were not dose-related. The highest incidence of mutations was found with the lowest test concentration in *S. typhimurium* TA1530. The values obtained with the higher concentrations and the other *S. typhimurium* strains were all similar. For *S. cerevisiae* D3 only one borderline positive result is mentioned for the highest dose. There are no data given for range or for the incidence of spontaneous mutations (Simmon *et al.* 1979).

With *Drosophila melanogaster* several tests for sex-linked recessive lethal mutations and also a test for mutation in somatic cells ("wing spot test") were carried out. The latter yielded a positive result after exposure for 48 and 72 hours (Tripathy *et al.* 1990).

No evidence of genotoxic effects of amitrole was found with the following tests: two micronucleus tests with mice (two intraperitoneal doses of 500 mg/kg body weight; Tsuchimoto and Matter 1981, and one oral dose of 10000 mg/kg body weight; BAYER 1982) and two dominant lethal tests with mice (one oral dose of 1000 mg/kg body weight; BAYER 1977a, and doses of 100 mg/kg body weight and day with the diet for 49 days) (WHO 1994).

[1] DMSO: dimethyl sulfoxide; UDS: unscheduled DNA synthesis; SCE: sister chromatid exchange; CHO: Chinese hamster ovary; SHE: Syrian hamster embryo; HPRT: hypoxanthine guanine phosporibosyl transferase

4.5 Carcinogenicity

4.5.1 Short-term studies

In several studies amitrole caused cell transformation (Table 1).

Table 1. Cell transformation tests

Author:	Styles (1979)			
Substance:	amitrole (purity not specified)			
Cell type:	Syrian hamster, kidney cells (BHK cells)			
Metabolic activation:	+ S9			
Solvent:	not specified			
Concentration:	0–250 µg/ml			
Result:	positive			
Author:	Daniel and Dehnel (1981)			
Substance:	amitrole (purity not specified)			
Cell type:	Syrian hamster, kidney cells (BHK cells)			
Metabolic activation:	+/– S9			
Solvent:	DMSO			
Concentration:	0–1000 µg/ml			
Result:	negative			
Author:	Inoue *et al.* (1981)			
Substance:	amitrole (purity not specified)			
Cell type:	Syrian hamster, embryonal cells (SHE cells)			
Metabolic activation:	none			
Solvent:	DMSO			
Result:	positive			
Concentration (µg/ml)	0	10	50	100
Transformed cells/ total number of cells				
Experiment 1	0/535	1/504	1/474	0/442
Experiment 2	0/444	0/453	1/440	0/421
Experiment 3	0/553	0/531	0/532	1/535

Table 1. continued

Author:	Tsutsui *et al.* (1984)				
Substance:	amitrole (purity not specified)				
Cell type:	Syrian hamster, embryonal cells (SHE cells)				
Metabolic activation:	none				
Solvent:	culture medium				
Result:	positive				
Concentration (μg/ml)	0.3	1	3	10	
Transformed cells/ total number of cells	4/3378	4/3360	5/2980	7/3552	
Author:	Dunkel *et al.* (1981)				
Substance:	amitrole (purity not specified)				
Cell type:	Syrian hamster, embryonal cells (SHE cells)				
Metabolic activation:	none				
Solvent:	DMSO				
Result:	positive				
Concentration (μg/ml)	0	0.1	1	10	100
Transformed colonies/ surviving colonies	0/704	0/537	1/684	4/782	0/470
Author:	Dunkel *et al.* (1981)				
Substance:	amitrole (purity not specified)				
Cell type:	rat, embryonal cells infected with R-MuLV (R-MuLV-RE cells)				
Metabolic activation:	none				
Solvent:	DMSO				
Result:	positive				
Concentration (μg/ml)	0	10	80	100	1200
Transformed foci	–	–	+	+	+
Number of passages	5	5	5	5	4
Author:	Dunkel *et al.* (1988) (inter-laboratory evaluation)				
Substance:	amitrole (purity not specified)				
Cell type:	mouse, embryonal cells (C3H/10T1/2 cells)				
Metabolic activation:	none				
Solvent:	DMSO				

Table 1. continued

Result:	negative (laboratory A), positive (laboratory B)				
Laboratory A					
	Concentration (µg/ml)				
	0	2	10	50	250
Number of foci/plates with foci/total number of plates	1/1/20	1/1/21	0/0/21	0/0/21	0/0/21
Laboratory B					
	Concentration (µg/ml)				
	0	125	250	500	1000
Number of foci/plates with foci/total number of plates	2/2/10	19/9/10	8/7/10	4/2/9	4/4/9

4.5.2 Long-term studies

In addition to the studies of the carcinogenic effects of amitrole described in the MAK documentation of 1983 further long-term studies have become available (Table 2). In no study were the results analysed statistically.

Table 2. Studies of the carcinogenic effects of amitrole

Author:	WHO (1994)			
Substance:	amitrole (purity not specified)			
Species:	rat (Wistar), 35 animals per sex and dose			
Exposure:	0, 10, 50, 100 mg amitrole per kg diet (corresponding to 0, 0.5, 2.5, 5 mg/kg body weight and day)			
Duration:	2 years			
Toxicity:	10 mg/kg diet: no thyroid hyperplasia 50 mg/kg diet: histopathological changes in the thyroid gland of one animal, thyroid hyperplasia in 3 animals 100 mg/kg diet: body weights decreased during the first 13 weeks (only ♂); relative thyroid weights increased, histopathological changes in the thyroid gland			
Tumours:	Concentration in diet (mg/kg)			
	0	10	50	100
thyroid tumours	0/5	1/10	2/15	15/27

Table 2. continued

Author	WHO (1994)		
Substance:	amitrole (purity 94.6 %)		
Species:	rat (Fischer 344), 75 animals per sex and dose		
Exposure:	0 to 100 mg amitrole per kg diet		
Duration:	115 (♂) and 119 (♀) weeks		
Toxicity:	no clinical symptoms, no pathological changes, no consistent effects on levels of T_3 and T_4, thyroid weights increased (groups B, E), follicular hyperplasia in the thyroid gland (groups B, D, E, less severe also in group C)		
Tumours:	weeks 1–39	weeks 40–115 (♂) or weeks 40–119 (♀)	thyroid tumours
Group A:	–	–	–
Group B:	5 mg/kg diet	100 mg/kg diet[a]	+ (♂, ♀)
Group C:	1 mg/kg diet	20 mg/kg diet[b]	–
Group D:	3 mg/kg diet	60 mg/kg diet[b]	+ (♂)
Group E:	10 mg/kg diet	100 mg/kg diet[b]	+ (♂, ♀)

[a] continuous administration
[b] intermittent administration in 4-week periods

Author:	Hiasa *et al.* (1982)			
Species:	rat (Wistar), 12 animals (♂) per group			
Exposure:	700 mg *N*-bis(2-hydroxypropyl) nitrosamine (DHPN)/kg body weight (subcutaneous); 2000 mg amitrole per kg diet			
Exposure:	20 weeks in total			
Tumours:	amitrole promoted the development of thyroid tumours			
	DHPN	amitrole	thyroid hyperplasia	thyroid tumours
Group 1	4 × in 3 weeks	12 weeks	3/11 (27 %)	9/11 (91 %)
Group 2	4 × in 3 weeks	–	0/12 (–)	0/12 (–)
Group 3	–	12 weeks	3/12 (25 %)	0/12 (–)
Group 4	8 × in 7 weeks	12 weeks	12/12 (100 %)	12/12 (100 %)
Group 5	8 × in 7 weeks	–	12/12 (100 %)	7/12 (58 %)
Group 6	–	12 weeks	12/12 (100 %)	0/12 (–)
Group 7	–	–	0/12 (–)	0/12 (–)

Author:	Sumi *et al.* (1985)
Species:	rat (Wistar), 5–20 animals (♂, castrated) per group
Exposure:	5 mg diethylstilbestrol (DES) pellets implanted every 2 months; 250 mg *N*-nitrosobutylurea (NBU)/l drinking water (≙ 5 mg/day); 1500 mg amitrole per l drinking water
Duration:	12 months in total

Table 2. continued

Tumours:	amitrole inhibits the formation of liver tumours induced by DES, or DES and NBU				
	DES	NBU	amitrole	liver tumours	pituitary tumours
Group 1:	–	–	–	0/5 (–)	0/5 (–)
Group 2:	–	–	11 months	0/9 (–)	0/9 (–)
Group 3:	12 months	–	–	4/9 (44 %)	7/9 (78 %)
Group 4:	12 months	–	11 months	2/11 (18 %)	8/11 (73 %)
Group 5:	12 months	30 days	–	15/17 (88 %)	12/17 (71 %)
Group 6:	12 months	30 days	11 months	3/14 (21 %)	10/14 (71 %)

Author:	Wynford-Thomas *et al.* (1983)	
Substance:	amitrole (purity not specified)	
Species:	rat (Wistar), 10 animals (♂), no controls	
Exposure:	0.1 % amitrole in drinking water	
Duration:	1 year, autopsy of subgroup after 7 months	
Toxicity:	inhibition of thyroid hormone synthesis, increase in the serum TSH levels	
Tumours:	0.1 % in drinking water	
	7 months	12 months
thyroid tumours	3 tumours in 1/5 animals	16 tumours in 4/5 animals

Author:	Doniach (1974)	
Substance:	amitrole (purity not specified)	
Species:	rat (Hooded Lister), 20 animals (♂) exposed, 26 controls (♂)	
Exposure:	1000 mg amitrole per l drinking water (0.1 %)	
Duration:	18–20 months	
Toxicity:	thyroid hyperplasia	
Tumours:	Concentration in drinking water (mg/l)	
	0	1000
thyroid adenomas	0/26	20/20
thyroid carcinomas	0/26	2/20

Table 2. continued

Author:	Vesselinovitch (1983)				
Substance:	amitrole (purity not specified)				
Species:	mouse ($B6C3F_1$)				
Exposure:	controls		(90 weeks)		
	exposed animals		500 mg amitrole per kg diet		
	group 1:		prenatal exposure (treatment of the dams from day 12 of gestation up to the birth of the offspring)		
	group 2:		preweaning exposure (treatment of the dams from the birth up to the weaning of the offspring)		
	group 3:		adult exposure (treatment of the offspring for 90 weeks after weaning)		
Duration:	90 weeks (group 3)				
Tumours:		controls	500 mg/kg diet		
			group 1 (prenatal)	group 2 (preweaning)	group 3 (adult)
hepatocellular adenomas	♂	2 %	4/74 (5 %)	6/45 (13 %)	9/55 (16 %)
	♀	0 %	0/83 (–)	0/55 (–)	5/49 (10 %)
hepatocellular carcinomas	♂	0 %	2/74 (3 %)	4/45 (9 %)	11/55 (20 %)
	♀	0 %	0/83 (–)	0/55 (–)	4/49 (8 %)

Author:	Mori *et al.* (1985)					
Substance:	amitrole (purity not specified)					
Species:	mouse (various strains: NOD, DS, ICR), ♀, no controls					
Exposure:	10000 mg amitrole per l drinking water (1 %)					
Duration:	6 months					
Toxicity:	NOD: marked fibrosis after 3 months; DS: slight fibrosis after 3 months, moderate fibrosis after 6 months; ICR: no fibrosis after 3 months, slight fibrosis after 6 months					
Tumours:	NOD		DS		ICR	
	3 months	6 months	3 months	6 months	3 months	6 months
hyperplastic nodes in the liver	15/19 (79 %)	19/19 (100 %)	3/5 (60 %)	18/18 (100 %)	0/5 (–)	17/19 (89 %)
hepatocellular carcinomas	0/19	1/19	0/5	0/18	0/5	0/19

It is confirmed that in rats the incidence of thyroid tumours is increased after long-term administration of amitrole in concentrations of more than 20 mg/kg diet (corresponding to about 1 mg/kg body weight and day) and from 1000 mg/l drinking water. At the same time hyperplasia of the thyroid gland is also observed. The thyroid tumours are regarded as an indirect consequence of the antithyroid effects of amitrole.

In two studies with mice, liver tumours were observed at the high amitrole doses administered (500 mg/kg diet and 10000 mg/l drinking water) (Mori *et al.* 1985, Vesselinovitch 1983). The reason for the liver tumours is unclear; reactive intermediates are perhaps involved.

5 Manifesto (MAK value/classification)

The main target organ of the toxic and carcinogenic effects of amitrole in the rat is the thyroid gland. Toxic changes in the thyroid gland were also found in the dog. The reason for the formation of the thyroid tumours observed in the rat is thought to be the anti-thyroid effect with subsequent disturbance of the hormone feedback system between the pituitary gland and the thyroid gland. As rodents and in particular the rat react very sensitively to disturbances in the growth regulation of the thyroid gland, the formation of thyroid tumours in the rat is regarded as a high dose effect in a particularly sensitive species, an effect which is not to be expected in man. The development of genotoxic effects as a result of the activation of amitrole by specific peroxidases cannot be excluded. The reason for the liver tumours observed in the mouse and the relevance of this finding for man is unclear. The available data suggest that amitrole could have carcinogenic activity but are not sufficient to permit a final classification of the substance. Further studies are necessary. Therefore amitrole is classified in Carcinogen category 3B in Section III of the *List of MAK and BAT Values*.

The MAK value valid since 1983 is 0.2 mg/m^3. Assuming complete absorption, this corresponds to a daily intake (10 m^3 inhaled air) of 2 mg or about 0.03 mg/kg body weight (for a body weight of 70 kg). In a study carried out by Astwood (1960, see documentation from 1983), in healthy persons given a single oral dose of 100 mg amitrole, the absorption of ^{131}iodine by the thyroid gland was inhibited; 10 mg had only a very slight effect and was regarded as the threshold dose. In rats the NOEL (no observed effect level) for thyroid changes was given as 0.5 mg/kg diet (about 0.025 mg/kg body weight and day; see documentation from 1983), in dogs as 0.3 mg/kg body weight and day. As the studies with man and animals show that the rat reacted more sensitively than man, the MAK value of 0.2 mg/m^3 has been retained. The substance is classified in Peak limitation category III.

The data for dermal absorption indicate that the substance need not be designated with an "H".

In animal experiments with Freund's adjuvant (maximization test), amitrole was found to have sensitizing effects on the skin. Sensitizing properties were not detectable without the use of adjuvant and with non-occlusive application (open epicutaneous test). With the exception of one case, there are no reports of sensitizing effects in man despite the use of the substance for decades as an industrial intermediate and its widespread use as a field herbicide. Therefore, and also taking into account the experiments with animals, amitrole is regarded as having a low sensitizing potential for the skin. Data for sensitizing effects on the respiratory passages are not available. Designation with an "Sh" or "Sa" is therefore not necessary.

The toxic effects of amitrole on reproduction have been investigated with rats and mice. No toxic effects on development were observed in rats given amitrole with the diet continuously over 2 generations in concentrations up to 100 ppm (about 7 mg/kg body weight and day) or gavage doses of amitrole up to 1000 mg/kg and day on days 6 to 15 or 7 to 15 of gestation. In mice, embryotoxic effects such as reduced foetal weights and delayed skeletal development occurred only in animals given maternally toxic concentrations of 1000 ppm and more in the drinking water (about 140 mg/kg body weight and day) on days 6 to 18 of gestation. The MAK value of 0.2 mg/m^3, corresponding to about 0.03 mg/kg body weight, is lower by a factor of more than 100. Amitrole is therefore classified in Pregnancy risk group C.

There is no evidence that amitrole has potential to cause germ cell mutations.

6 References

Andrae U (1996) Reactive intermediates of xenobiotics in thyroid. *Adv Exp Med Biol 387*: 213–219

Astwood EB (1960) Cranberries, turnips, and goiter. *J Am Med Ass 172*: 1319–1320

BAYER (1977a) *Amitrole. Dominant-Letal-Test an der männlichen Maus zur Prüfung auf mutagene Wirkung* (Amitrole. Dominant lethal test for mutagenic effects in the male mouse) (German). Bayer AG, Wuppertal, No. 6679, unpublished report

Bayer AG (1977b) *Amitrol: Untersuchung auf embryotoxische und teratogene Wirkung nach oraler Verabreichung an der Ratte* (Amitrole. Embryotoxic and teratogenic effects after oral administration to the rat) (German). Bayer AG, Institut für Toxikologie, No. 6634, unpublished report

BAYER (1982) *Amitrol. Aminotriazol-Wirkstoff. Mikronukleus-Test an der Maus zur Prüfung auf mutagene Wirkung* (Amitrole. Micronucleus test for mutagenic effects in the mouse) (German). Bayer AG, Wuppertal, No. 11139, unpublished report

BAYER (1983) *Produktion von Aminotriazol: Gesundheitsüberwachung* (Aminotriazole production: health surveillance) (German). Letter of 22.04.1983 from Dr. Miksche, medical department, Bayer AG, Leverkusen, unpublished

BAYER (1984a) *Aminotriazol. Subakuter dermaler Toxizitätsversuch an Kaninchen* (Aminotriazole. Subacute dermal toxicity in the rabbit) (German). Bayer AG, Wuppertal, No. 12369, unpublished report

BAYER (1984b) *Aminotriazol. Untersuchungen auf hautsensibilisierende Wirkung bei Meerschweinchen* (Aminotriazole. Skin sensitization in the guinea pig) (German). Bayer AG, Wuppertal, No. 12607, unpublished report

BAYER (1985) *Aminotriazol (c.n.: Amitrol). Untersuchung auf hautsensibilisierende Wirkung im offenen Epicutan-Test an Meerschweinchen* (Aminotriazole (Amitrole). Skin sensitization in the open epicutaneous test in the guinea pig) (German). Bayer AG, Wuppertal, No. 13775, unpublished report

BAYER (1994) *Technical grade amitrole: a chronic toxicity feeding study in the Beagle dog.* Miles Inc. Agriculture Division Toxicology, South Metcalf, Stilwell, Kansas, No. R6321, unpublished report

Capen CC, Martin SL (1989) The effects of xenobiotics on the structure and function of thyroid follicular and C-cells. *Toxicol Pathol 17*: 266–293

Curran PG, DeGroot LJ (1991) The effect of hepatic enzyme-induced drugs on thyroid hormones and the thyroid gland. *Endocrine Rev 12*: 135–150

Daniel MR, Dehnel JM (1981) Cell transformation test with baby hamster kidney cells. in: de Serres FS, Ashby J (Eds) (1981) *Evaluation of short-term tests for carcinogens. Progress in mutation research*, Vol 1, Elsevier, North-Holland, 626–637

Doerge DR, Niemczura WP (1989) Suicide inactivation of lactoperoxidase by 3-amino-1,2,4-triazole. *Chem Res Toxicol 2*: 100–103

Doniach I (1974) Carcinogenic effect of 100, 250 and 500 rad x-rays on the rat thyroid gland. *Br J Cancer 30*: 487–495

Dunkel VC, Pienta RJ, Sivak A, Traul KA (1981) Comparative neoplastic transformation responses of Balb/3T3 cells, Syrian hamster embryo cells, and Rauscher murine leukemia virus infected Fischer 344 rat embryo cells to chemical carcinogens. *J Natl Cancer Inst 67*: 1303–1315

Dunkel VC, Schechtman LM, Tu AS, Sivak A, Lubet RA, Cameron TP (1988) Interlaboratory evaluation of the C3H/10T$^1/_2$ cell transformation assay. *Environ Mol Mutagen 12*: 21–31

Eling TE, Petry TW, Hughes MF, Krauss RS (1988) Aromatic amine metabolism catalyzed by prostaglandin H synthase. in: King CM, Romano LJ, Schuetzle D (Eds) *Carcinogenic and mutagenic responses to aromatic amines and nitroarenes*, Elsevier, 161–172

English JSC, Rycroft RJG, Calnan CD (1986) Allergic contact dermatitis from aminotriazole. *Contact Dermatitis 14*: 155–156

Gaines TB, Kimbrough RD, Linder RE (1973) The toxicity of amitrole in the rat. *Toxicol Appl Pharmacol 26*: 118–129

Hapke H-J (1967) Die Giftigkeit von Aminotriazolen für Haustiere (The toxicity of aminotriazoles for pets) (German). *Zentralbl Veterinaermed, Reihe A 14*: 469–486

Hayes S, Gordon A, Sadowski I, Hayes C (1984) RK bacterial test for independently measuring chemical toxicity and mutagenicity: short-term forward selection assay. *Mutat Res 130*: 97–106

Hiasa Y, Ohshima M, Kitahori Y, Yuasa T, Fujita T, Iwata C (1982) Promoting effects of 3-amino-1,2,4-triazole on the development of thyroid tumors in rats treated with *N*-bis(2-hydroxypropyl)nitrosamine. *Carcinogenesis 3*: 381–384

IARC (International Agency for Research on Cancer) (1986) *Some halogenated hydrocarbons and pesticide exposures. IARC monographs on the evaluation of the carcinogenic risk of chemicals to humans*, Vol 41, IARC, Lyon, 293–317

Inoue K, Katoh Y, Takayama S (1981) *In vitro* transformation of hamster embryo cells by 3-(*N*-salicyloyl)amino-1,2,4-triazole. *Toxicol Lett 7*: 211–215

Kada T (1981) The DNA-damaging activity of 42 coded compounds in the rec-assay. in: de Serres FS, Ashby J (Eds) (1981) *Evaluation of short-term tests for carcinogens. Progress in mutation research*, Vol 1, Elsevier, North-Holland, 171–182

Krauss RS, Eling TE (1987) Macromolecular binding of the thyroid carcinogen 3-amino-1,2,4-triazole (amitrole) catalyzed by prostaglandin H synthase, lactoperoxidase and thyroid peroxidase. *Carcinogenesis 8*: 659–664

Martin CN, McDermid AC (1981) Testing of 42 coded compounds for their ability to induce unscheduled DNA repair synthesis in HeLa cells. in: de Serres FS, Ashby J (Eds) (1981) *Evaluation of short-term tests for carcinogens. Progress in mutation research*, Vol 1, Elsevier, North-Holland, 533–537

Mori S, Takeuchi Y, Toyama M, Makino S, Ohhara T, Tochino Y, Hayashi Y (1985) Amitrole: strain differences in morphological response of the liver following subchronic administration to mice. *Toxicol Lett 29*: 145–152

Morpurgo G, Bellincampi D, Gualandi G, Baldinelli L, Serlupi Crescenzi O (1979) Analysis of mitotic nondisjunction with *Aspergillus nidulans. Environ Health Perspect 31*: 81–95

Perry PE, Thomson JE (1981) Evaluation of the sister chromatid exchange method in mammalian cells as a screening system for carcinogens. in: de Serres FS, Ashby J (Eds) (1981) *Evaluation of short-term tests for carcinogens. Progress in mutation research*, Vol 1, Elsevier, North-Holland, 560–569

Seidenberg JM, Anderson DG, Becker RA (1986) Validation of an *in vivo* development toxicity screen in the mouse. *Teratogen Carcinogen Mutagen 6*: 361–374

Sharp DC, Parry JM (1981a) Use of repair-deficient strains of yeast to assay the activity of 40 coded compounds. in: de Serres FS, Ashby J (Eds) (1981) *Evaluation of short-term tests for carcinogens. Progress in mutation research*, Vol 1, Elsevier, North-Holland, 502–516

Sharp DC, Parry JM (1981b) Induction of mitotic gene conversion by 41 coded compounds using the yeast culture JD1. in: de Serres FS, Ashby J (Eds) (1981) *Evaluation of short-term tests for carcinogens. Progress in mutation research*, Vol 1, Elsevier, North-Holland, 491–501

Simmon VF, Rosenkranz HS, Zeiger E, Poirier LA (1979) Mutagenic activity of chemical carcinogens and related compounds in the intraperitoneal host-mediated assay. *J Natl Cancer Inst 62*: 911–918

Styles JA (1979) Studies on the detection of carcinogens using a mammalian cell transformation assay with liver homogenate activation. in: Norpoth K, Garner RC (Eds) *Short-term mutagenicity test systems for detecting carcinogens*, Springer-Verlag, Berlin, Heidelberg, New York, 226–238

Sumi C, Yokoro K, Matsushima R (1985) Inhibition by 3-amino-1*H*-1,2,4-triazole of hepatic tumorigenesis induced by diethylstilbestrol alone or in combination with *N*-nitrosobutylurea in WF rats. *J Natl Cancer Inst 74*: 1329–1334

Tjälve H (1974) Fetal uptake and embryogenetic effects of aminotriazole in mice. *Arch Toxicol 33*: 41–48

Tripathy NK, Würgler FE, Frei H (1990) Genetic toxicity of six carcinogens and six non-carcinogens in the *Drosophila* wing spot test. *Mutat Res 242*: 169–180

Tsuchimoto T, Matter BE (1981) Activity of coded compounds in the micronucleus test. in: de Serres FS, Ashby J (Eds) (1981) *Evaluation of short-term tests for carcinogens. Progress in mutation research*, Vol 1, Elsevier, North-Holland, 705–711

Tsutsui T, Maizumi H, Barrett JC (1984) Amitrole-induced cell transformation and gene mutations in Syrian hamster embryo cells in culture. *Mutat Res 140*: 205–207

Venitt S, Crofton-Sleigh S (1981) Mutagenicity of 42 coded compounds in a bacterial assay using *Escherichia coli* and *Salmonella typhimurium*. in: de Serres FS, Ashby J (Eds) (1981) *Evaluation of short-term tests for carcinogens. Progress in mutation research*, Vol 1, Elsevier, North-Holland, 351–360

Vesselinovitch SD (1983) Perinatal hepatocarcinogenesis. *Biol Res Pregnancy Perinatol 4*: 22–25

Wynford-Thomas D, Stringer BMJ, Gomez Morales M, Williams ED (1983) Vascular changes in early TSH-induced thyroid tumours in the rat. *Br J Cancer 47*: 861–865

WHO (World Health Organization) (1994) *Amitrole. Environmental health criteria 158, International Programme on Chemical Safety – IPCS*, World Health Organisation, Genf

completed 31.03.1998, reproductive toxicity completed 15.01.1999

Carbon black in the form of inhalable dust

MAK value	–
Peak limitation	–
Absorption through the skin	–
Sensitization	–
Carcinogenicity (1999)	**Category 3B**
Prenatal toxicity	–
Germ cell mutagenicity (1998)	–
EKA	–
Synonyms	–
Chemical name (CAS)	carbon black
CAS number	1333-86-4
Density at 20°C	1.80 g/cm^3

1 Chemical and Physical Properties, Production, Uses

1.1 Introduction

Carbon black is produced from liquid hydrocarbons by controlled gas phase pyrolysis and, unlike soot from factory chimneys and diesel engines, is a substance produced under defined conditions.

1.2 Physicochemical characterization

Carbon black consists mainly (98.0–99.8 % w/w) of carbon. In addition, it contains hydrogen, oxygen, sulfur, extractable organic materials (including polycyclic aromatic hydrocarbons (PAH)) and ash. The hydrogen is present in hydrocarbons, the oxygen in surface-bound functional groups (carbonyl, carboxyl, pyrone, phenol, quinone, lactol and ether groups). By heating the carbon black (to about 950°C) in the presence of oxygen, the oxygen-containing groups and the PAH can be more or less completely volatilized. Sulfur can be detected as the bound element and in oxidized form. Analysis of carbon black also reveals traces of metals (generally < 10 ppm; iron up to 60 ppm). When nitrogen is present, it is generally incorporated into the graphite lattice. Different types of carbon black contain, depending on the production process, low levels of PAH firmly bound by surface adsorption. They may be extracted with organic solvents and are designated as extractable organic material (EOM). The level of PAH is mostly markedly less than 0.1 % (Degussa 1992). The benzo[*a*]pyrene (BaP) levels (extractable in benzene) in 9 types of carbon black which were produced by the furnace combustion process were in the range between 0.14 and 5.47 mg/kg and that in a type of carbon black produced by the thermal process was 35 mg/kg (Taylor *et al.* 1980, Zoccolillo *et al.* 1984). The BaP levels in 7 different types of carbon black which were used in Poland for tyre production were determined after extraction with toluene to be 1.44 to 3.07 mg/kg (IARC 1996). In a Swedish carbon black produced by the furnace combustion process, a BaP level of 0.9 to 28 mg/kg was determined after extraction with benzene (Agurell and Löfroth 1993). The PAH levels in one kind of carbon black (Printex 90) were determined (Degussa 1998c) with the FDA method (FDA 1994):

naphthalene	0.661 mg/kg
acenaphthylene	0.008 mg/kg
acenaphthene	0.035 mg/kg
fluorene	0.011 mg/kg
phenanthrene	0.051 mg/kg
anthracene	0.004 mg/kg
fluoranthene	0.015 mg/kg
pyrene	0.020 mg/kg
benzo[*g,h,i*]fluoranthene	0.002 mg/kg
benz[*a*]anthracene	0.005 mg/kg
cyclopenta[*c,d*]pyrene	0.001 mg/kg
chrysene [a]	0.003 mg/kg
benzo[*b/j*]fluoranthene [b]	0.002 mg/kg
benzo[*k/j*]fluoranthene [c]	0.002 mg/kg
benzo[*e*]pyrene	0.002 mg/kg
benzo[*a*]pyrene	0.001 mg/kg
perylene	0.001 mg/kg
dibenz[*a,h*]anthracene [d]	<0.001 mg/kg
benzo[*g,h,i*]perylene [e]	<0.001 mg/kg
indeno[1.2.3-*cd*]pyrene	0.001 mg/kg
anthanthrene	<0.001 mg/kg
coronene	0.009 mg/kg

[a] mixture of chrysene and triphenylene isomers
[b] mixture of benzo[*b*]fluoranthene and benzo[*j*]fluoranthene isomers
[c] mixture of benzo[*k*]fluoranthene and benzo[*j*]fluoranthene isomers
[d] mixture of dibenz[*a,c*]anthracene and dibenz[*a,h*]anthracene isomers
[e] also called 1,12-benzoperylene

The heavy metal content of this type of carbon black (Printex 90) was given as follows (Degussa 1998c):

arsenic (As)	< 2 mg/kg
cadmium (Cd)	< 0.4 mg/kg
cobalt (Co)	< 1 mg/kg
chromium (Cr)	< 1 mg/kg
copper (Cu)	< 1 mg/kg
iron (Fe)	11 mg/kg
nickel (Ni)	< 2 mg
mercury (Hg)	< 0.2 mg/kg
selenium (Se)	< 10 mg/kg
vanadium (V)	< 1 mg/kg
zinc (Zn)	< 2 mg/kg

Carbon black forms chain-shaped or clustered, branched aggregates of practically spherical particles, the so-called primary particles which are also called ultrafine particles (see "Aerosols – Ultrafine particles, their agglomerates and aggregates", in Volume 16 of the present series). The average size of the primary particles depends on the production process and is in the range between 5 and 500 nm (see Table 1). The level of linkage or branching of the aggregates of primary particles is designated as the structure. The intergrown aggregates associate to form agglomerates (Degussa 1992, Vohler *et al.* 1998).

Table 1. Data for carbon black types from different production processes (Degussa 1992)

Chemical process	Thermal-oxidative cracking		Thermal cracking	
Production process	Lampblack process	Furnace process	Thermal process	Acetylene black process
nitrogen adsorption surface (BET*) [m^2/g]	16–24	15–450	6–15	about 64
adsorption of iodine [mg/g]	23–33	15–450	6–10	about 100
primary particle diameter (arithmetic mean) [nm]	110–120	10–80	120–500	32–42
volatile components [a] [%]	1–2.5	0.5–6	0.5–1.0	0.5–2.0
pH value	6–9	7–10	7–9	5–8

[a] DIN 53 552/ISO 1126 (950°C, 7 minutes)

Degussa AG carried out studies of particle size distribution and stability of aggregates and agglomerates of carbon black particles. Two types of carbon black (Printex G: average primary particle diameter 51 nm, BET* surface area 30 m^2/g; Printex 90: average primary particle diameter 14 nm, BET surface area 300 m^2/g) were studied with the laser diffraction method (apparatus Coulter LS230) to determine the distribution function of their geometric particle diameters both after dry dispersion in a stream of air and after

* determined by the method of Brunauer, Emmett and Teller

dispersion in water by ultrasonic treatment. From the results it was concluded that even after intensive dispersion, particles of less than 1 μm diameter are not formed by these two types of carbon black. Whereas after dry dispersion and after wet dispersion without ultrasonic treatment, larger unstable agglomerates are still observed, after ultrasonic treatment the samples no longer differed significantly. They appear to contain only aggregates or stable agglomerates with particle sizes > 1 μm. Although the BET surface areas of the two types of carbon black differ by a factor of 10 and the average diameters of the primary particles by a factor of about 3 to 4, the distribution functions for their aggregate sizes (after dispersion) are identical (Degussa 1998c). In other studies, average aggregate diameters in the range from 80 to 500 nm were reported for various types of carbon black (IARC 1996).

1.3 Production

For the production of carbon black, a variety of processes and a variety of starting materials are used (see Table 2). More than 98 % of the world production of carbon black for use in rubber and pigments, about 6 million tons per year, is furnace black. The total capacity of the plants in Germany for production of carbon black for use in rubber and pigments is about 370000 tons per year and in the whole world about 8 million tons per year. For the production of carbon black in Germany, black coal tar oils (anthracene oils) or oils refined from petroleum (pyrolysis and crack oils) are used (Degussa 1992, IARC 1996, Vohler *et al.* 1998).

Table 2. Processes and raw materials for the production of various types of carbon black (Degussa 1992)

Chemical process	Production process	Main raw materials
thermal-oxidative cracking	lampblack process	aromatic oils from black coal or petroleum
	furnace process	aromatic oils from black coal or petroleum, natural gas
	Degussa gas black process	black coal tar distillates
thermal cracking	thermal process	natural gas or oils
	acetylene black process	acetylene

1.4 Uses

More than 90 % of the world production of carbon black is for the rubber industry (see Table 3). Carbon black for use in rubber is processed as pellets. About 3 % is used in printing inks, less than 1.5 % in paints and coatings. Less than 5 % is used in the plastics, battery and metallurgical industries and for isolated special purposes. Less than 0.1 % of the total world production is delivered in powdered form.

Table 3. The main uses of carbon black (Degussa 1992)

Sector	Use
rubber	in tyres and industrial rubber products to increase durability
printer's inks	pigmentation, rheology
paints	black and grey pigmentation, tinting
plastics	black and grey pigmentation, tinting, UV stabilization of polyolefins
fibres	pigmentation
paper	black and grey colouring, light-excluding papers for photography
building sector	colouring cement
electrical industry	electrodes, batteries

2 Exposure

Occupational exposure to carbon black occurs mainly during production and affects all workers who come into direct contact with the material, for example, production workers, mechanics, packers and the persons cleaning the production plants. The exposure levels cover a wide range both within one plant (workplace-specific) and between plants. In plants which process carbon black in the production of rubber and pigments, closed systems are used increasingly in storage and processing.

As recently as during the 1960s, high levels of carbon black of up to 1000 mg/m^3 were determined in the air of production plants. In a plant producing carbon black in Zagreb in the years 1964 and 1971, the geometric mean concentration of carbon black as respirable dust was 7.2 and 7.9 mg/m^3 (Valic *et al.* 1975). As the dust was sampled with a 2-step Hexleth dust collector, the actual dust concentration was probably higher. Extensive studies carried out in western European and American production plants at the end of the 1980s and beginning of the 1990s demonstrated that technological improvements had reduced the levels of inhalable carbon black to an average of less than 1 mg/m^3 (IARC 1996). From the DEGUSSA plant in Kalscheuren (Germany) concentrations of respirable carbon black dust of 0.01 to 9.14 mg/m^3 and of inhalable carbon black dust of 1.08 to 19.95 mg/m^3 have been reported (Küpper *et al.* 1996). The concentrations were determined after personal sampling with an Institute of Occupational Medicine sampler (arithmetic mean sampling period 327 min.). These data (Küpper *et al.* 1996) have also been published elsewhere (Gardiner *et al.* 1992) together with data from other plants producing carbon black. For occasional static sampling, a VC 25 sampler was used (measuring periods about 2 hours); the amount of data collected by this method (Küpper *et al.* 1996) is not comparable with that obtained in the Institute of Occupational Health study (Gardiner *et al.* 1992). In 1998, samples were collected in the Kalscheuren carbon black plant of the DEGUSSA AG with an 11-stage impactor LPI 25 (Haucke, Gmünden, Austria) set up about 2 m in front of the packing station (filling 10 kg sacks with carbon black powder). After both 180 minutes and 960 minutes sampling time, the mass median aerodynamic diameter (MMAD) was 5.6 µm. Increasing the sampling time from 180 to 960 minutes produced a marked increase in particle mass also on the impactor stage for particle sizes in the range from 0.138 to 0.272 µm. The particle concentration estimated from the impactor data was about 0.2 mg/m^3. To what extent corrections are necessary because of overloading of impactor stages and background dust levels is still to be determined in future studies (Degussa 1998c).

3 Toxic Effects and Mode of Action

In persons exposed for long periods to carbon black, impairment of lung function and the development of pneumoconiosis, inflammation and fibrosis of the lungs has been seen.

Inhaled carbon black can be deposited in the lungs of experimental animals and persons exposed at work. In experimental animals exposed to carbon black by inhalation or intratracheal instillation, when the lung load of carbon black particles reaches a certain level the alveolar clearance is impaired because of reduced macrophage motility. As with other poorly soluble dusts, the impairment of alveolar particle clearance by carbon black can result in the accumulation of particles in the lungs which can in turn lead to inflammatory, proliferative and fibrotic reactions. Long-term exposures which caused marked accumulation of carbon black in the lungs induced benign and malignant lung tumours in female rats. In one experiment with male rats, however, the incidence of lung tumours was not increased significantly. After intratracheal instillation of two different types of carbon black into female rats, a significantly increased tumour incidence was seen only in the animals treated with the carbon black type with very fine primary particles (particle diameter 14 nm, surface area 270 m^2/g). Thus the carcinogenic potency of carbon black under the conditions in this experiment was seen to be dependent on the primary particle size and the specific surface area of the particles. With another poorly soluble fine dust (diesel soot) the incidence of induced lung tumours was also higher in female rats than in the male animals; in this case, however, the incidence of tumours in the males was also increased significantly (Mauderly *et al.* 1987).

In addition to adenomas, adenocarcinomas and squamous cell carcinomas, carbon black and other poorly soluble fine dusts induce non-neoplastic keratin cysts, benign cystic keratinizing squamous cell tumours and cystic keratinizing squamous cell carcinomas (Dungworth *et al.* 1994, Boorman *et al.* 1996). The mechanism of induction of tumours by particles is thought to involve the marked persistent inflammatory reaction and the increased levels of reactive oxygen species which it generates. An increased incidence of mutations in pulmonary epithelial cells exposed to carbon black has been demonstrated *in vitro* and *in vivo*. The addition of antioxidative enzymes prevented this effect (Driscoll *et al.* 1997). The cytotoxic effects and the genotoxic effects caused by the inflammatory reaction are more marked with ultrafine carbon black than with fine carbon black (diameter of the primary particles < 14 nm and > 60 nm, respectively) (Stone *et al.* 1998).

Of great importance for the toxic effects of carbon black is the very small size of the primary particles and the large specific surface area of these particles as well as the formation of aggregates and agglomerates. The specific surface areas (BET) of different types of carbon black can differ by almost two orders of magnitude and the size of the primary particles can range from 5 to 500 nm. It is not known to what extent the agglomerates break down into aggregates in the lungs and what size and surface area these aggregates can have. Since the toxic effects of poorly soluble dusts in the lungs do not correlate as well with the particle concentration expressed in terms of weight but correlate better with the concentration expressed in terms of particle number or surface area, it is not possible to estimate the severity of toxic effects of carbon black in the respiratory tract because of lack of data for doses or concentrations expressed in terms of particle counts or surface areas. Data available for TiO_2 particles of differing sizes suggest that the toxic effects are a function of the particle diameter (Oberdörster *et al.* 1994; see also "Aerosols – Ultrafine aerosol particles, their agglomerates and aggregates", in Volume 16 of the present series).

4 Toxicokinetics

The mechanisms of deposition and clearance of ultrafine particles, agglomerates and aggregates of the kind found in carbon black are described in the chapter "Aerosols – Ultrafine aerosol particles, their agglomerates and aggregates" in Volume 16 of the present series. Studies with mice and more particularly those with rats have investigated deposition and retention of carbon black after administration by inhalation and by intratracheal instillation. The results of the more important studies of the retention kinetics of carbon black are shown in Table 4. To summarize, the studies show that exposure of rats to carbon black in doses resulting in a lung load of 0.5 to 1 mg/lung causes a significant increase in the retention times of the particles in the lungs, in the same way as does exposure to other poorly soluble dusts. In addition, the clearance function of the alveolar macrophages is impaired also for other dusts. A study which compared the retention kinetics of inhaled carbon black in the rat and mouse revealed similar retention behaviour in the two species.

In the derivation of the general threshold limit for dust, it was assumed that the aerodynamic diameters of particles deposited in the alveoli in man are in the range between 0.5 and 5 µm. The half-time for the alveolar clearance of particles in this size range in man is about 400 days (see "General Threshold Limit Value for Dust" in Volume 12 of the present series). This value cannot be used for carbon black because of its ultrafine primary particles and its tendency to form agglomerates and aggregates, and because of the marked variation in specific surface area of different types of carbon black (see Table 1). It may be assumed that carbon black, like the ultrafine particles of other poorly soluble dusts, causes a reduction in the rate of lung clearance (Oberdörster *et al.* 1994; see also "Aerosols – Ultrafine aerosol particles, their agglomerates and aggregates", in Volume 16 of the present series).

Table 4. Retention kinetics of carbon black in animal studies (from IARC 1996)

Type of carbon black	Particle diameter, Specific surface area	Species Sex	Exposure conditions Dose or concentration Duration of study	Results Comments	References
colloidal carbon	30 nm (primary), n. s.	Swiss mouse n. s.	intratracheal instillation; 4 mg, once; 6 months	mainly mucociliary clearance, practically no transepithelial passage of particles; alveolar macrophages heavily loaded with particles until the end of the study; effects not quantified; observations from histological examination	Bowden and Adamson 1984
RCF-7, furnace black	37 nm (primary), 0.22 µm (MMAD) n.s.	Fischer 344 rat n.s.	inhalation, 6.6 mg/m^3 20 h/day, 7 days/week, 1–11 weeks 1 year (recovery period)	linear increase in lung load of carbon black with exposure duration; $t_{½}$ significantly increased as a function of lung load of carbon black	Lee *et al.* 1987
Elftex 12, furnace black	0.24 µm (MMAD) n.s.	Fischer 344 rat n.s.	inhalation, 7 mg/m^3 20 h/day, 7 days/week, 1, 3, 6 weeks up to 1 year (recovery period)	lung load and retention of carbon black increased with the exposure period; doubling of the normal $t_{½}$ of about 50 days at a lung load of about 0.8 mg/g lung	Strom *et al.* 1989
Printex 90, furnace black	14 nm (primary), 0.64 µm (MMAD), n.s.	Wistar rat n.s.	inhalation, 7.4 mg/m^3 95 h/week, 4.5 months	longer $t_{½}$ at lung loads of carbon black of about 0.5 mg/g lung; inhibition of the alveolar macrophage-mediated clearance of other particles	Muhle *et al.* 1990
Printex 90, furnace black	0.64 µm (MMAD), n.s.	Wistar rat ♀	inhalation, 12 mg/m^3 19 h/day, 5 days/week, 24 months	longer $t_{½}$ of carbon black; inhibition of alveolar macrophage-mediated clearance of other particles	Creutzenberg *et al.* 1990, Muhle *et al.* 1994
Elftex 12, furnace black	2–2.4 µm (MMAD), n.s.	Fischer 344 rat ♀	inhalation, (I) 3.5 mg/m^3, 16 h/day, 7 days/week (II) 13 mg/m^3, 6 h/day, 5 days/week (III) 98 mg/m^3, 4 h/day, 1 day/week in each case for 12 weeks, 24 weeks recovery period	lung load of carbon black 3–4 mg/g lung in all 3 groups, $t_{½}$ comparable in all 3 groups	Henderson *et al.* 1992

Table 4. continued

Type of carbon black	Particle diameter, Specific surface area	Species Sex	Exposure conditions Dose or concentration Duration of study	Results Comments	References
Elftex 12, furnace black	20 µm (MMAD) 43 m^2/g	Fischer 344 rat ♂ and ♀	inhalation, 2.5 and 6.5 mg/m^3 16 h/day, 5 days/week, 23 months	retention determined with radioactively labelled carbon black, clearance obeys a 2-phase model; in comparison with controls, animals treated with carbon black reveal little or no clearance in the second phase	Mauderly *et al.* 1994
Printex 90, furnace black	0.64 µm (MMAD) 227 m^2/g	Wistar rat ♀	inhalation, 11.6 mg/m^3, 18 h/day, 5 days/week, 24 months 6 months recovery period	lung loads of carbon black similar in the two species (after exposure for 1 year)	Heinrich *et al.* 1995
		C57Bl/6N mouse, ♀	18 h/day, 5 days/week, 13.5 months 9.5 months recovery period		

MMAD mass median aerodynamic diameter, $t_{½}$ retention half-time

Table 5. Retention kinetics and effects of PAH adsorbed on carbon black (from IARC 1996)

Type of carbon black Particle diameter Specific surface area	Adsorbed substance	Species	Exposure conditions Dose or concentration Duration of study	Results	References
furnace black 26–160 nm n.s.	benzo[*a*]pyrene (radioactive)	golden Syrian hamster	intratracheal instillation dose n.s. 21 days	adsorbed benzo[*a*]pyrene retained longer than non-adsorbed; increased numbers of macrophages after instillation of carbon black with benzo[*a*]pyrene	Pylev *et al.* 1970
Elftex 12, furnace black 37 nm 43 m^2/g	benzo[*a*]pyrene (radioactive)	Fischer 344/N rat	inhalation: carbon black: 100 mg/m^3 with benzo[*a*]pyrene: 0.2 %, 2 %, 20 % intratracheal instillation: carbon black: 500 μg with benzo[*a*]pyrene: 10, 100 μg benzo[*a*]pyrene alone: 2, 20 mg/m^3 30 days	2-phase lung retention; long term retention of benzo[*a*]pyrene after adsorption to carbon black 16–60 times that without carbon black, effect more marked after intratracheal instillation than after inhalation; more covalent binding to lung macromolecules with benzo[*a*]pyrene adsorbed to carbon black than with benzo[*a*]pyrene alone	Sun *et al.* 1989
Elftex 12, furnace black 37 nm 43 m^2/g	nitropyrene (radioactive)	Fischer 344/N rat	inhalation: carbon black: 98 mg/m^3 with nitropyrene: 2 mg/m^3 nitropyrene alone 30 days	2-phase lung retention; adsorbed nitropyrene is retained longer than non-adsorbed; with nitropyrene adsorbed on carbon black covalent binding to lung macromolecules is 10 times that with nitropyrene alone	Wolff *et al.* 1989

n.s. not specified, PAH polycyclic aromatic hydrocarbons

Studies with rats and hamsters exposed by inhalation or intratracheal instillation have shown that PAH adsorbed onto carbon black remain in the lung for longer periods than do PAH which are not adsorbed onto particles (depot effect). The results of these experiments are shown in Table 5.

5 Effects in Man

There are no data available for the effects in man of single exposures to carbon black.

5.1 Repeated exposures

In lung tissue samples from persons employed in the carbon black production industry, marked deposition of carbon black was observed. Radiological examination of persons exposed to carbon black at work revealed changes of the kind seen in pneumoconiosis. Such results were first published in 1951 and were confirmed in later studies. Exposure to high levels of dust can cause unspecific inflammatory reactions in the lungs. In some studies, fibrotic tissue reactions localized specifically around the carbon black deposits were also detected in the lung parenchyma.

As well as pneumoconiosis, chronic bronchitis and slight impairment of lung function (effects on spirometric parameters) were diagnosed in workers exposed to carbon black (IARC 1996).

The results of the most important studies of effects of carbon black on the lungs are summarized in Table 6. They are either case reports or workplace studies which recorded incidences. Most of the studies are inadequately documented. Often important data are missing, for example, there are no data for exposure concentrations or smoking habits, or the selection criteria for the workers studied are not adequately defined (see Table 6). Particularly in the earlier studies, the diagnosis of pneumoconiosis was not based on the standardized classification criteria of the International Labour Organization (ILO). The more recent studies are discussed in detail below.

In a European multicentre study (Gardiner *et al.* 1993), 1742 exposed workers from a total of 15 plants producing carbon black were examined for respiratory symptoms by means of questionnaires and spirometric measurements. In 10 of the plants, a total of 1096 lung roentgenograms were also evaluated. The average exposure period was 14.2 years. In 1317 air samples from the 15 plants, the concentration of inhalable dust was 0.57 mg/m^3 (geometric mean value; geometric standard deviation 4.0) and in 1298 samples the concentration of respirable dust was found to be 0.21 mg/m^3 (geometric mean value; geometric standard deviation 2.7). Exposure-related coughing, sputum formation and symptoms of bronchitis were reported frequently. Lung function tests revealed slight exposure-related impairment of lung function (forced vital capacity (FVC) and forced expiratory volume in one second (FEV_1)) in smokers and non-smokers. Evaluation of the roentgenograms according to the ILO system for classifying pneumoconiosis findings revealed that 24.4 % had lung shadows with scores > 0/1, 9.9 % with scores > 1/0 and 4.7 % with scores > 1/1; the incidence was said to be dependent on the cumulative exposure. Lung shadows provide unequivocal evidence of the early phases of pneumoconiosis only when the ILO score is > 1/1 (see "Coal mine dust (black coal mines)", this volume).

In 913 employees from 6 plants producing carbon black, no association between lung function (FVC and FEV_1) and cumulative dust exposure was found when age and smoking habits were taken into account (IARC 1996, Robertson 1996). Three years later during repeat examinations of 697 (76 %) of these employees, a non-significant impairment of lung function (FVC and FEV_1) was found. For the subgroups of non-smokers, ex-smokers and smokers, these parameters were changed significantly; the effects were smallest in the smokers. The observed changes were put down to the variability of lung function measurements and thus considered to be questionable (Robertson 1996).

In a plant producing carbon black, 677 employees were examined for effects on lung function by means of spirometric measurements and a health questionnaire (Küpper *et al.* 1996). Workplace analyses for 99 of these workers (14.7 %) could exclude exposure to carbon black almost completely; they served as the control collective. The concentrations of inhalable carbon black dust were 1.08–19.95 mg/m^3 and of respirable carbon black dust 0.01–9.14 mg/m^3 (average 0.58 mg/m^3). In smokers a slight but not significant effect of cumulative dust exposure on lung function was observed; smoking itself changed various parameters significantly. However, the proportion of exposed smokers with obstructive respiratory disease (7.3 %) was higher than the proportion of exposed non-smokers (3.9 %) with this disorder.

In persons active in carbon black production, in addition to the above-mentioned toxic effects of carbon black on the lung, symptoms in the nasal region including hyposmia (diminished acuteness of the sense of smell; concentration of inhalable dust 4.4–13.4 mg/m^3) and dermatological changes ("carbon black tattoos") were reported (IARC 1996).

5.2 Carcinogenicity

The studies relevant for an assessment of the carcinogenic effect of carbon black in man are primarily epidemiological studies in which the specific exposure to carbon black was recorded. They include studies which investigated the workplace exposure in plants producing and processing carbon black; the exposure levels are expected to be higher in the producing plants than in those processing carbon black. Both cohort studies and case-control studies are available (review in Tables 7 and 8).

Table 6. Effects in man of repeated exposure to carbon black (from IARC 1996)

Study population Carbon black production	Exposure Exposure assessment	Results Comments	References
189 workers in a German carbon black production plant; carbon black from anthracene	n.s.	in 31 of 161 workers, pneumoconiosis-like radiological changes: 18 with slight and 13 with marked changes standard classification criteria not used	Gärtner and Brauss 1951
56 workers in two German carbon black production plants, 52 controls; carbon black produced by burning oil or acetylene	16 workers exposed for more than 10 years; no data for concentration	radiological changes in two of the exposed workers (exposed for 14 and 17 years); selection criteria not given; standard classification criteria not used	Mai 1966
12 workers in a Czech carbon black production plant; carbon black from anthracene	exposure period > 15 years; no data for concentration	no symptoms of impaired lung function; in 1 worker (exposed for 21 years) pneumoconiosis-like radiological changes, in 1 worker (exposed for 27 years) lung biopsy subjected to histological examination: extensive deposits of carbon black, reticular fibrosis, emphysema; co-exposure to quartz (lung ash 5 % $SiO2$) selection criteria unclear, standard classification criteria not used, smoking habits not taken into account	Rosmanith *et al.* 1969
143 workers; furnace process	average exposure period 19 years; no data for concentration	pneumoconiosis: in 29 (20.3 %) ILO ≥ 1/0, in 26 (18 %) ILO ≥ 1/1; often accompanied by generalized or local emphysema and lung perfusion disorders; serum IgA and IgM increased	Cocarla *et al.* 1976

Table 6. continued

Study population Carbon black production	Exposure Exposure assessment	Results Comments	References
125 Nigerian workers in carbon black processing: 20 in tyre production, 105 in battery (dry cell) production; 145 controls	tyre production: average exposure time 3.6 years battery production: exposure period 1–11 years average particle concentration[1] 9.7–31 mg/m^3	significant impairment of lung function (FVC, FEV_1), greater in exposed smokers than in exposed non-smokers, most frequent symptoms coughing and mucous, no radiological changes selection criteria unclear, exposure to mixtures of substances	Oleru *et al.* 1983
35 Croatian workers exposed to carbon black (no details), 35 controls	average exposure period 12.9 years; inhalable dust: 8.5 mg/m^3 (1964), 8.2 mg/m^3 (1971) (g-mean, g-SD = 1.6) respirable dust: 7.2 mg/m^3 (1964), 7.9 mg/m^3 (1971) (g-mean, g-SD = 1.7) observation period 1964–1971	1964: no significant changes in FVC, FEV_1 1971: significant impairment of lung function (FVC, FEV_1) for all exposed persons and for exposed smokers but not for exposed non-smokers; impairment of lung function progressive during the 6 year study period in 6 workers (17.1 %) radiological changes; mild interstitial fibrosis which became progressively worse in the course of the study; selection criteria unclear, controls matched for age, body size and smoking habits	Valić *et al.* 1975
83 active workers (aCb) and 46 ex-workers (eCb) in carbon black production in Germany, 144 controls (C) no details of production process	exposure period 2–30 years no data for concentration	in carbon black workers incidence of respiratory diseases significantly increased: chronic bronchitis (aCb 60 %, eCb 39 %, C 19 %), obstructive ventilation disorders diagnosed in spirometric examination (aCb 24 %, eCb 26 %, C 6 %), non-specific bronchial hyperreactivity (aCb 28 %, eCb 17 %, C 3 %), FVC and FEV_1 unaffected in 3 workers pneumoconiosis-like radiological changes, standard classification criteria not used selection criteria inadequately defined	Kandt 1985

[1] "total suspended particulate level"

g-mean geometric mean value
g-SD geometric standard deviation
FEV_1 forced expiratory volume in one second
FVC forced vital capacity

Table 6. continued

Study population Carbon black production	Exposure Exposure assessment	Results Comments	References
3027 exposed workers (92 % male, 8 % female) from 18 European and 1 American carbon black production plants no details of production process	average exposure period 10.9 years (1–35 years); exposure estimated from description of workplace	in carbon black workers chronic coughing and mucous formation significantly increased in an exposure-related manner, significant impairment of lung function (FVC, FEV_1) X-ray examination of 935 workers from 11 firms (396 workers exposed for > 10 years): pneumoconiosis-like changes in 6 workers (all exposed for > 10 years), 2 with ILO 0/1, 3 with ILO 1/1, 1 with ILO 1/2 26 % of the roentgenograms not suitable for the detection of slight lung changes (IARC 1996)	Crosbie 1986
1742 workers (92 % male, 8 % female) from 15 carbon black production plants (European multicentre study) no details of production process	average exposure period 14.2 years (10.3–17.7 years); respirable dust: 0.21 mg/m^3 (g-mean), g-SD 2.7 dust samples collected at the workplace	incidence of coughing and mucous formation, symptoms of chronic bronchitis, and slight impairment of lung function (FVC, FEV_1) increased in an exposure-dependent manner pneumoconiosis in 1096 workers, 24.4 % ILO ≥ 0/1, 9.9 % ILO ≥ 1/0, 4.7 % ILO ≥ 1/1; incidence dependent on cumulative exposure; preliminary analysis revealed association between radiological findings and results of lung function tests study population almost identical with that of Crosbie (1986)	Gardiner *et al.* 1993
913 workers from 6 USA carbon black producers no details of production process	inhalable dust 0–2.0 mg/m^3 (TWA) dust samples collected at the individual workplaces	no relationship between lung function (FVC, FEV_1) and cumulative dose (< 50 to ≥ 200 mg/m^3 × months); at follow-up examination of 697 (76 %) of the workers 3 years later, lung function (FVC, FEV_1) was impaired but not significantly; high drop-out rate (804 of 1717: 47 %) makes evaluation difficult	IARC 1996, Robertson 1996
578 workers and 99 controls in a German carbon black production plant; no details of production process	inhalable dust: 1.08–19.95 mg/m^3; respirable dust: 0.01–9.14 mg/m^3 dust samples collected at the individual workplaces	in smokers slight, not significant effect of cumulative dust exposure on lung function (synergistic effect of inhalative smoking and dust exposure); obstructive respiratory disorders in persons exposed to carbon black: smokers 7.3 %, non-smokers 3.9 %	Küpper *et al.* 1996

TWA time-weighted average value

Table 7. Cohort studies of the carcinogenicity of carbon black in exposed workers (from IARC 1996)

Population Country	Study period	Exposure	Cause of death Tumour location	Deaths obs/exp	SMR	95 % CI	Comments	References
1422 male employees in 5 carbon black production plants, England	1947–1980	19266 person-years	all causes of death	129/155.6	0.83	0.7–1.0[1]	age-adjusted comparison with regional population, no data for smoking habits, only 4 % of the person-years at risk from workers > 65 years	Hodgson and Jones 1985
			all tumours	42/44.1	0.95	0.7–1.3[1]		
			lungs	25/16.5	1.52	1.0–2.2[1]		
			urinary bladder	3/1.2	2.50	0.5–7.3[1]		
all employees (male, number not specified) of 4 carbon black production firms, southern USA	1935–1974	34729 person-years	all causes of death	190/227	0.84	0.72–0.95	comparison with regional population adjusted for age and ethnic group, no data for smoking habits, only 2% of the person-years at risk from workers > 65 years	Robertson and Ingalls 1980
			all tumours	29/41	0.71	0.47–1.99		
			gastrointestinal tract	6/9.6	0.62	0.12–1.23		
			respiratory tract	13/14.7	0.88	0.47–1.43		
		subcohort ≥ 20 years	all tumours	15/17.9	0.8	n.s.		
employees (male, number not specified) of 3 carbon black production firms, southern USA	1935–1994	54784 person-years	all causes of death	377/540	0.70	0.63–0.77	comparison with regional population adjusted for age and ethnic group, no data for smoking habits, 20% of the person-years at risk from workers > 65 years; cohort largely the same as that of Robertson and Ingalls (1980)	Robertson and Inman 1996
			all tumours	79/110	0.72	0.57–0.89		
			gastrointestinal tract	12/25	0.48	0.25–0.79		
			respiratory tract	34/41	0.84	0.58–1.14		
26561 men employed in formaldehyde production, USA	1966–1980	carbon black	lungs	20/15.6	1.3	0.8–2.0[1]	age-adjusted comparison with the population of the USA no data for smoking habits; some persons exposed not only to formaldehyde but also to 17 other substances including carbon black	Blair *et al.* 1990, IARC 1996
		subcohort: ≥ 20 years,	lungs	6	2.4	0.9–5.2[1]		
		latency period: ≥ 20 years	lungs	11	1.4	0.7–2.5[1]		

95 % CI: 95 % confidence interval, SMR: standardized mortality ratio
[1] 95 % confidence interval from IARC 1996

Table 8. Case-control studies of the cancer risk associated with exposure to carbon black (from IARC 1996)

Population Country	Cases	Controls	Exposed persons cases/ controls	Assessment of carbon black exposure	Tumour location	Risk	95 % CI	Comments	References
all employees (male) of 7 carbon black producers (active in 1980) USA	n = 24 employees with diagnosed skin tumours reported as suspected occupational disease	n = 48 1 matched for age and 1 matched for age and job	24/n.s.	cumulative carbon black exposure, workplace concentrations measured, retrospective grouping	skin	OR 0.9	0.3–3.2	questionable whether incidence among cases representative; unclear assignment to exposure groups	IARC 1996, Robertson and Ingalls 1989
employees (male) in rubber production (active 1964 or earlier) USA	n = 65 employees with squamous cell carcinoma of the skin diagnosed in local hospitals	n = 254 4 for each case from the same plant and matched for year of birth and year of starting job		estimated dermal exposure to carbon black	skin, squamous cell carcinoma	RR		estimation of exposure to carbon black included concentration and frequency of contact; no dose-response relationship	Bourguet *et al.* 1987
			14/47	low		0.7	n.s.		
			14/49	medium		1.2	n.s.		
			8/48	high		0.7	n.s.		
general male population of Stockholm Sweden	n = 254 cases of urothelial tumours diagnosed 1985–1987	n = 287 matched for sex and year of birth	14/9	all possible sources of carbon black exposure including printer's inks	urothelial tumours	RR 2.0	0.8–4.9	adjusted for year of birth and smoking	Steineck *et al.* 1990

Table 8. continued

Population Country	Cases	Controls	Exposed persons cases/ controls	Assessment of carbon black exposure	Tumour location	Risk	95 % CI	Comments	References
general male population of Montreal Canada	tumours diagnosed 1979–1985:	cancer cases, not matched	227/n.s.	exposure to carbon black estimated from job description and cumulative exposure index (n = 227)		OR		adjusted for age, social status, ethnic group and smoking	IARC 1996, Parent *et al.* 1996
	oesophagus n = 99	2546	11/n.s.		oesophagus	2.2	1.1–4.4		
	stomach n = 251	2397	9/n.s.		stomach	0.8	0.4–1.4		
	large intestine n = 497	2056	17/n.s.		large intestine	0.7	0.5–1.1		
	rectum n = 257	1299	10/n.s.		rectum	0.7	0.4–1.3		
	pancreas n = 116	2454	3/n.s.		pancreas	0.7	0.2–2.2		
	prostate n = 449	1550	25/n.s.		prostate	1.2	0.7–1.9		
	bladder n = 484	1879	26/n.s.		bladder	1.2	0.7–1.9		
	kidney n = 177	2481	14/n.s.		kidney	1.9	1.1–3.3		
	skin* n = 103	2525	2/n.s.		skin	0.4	0.1–1.8		
	lymphoma* n = 215	2357	9/n.s.		lymphoma	0.9	0.5–1.8		
	lung n = 857	1360 with cancer	52/58		lung (all tumours)	1.6	1.1–2.3		
		1360 with cancer	18/11	only high exposure	lung (all tumours)	2.2	1.0–4.9		
			7/n.s.		br. carcinoma	5.1	1.7–14.9		
			4/n.s.		squ. carcinoma	0.8	0.2–2.8		
			3/n.s.		adenocarcinoma	2.1	0.5–8.2		
		533 general population		only high exposure	lung (all tumours)	1.5	0.6–4.0		
					br. carcinoma	4.8	1.4–17.0		
					squ. carcinoma	0.4	0.1–1.6		
					adenocarcinoma	1.8	0.4–7.7		

n.s.: not specified, OR: odds ratio, RR: relative risk, skin*: skin melanoma, lymphoma* non-Hodgkin's lymphoma, br. carcinoma: small cell bronchial carcinoma, squ. carcinoma: squamous cell carcinoma

5.2.1 Cohort studies

In a cohort study in England, only male employees who had been exposed to carbon black for at least one year between 1947 and 1974 in one of five carbon black producing plants were included in the study population which finally amounted to 1422 persons (Hodgson and Jones 1985). The central death register listed 129 deaths (9 % of the study population) and their causes of death coded according to ICD 8. In 1976, dust concentrations (total dust) were measured at various workplaces. They revealed high levels of dust (up to 79 mg/m^3); about 50 % of the values were above the time-weighted average level of 3.5 mg/m^3. The authors remarked that these samples could be considered to be representative for the whole study period. The person-years at risk were calculated from one year after the beginning of exposure until death or emigration of the person or the end of the study period (31.12.1980). In 2 of the 5 plants, the personnel data were incomplete and the calculations could not begin until 1967 and 1968, respectively. The person-years at risk were divided into 5 year intervals for the comparison with cohorts of corresponding age from the male population of England and Wales from which the expected age-adjusted death rates were calculated. In addition, to take into account regional effects, regional expected values were calculated for 10 year periods for each plant. For workers with lung cancer, the period of employment was compared with that of matched control persons with the same year of birth and from the same plant.

The comparison with national data revealed no increases in mortality for all causes of death or for all kinds of cancer but did reveal increases for lung cancer and bladder cancer (see Table 7). The standardized mortality ratio (SMR) for stomach cancer and non-malignant lung disorders (SMR 0.36) was significantly reduced, also in the 3 plants for which the personnel data were complete. Comparison with regional data also revealed an excess of deaths from lung cancer and bladder cancer but, as in the national comparison, the differences were not significant. In the three plants for which the personnel data were complete, a "healthy worker effect" was found during the first ten years of the study period. The non-significant excess of deaths from lung tumours began after the 10th year of the study. The authors concluded that their results are difficult to interpret because of the incomplete exposure duration data. As only 9 % of the study population had died by 1980, reanalysis of the cohort at a later time would be expected to yield clearer results.

A series of epidemiological studies carried out in North American carbon black production plants and covering the years between 1950 and 1994 were described in a review (Robertson 1996). In 1950, Ingalls presented the results of a retrospective cohort study carried out in a plant producing carbon black in which deaths in the years 1940 to 1949 and illnesses in 1944 to 1949 among persons exposed to carbon black and control persons (mainly working in gas and oil production) were reported. The observed mortality (compared with the age-specific national death rates in the USA in 1940) was reduced and the observed morbidity (compared with age-specific regional cancer incidence) was slightly increased. Because of the short observation period and the small size of the study population, only very few cases of cancer were observed. Two cases of stomach cancer were seen among the workers exposed to carbon black and two lung tumours were diagnosed in the non-exposed control group. Two skin tumours were reported in the carbon black group against one skin tumour among the controls.

The same cohorts (see above) were followed up until the end of 1956 (total observation period 17 years). The observed mortality and morbidity in both the group exposed to carbon black and the control group were below the expected values (questionable "healthy worker effect"). In the carbon black group there were 6 digestive tract tumours, in the control group one tumour. The only two lung tumours observed were in the control group. There were four skin tumours in the group exposed to carbon black and three in the control group. When the observation period was extended from the year 1957 to 1974 in a subsequent study in which the control population was abandoned, the mortality in the group exposed to carbon black (50 deaths) was the same as the expected value; the SMR values for lung cancer (SMR 2.94; 95 % CI 0.93–6.08) and for leukaemia (SMR 4.00; 95 % CI 1.04–8.88) were increased (Robertson 1996).

The earlier retrospective cohort study of the employees in a single plant was extended to include the workers from three other plants producing carbon black for the years between 1935 and 1974; the same methods were used (Robertson 1996, Robertson and Ingalls 1980). This extension of the cohort yielded a total of almost 35000 observed person-years (not including retired workers or those who had left the plants before retirement). The mortality (190 deaths) was clearly lower than the expected value of 277 (SMR 0.84) as was the observed number of 29 deaths from cancer (41 expected, SMR 0.71; see Table 7). There were 89 observed deaths from cardiac disorders where 85.5 were expected. A subgroup analysis revealed no association between duration of exposure (5-year intervals) and death from cancer or cardiac disorders.

In a short communication, results obtained for a cohort of employees from three carbon black producing concerns were reported. Two of these concerns, with a total of 9 of the 20 carbon black production plants in the USA, were among those included in the earlier cohort studies of Robertson (see above). Employees of one other concern were also included. The retrospective analysis covered the years 1975 to 1994 with 54784 person-years at risk. The death certificates were available for 354 of the 377 deaths (94 %). The observed numbers of deaths from all causes, from all cancers and from tumours of the respiratory tract or digestive system were lower than the expected values (SMR 0.48–0.72; see Table 7, Robertson and Inman 1996).

From the only prospective cohort study currently in progress in the USA (6 carbon black producing concerns, begun 1978–1980), morbidity and mortality data are not yet available (Robertson 1996).

In a retrospective cohort study of 26561 workers employed by ten American formaldehyde producers, some of the workers were exposed to 17 other substances including carbon black (Blair *et al.* 1990). For carbon black, neither exposure concentrations nor the number of persons exposed is recorded; the exposure periods are given only in categories of < 1, 1–9, 10–19, > 20 years. The authors claim that there was an increase in the standardized lung cancer incidence in workers exposed to carbon black and that the risk (SMR) relative to that for the control population was > 1.5. There are no data to document this claim. To investigate the possibility of an interaction between formaldehyde and carbon black exposure, the relative lung cancer risk for persons exposed only to formaldehyde (cumulative exposure) was compared with that for persons also exposed to carbon black. Only when carbon black and formaldehyde occurred together was the SMR for lung cancer increased; for persons exposed to formaldehyde alone the SMR was 1.0.

In all, 20 cases of lung cancer were recorded in the group exposed to both substances (expected value 15.6; $p > 0.05$). The SMR did not increase with increasing exposure period. Adequate correction for smoking habits was not possible because of lack of data.

5.2.2 Case-control studies

A case-control study investigated the employees of seven American carbon black producers (Robertson 1996, Robertson and Ingalls 1989). Workers who had lodged a claim with their insurers for compensation for a malignant tumour or cardiovascular or lung disease were each compared with two control persons of the same age or the same age and same period of employment. There were 36 cases of cancer (ICD 9) including 3 lung tumours and 24 skin tumours, 66 deaths from cardiovascular disease and 38 non-malignant lung disorders. Comparison of the median cumulative dust doses for the cases ($mg/m^3 \times$ months) with those for the controls revealed no association between specific mortality risks and exposure to carbon black dust. The cumulative dust doses for workers with cardiovascular disease were significantly lower than those for the control persons.

In Canada, a case-control study of cancer in the general population was carried out (IARC 1996, Parent *et al.* 1996). In the study were included men aged 37 to 70 years who were living in Montreal and the immediate vicinity and for whom a tumour had been demonstrated histologically in a hospital in the region during a period of 6 years (1979–1985). Interviews were carried out with 3730 patients (82 % of the cancer cases) to establish their occupations and possible confounders. From the workplace descriptions, a team of chemists and occupational physicians established a cumulative exposure index based on exposure duration, frequency and concentration so that the persons could be classified into high or low exposure groups for each of 293 substances including carbon black. For 5 % of the study population (227 patients) including painters (26 %), printers (17 %), car mechanics (8 %) and factory workers producing rubber and plastic products (6 %), occupational exposure to carbon black was assumed. These persons were, however, also exposed to other gaseous substances and aerosols. For each of the tumour locations listed below two control persons were selected, a so-called cancer control from the cancer patients in the study population with tumours in other locations, and a control group consisting of 533 age-stratified persons from the general population. The published results for relative cancer risk are based mainly on the comparison with the cancer controls. For tumours of the stomach, large intestine, rectum, pancreas, prostate, bladder and skin (melanomas) and for non-Hodgkin's lymphoma, carbon black exposure was not associated with an increase in the odds ratio (OR). For tumours of the oesophagus (OR 2.2), kidneys (OR 1.9) and the lungs (OR 1.6) there was evidence of an increased relative risk (see Table 8; IARC 1996). Detailed retrospective workplace descriptions for 857 persons with histologically verified lung cancer were compared with those for 1360 cancer controls and 533 age-stratified controls from the regional population. With logistic regression analysis (with adjustment for age, ethnic group, social status, smoking habits and other occupational and non-occupational confounders such as exposure to asbestos or chromium compounds) it was investigated whether there is an association between exposure to carbon black and lung cancer. After relatively low level exposure to carbon black there was no increase in OR. Comparison with the cancer control group revealed an increased risk (OR 2.17) for workers exposed to relatively high carbon black concentrations at the workplace. This association was also revealed in the comparison with the second control group from the general population (OR 1.52). However, the confidence intervals overlap entirely (0.95–4.91 and 0.58–3.97). The association between high level carbon black exposure and lung cancer is particularly marked for small cell bronchial carcinoma ("oat-cell carcinoma") with OR values of 5.05 (cancer control) and 4.82 (general population). The authors conclude that the results of their analysis provide evidence for an association between occupational exposure to carbon black and the development of lung cancer. However, as the exposure levels were all estimates, it remains unclear to what extent the cases were really exposed to high concentrations of carbon black.

5.2.3 Summary

The cohort studies of two populations of persons in carbon black production in England and the USA and of formaldehyde production workers did not yield consistent results. In carbon black production in the USA, the relative risk of lung carcinoma was not increased (SMR below 1.0). The other two studies yielded increased relative lung cancer risks but the increases were not significant. In one study (carbon black production in England) the SMR for bladder cancer was also increased but not significantly. Case-control studies recorded data for several tumour locations (lungs, skin, urothelium). In a case-control study of Canadian cancer patients, the relationship between various tumours and carbon black exposure was analysed. Noteworthy is the significantly increased OR for lung cancer at high level carbon black exposure. However, in this last study the question of the actual level of exposure to carbon black was not answered.

In view of the methodical problems in some of the studies and the inconsistency of the results, it is not possible to derive adequate evidence for a carcinogenic effect of carbon black in man from these epidemiological studies.

6 Animal Experiments and *in vitro* Studies

There are no studies available of the acute toxicity of carbon black in animals.

6.1 Subacute, subchronic and chronic toxicity

6.1.1 Inhalation

The results of studies with carbon black which appear to be adequate in their planning, implementation and documentation are summarized in Table 9.

Male Fischer 344/N rats inhaled carbon black (furnace black, aerodynamic diameter 2.0 µm) in a concentration of 10 mg/m³, 7 hours daily, 5 days weekly for 12 weeks (Wolff *et al.* 1990). At the end of the study the lung load of carbon black was 2.2 mg. There were inflammatory reactions in the lung recognized by increases in the count of polymorphonuclear neutrophils and increased protein concentration in the broncho-alveolar lavage fluid. Histopathological examination of the lungs revealed hyperplasia of type II cells and aggregates of macrophages loaded with carbon black particles in the terminal bronchioles and the alveolar passages.

The effect of the exposure conditions on the toxicity of carbon black (furnace black, aerodynamic diameter: 2.0–2.4 µm) was investigated in a study with female Fischer 344/N rats (Henderson *et al.* 1992). Three groups of animals were exposed for a period of 12 weeks to the same weekly carbon black dose of 390 mg/m³ × h but this weekly dose was achieved with various exposure concentrations and times: 3.5 mg/m³, 16 h/day, 7 days/week; 13 mg/m³, 6 h/day, 5 days/week; 98 mg/m³, 4 h/day, 1 day/week. The recovery period was 24 weeks. At the end of exposure the lung load of carbon black was 3 to 4 mg/lung. In all groups there were increases in lung weights and inflammatory reactions were detected (increases in the polymorphonuclear neutrophil count and in biochemical parameters such as protein, lactate dehydrogenase (LDH) and β-glucuronidase in the bronchoalveolar lavage fluid were detectable after as little as 6 weeks and were more pronounced after 12 weeks exposure). The described morphological changes were a slight thickening of the alveolar septa with hypertrophy of the epithelial cells, and inflammatory cells around deposits of particles. Under the conditions of this study, the different exposure patterns had no effect on particle accumulation in the lungs or on the inflammatory reactions which developed.

Male and female Fischer 344/N rats were exposed to carbon black (furnace black, aerodynamic diameter 2.0 µm) in concentrations of 2.5 and 6.5 mg/m³ for 24 months (16 h/day, 5 days/week) and then observed for the rest of their lives (Mauderly *et al.* 1994, Nikula *et al.* 1995). Survival after 23 months was 13.8 % for the males and 35.6 % for the females in the control group, 4.3 % and 40.4 % in the 2.5 mg/m³ group and 5.8 % and 25.9 % in the 6.5 mg/m³ group. After about 22 months, the body weights of the exposed animals were significantly lower than the control values. Progressive loading of the lungs and the adjacent lymph nodes with carbon black was observed in the course of the exposure. After 23 months, the lung loads in the 2.5 mg/m³ group were 24.7 mg/lung in the males and 17.3 mg/lung in the females, in the 6.5 mg/m³ group 40.1 mg/lung and 36.9 mg/lung, respectively. The lung weights were significantly higher than the control values in the male animals of both exposed groups after 12 months and in the female animals of the 2.5 mg/m³ group after 6 months and the 6.5 mg/m³ group after 3 months. After 3 and 18 months the rate of alveolar clearance was determined with radioactively labelled carbon black. It was reduced in the 6.5 mg/m³ group after as little as 3 months and to the same extent in the 2.5 mg/m³ group after 18 months. In the bronchoalveolar lavage fluid, signs of an inflammatory reaction were detected in the form of a significant dose-dependent increase in the polymorphonuclear neutrophil count and in the biochemical parameters, protein, LDH, β-glucuronidase. Most effects were more pronounced in the female rats than in the males.

Female rats inhaled carbon black (furnace black, aerodynamic diameter 0.64 µm) in a concentration of 6 mg/m^3, 17 h/day, 5 days/week for 10 months and were then observed for another 20 months or were exposed for 20 months followed by a recovery period of 10 months (Dungworth *et al.* 1994). Moderate to severe inflammation was observed in the lungs together with bronchoalveolar hyperplasia, alveolar histiocytosis, lipoproteinosis and squamous cell metaplasia.

Male Fischer 344 rats were exposed to carbon black (furnace black, aerodynamic diameter 0.88 µm, primary particle diameter 16 nm) in concentrations of 1.1, 7.1 and 52.8 mg/m^3, 6 h/day, 5 days/week for 13 weeks (Driscoll *et al.* 1996). At the end of the exposure period, the lung loads of carbon black were 0.35, 1.8 and 7.9 mg/lung. Whereas 3 and 8 months after the end of exposure, the lung load of carbon black had decreased by 33 % and 64 % in the lowest dose group, in both higher dose groups after 3 months no particle clearance had taken place and after 8 months the particle load had decreased by only 35 % and 25 %, respectively. In the bronchoalveolar lavage fluid there were changes in the total number of cells, the differential cell count, the level of LDH, β-glucuronidase and total protein in the medium and high dose groups. Likewise in the histopathological examination, inflammatory, proliferative and fibrotic changes were detected only in the two higher dose groups. An increased incidence of mutation in type II alveolar cells was detected in the HPRT (hypoxanthine guanine phosphoribosyl transferase) test after exposure to carbon black concentrations of 7.1 and 52.8 mg/m^3. After a recovery period of 8 months, the described effects were still significant only in the highest dose group.

The toxic effects of carbon black (aerodynamic diameter 0.64 µm) were studied with female Wistar rats and female NMRI mice (Heinrich *et al.* 1995). The mice inhaled carbon black in a concentration of 11.6 mg/m^3, 18 h/day, 5 days/week for 13.5 months, and were observed for another 9.5 months. After 12 months the lung load of carbon black was 7.4 mg/lung. Reductions in body weight and increases in lung weights were reported. The rats inhaled carbon black in a concentration of 11.6 mg/m^3, 18 h/day, 5 days/week for 24 months, and were observed for another 6 months. After 24 months the lung load of carbon black was 44 mg/lung. Carbon black also accumulated in the lymph nodes closest to the lung (6.7 mg/animal). The exposure to carbon black resulted in reduction of body weights and increase of lung weights. In the bronchoalveolar lavage fluid, signs of an inflammatory reaction were detected in the form of an increase in the polymorphonuclear neutrophil count and in biochemical parameters such as protein, LDH and β-glucuronidase. The histological examination revealed moderate to severe bronchoalveolar hyperplasia and slight to moderate interstitial fibrosis.

6.1.2 Intratracheal administration

The results of studies with intratracheal administration of carbon black which appear to be adequate in their planning, implementation and documentation are summarized in Table 10.

Table 9. Toxicity of inhaled carbon black in animal studies (from IARC 1996)

Carbon black type, particle size, specific surface area	Species, strain, sex, age	Exposure	Results Comments	References
thermal black 0.15–0.2 μm (primary), 10–15 m^2/g	golden Syrian hamster, n.s.	about 56 mg/m^3 236 days, 5 days/week, 6 h/day about 105 mg/m^3 53 or 172 days, 5 days/week, 6 h/day	only larynx and trachea were examined: oedema of the laryngeal fold, no morphological changes	IARC 1996
Elftex 12, furnace black bimodal: 2.0 μm (MMAD), 0.02–0.12 μm (MMDD), 43 m^2/g	Fischer 344/N rat ♂ 14–15 weeks	10 mg/m^3 7 h/day, 5 days/week, 12 weeks	lung load of carbon black: 2.2 mg; BAL: PMN and protein increased; in terminal bronchioles and alveolar passages macrophage aggregates loaded with carbon black; hyperplasia of type II pneumocytes	Wolff *et al.* 1990
Elftex 12, furnace black 2–2.4 μm (MMAD), n.s.	Fischer 344/N rat ♀ 11–15 weeks	3.5 mg/m^3 16 h/day, 7 days/week, 12 weeks 13 mg/m^3 6 h/day, 5 days/week, 12 weeks 98 mg/m^3 4 h/day, 1 day/week, 12 weeks recovery period 24 weeks	lung load of carbon black: 3–4 mg; lung weights increased; BAL: PMN, protein, LDH and β-glucuronidase increased; thickening of the alveolar septa with hypertrophy of the epithelial cells; different exposure conditions did not change the results	Henderson *et al.* 1992

Table 9. continued

Carbon black type, particle size, specific surface area	Species, strain, sex, age	Exposure	Results Comments	References
Elftex 12, bimodal: 2.0 µm (MMAD), 0.1 µm (MMDD), 43 m^2/g	Fischer 344/N rat *c* 100 ♂ and 100 ♀/group 7–9 weeks	(I) 2.5 mg/m^3 (II) 6.5 mg/m^3 16 h/day, 5 days/week, 24 months	body weights decreased; lung weights increased with the dose; lung load of carbon black (after 23 months): (I): ♀ 17.3 mg; ♂ 24.7 mg (II): ♀ 36.9 mg; ♂ 40.1 mg accumulation of carbon black in the lymph nodes BAL: PMN, protein, LDH and β-glucuronidase increased; *histology*: chronic inflammation, alveolar epithelial hyperplasia, septum fibrosis, alveolar proteinosis, bronchoalveolar metaplasia, focal fibrosis with epithelial hyperplasia, squamous cell metaplasia; delayed alveolar clearance of radioactive carbon black after 3 and 18 months exposure; accumulation of fluorescent microspheres in macrophage aggregates; effects mostly more severe in ♀ than in ♂	Mauderly *et al.* 1994, Nikula *et al.* 1995
Printex 90, furnace black 0.64 µm (MMAD), n.s.	Wistar rat 72 ♀/group 7 weeks	6 mg/m^3 17 h/day, 5 days/week,10 months exposure, 20 months recovery or 20 months exposure, 10 months recovery 11.3 mg/m^3 18 h/day, 5 days/week, 24 months exposure, 6 months recovery	moderate to severe inflammation, bronchoalveolar hyperplasia, alveolar histiocytosis, lipoproteinosis, squamous cell metaplasia	Dungworth *et al.* 1994

Table 9. continued

Carbon black type, particle size, specific surface area	Species, strain, sex, age	Exposure	Results Comments	References
Monarch 880, furnace black 16 nm (primary), 0.88 µm (MMAD), 220 m^2/g	Fischer 344/N rat ♂ n.s.	1.1, 7.1, 52.8 mg/m^3 6 h/day, 5 days/week, 13 weeks, recovery period 3 or 8 months	lung load of carbon black (at end of exposure): 0.35, 1.8, 7.8 mg; all effects seen at 7.1 mg/m^3 or more: delayed lung clearance of carbon black, BAL: PMN, LDH, β-glucuronidase, protein increased with the dose; *histology*: epithelial hyperplasia, pulmonary fibrosis, increased chemokine expression, alveolar cell proliferation increased with the dose	Driscoll *et al.* 1996
Printex 90, furnace black 0.64 µm (MMAD), 227 m^2/g	Wistar rat 100 ♀/ group, 217 controls, 7 weeks	11.6 mg/m^3 (mean value) 18 h/day, 5 days/week, 24 months, recovery 6 months	body weights decreased, lung weights increased, load of carbon black (end of exposure): lung 44 mg/animal, lymph nodes 6.7 mg/animal; delayed alveolar clearance; BAL: PMN, LDH, β-glucuronidase, protein increased; moderate to severe bronchoalveolar hyperplasia; mild to moderate interstitial fibrosis	Heinrich *et al.* 1995
	NMRI mouse 80 ♀/ group 7 weeks	11.6 mg/m^3 (mean value) 18 h/day, 5 days/week, 13.5 months, recovery 9.5 months	body weights decreased, lung weights increased, lung load of carbon black (after 12 months) 7.4 mg	

BAL: bronchoalveolar lavage
LDH: lactate dehydrogenase
MMAD: mass median aerodynamic diameter
MMDD: mass median diffusion equivalent diameter
PMN: polymorphonuclear neutrophil count

Table 10. Toxicity of single doses of carbon black administered to experimental animals by intratracheal instillation (from IARC 1996)

Type of carbon black, primary particle size	Species, strain, sex	Dose, Recovery period	Results Comments	References
Indian ink 30–40 nm	Swiss mouse, n. s.	4 mg/animal, up to 28 days	two-phase macrophage reaction: first phase without mitotic activity, second phase mitosis of the interstitial macrophages; BAL: initial marked increase in PMN; some penetration of carbon black particles through the epithelium into peribronchial and perivascular regions	IARC 1996
	Swiss mouse, n. s.	4 mg/animal, up to 7 weeks	rapid migration of monocytes from the blood into the alveoli, rapid formation of monocytes in the bone marrow after administration of carbon black	
	Swiss mouse, ♂	0.1, 1, 2, 4, 8 mg/animal, up to 14 days	dose-dependent increase in alveolar macrophages	
Regal 660, furnace black *c* 20 nm	Fischer 344 rat, ♂	0.5 mg/animal, 24 h	inflammatory reaction: PMN and protein increased in BAL; some carbon black particles penetrated into the interstitium (to a lesser extent than ultrafine TiO_2)	Oberdörster *et al.* 1992
carbon black (no other details)	Wistar rat, ♀	0.5 mg/animal for life (average survival 73–109 weeks)	deposits of carbon black in the lungs, mostly in the alveolar macrophages; metaplasia of the alveolar epithelium	IARC 1996
Monarch 900, furnace black 15 nm	Fischer 344 rat, ♀	10, 100 mg/kg, 15 months	inflammatory reaction: PMN, macrophages and lymphocytes significantly increased in BAL; histopathology: minimal to mild centroacinar alveolitis, bronchiolitis, hyperplasia of the type II pneumocytes in the bronchoalveolar region	Driscoll *et al.* 1997

BAL: bronchoalveolar lavage, PMN: polymorphonuclear neutrophils

24 hours after administration to male Fischer 344 rats of a single dose of 0.5 mg carbon black (primary particle diameter 20 nm) by intratracheal instillation, signs of an inflammatory reaction were seen in an increase in the polymorphonuclear neutrophil count and in the protein concentration in the bronchoalveolar lavage fluid (Oberdörster *et al.* 1992).

Driscoll *et al.* (1997) administered single doses of carbon black (primary particle diameter 15 nm) of 10 and 100 mg/kg body weight to female Fischer 344 rats; 15 months after dosing, neutrophilic inflammation and epithelial hyperplasia in the lungs of the animals were described.

6.1.3 Summary

It may be concluded from the animal studies that high lung loads of carbon black result in delayed alveolar lung clearance (see Section 4) and in inflammation (increase in cellular and biochemical parameters of inflammation in the bronchoalveolar lavage fluid) and fibrosis. It is assumed that the large specific surface area of carbon black is of particular relevance for the effects. Rats seem to be more sensitive than mice to the toxic effects of carbon black in the lung. Female rats produce more pronounced reactions than male rats, that is, the effects occur with higher incidence and are more severe in females. There are species-specific and sex-specific differences in the reactions to exposure to carbon black and other respirable dusts, that is, differences in the severity of the inflammatory reactions with release of cytokines and growth factors and the ability of the granulocytes and macrophages to function as radical traps. In addition, the differences in respiratory physiology of the different animal species must be taken into account (Heinrich 1996).

6.2 Genotoxicity

Various samples of carbon black and extracts have been tested *in vitro* and *in vivo* for genotoxic activity. The inconsistent, mostly negative results are summarized in Tables 11 and 12. In tests with particles *in vitro*, the test conditions such as presence of serum, concentration of DMSO or other solvents and the exposure period play an important role. In studies of carbon black extracts, the extraction conditions (extraction solvent, duration, temperature) have significant effects on the biological activity of the extract. Extraction with organic solvents probably dissolves more substances from the EOM than are dissolved in biological media (IARC 1996).

Table 11. Genotoxicity of carbon black and extracts of carbon black *in vitro* (from IARC 1996)

Test substance Specific surface area or extraction procedure	Test system	Test concentration Solvent[a]	Metabolic activation	Result	References
furnace black N339 100 m^2/g	*Salmonella* microsome test *S. typhimurium* TA98, TA100, TA1535, TA1537, TA1538	up to 7.5 mg/plate DMSO (suspension of carbon black particles)	–/+	–/–	Kirwin *et al.* 1981
extracts of 20 samples of carbon black, furnace black N330 Soxhlet extraction of 1 g carbon black with 200 ml benzene for 16 h	*Salmonella* microsome test *S. typhimurium* TA98, TA100 TA98NR, TA98/1,8-DNP_6	DMSO (substitution for benzene)	–/+ – –	+/+ + +	Agurell and Löfroth 1993
7 carbon black pastes for colouring leather n.s.	*Salmonella* microsome test *S. typhimurium* TA98, TA100	50 µg/ml (HID) n.s.	–/+	–/–	Venier *et al.* 1987
extracts of 7 carbon black pastes for colouring leather Soxhlet extraction of 4 g carbon black with 50 ml toluene for 48 h	*Salmonella* microsome test *S. typhimurium* TA98 TA100	2.5 µg/ml (LED) DMSO (substitution for toluene)	–/+ –/+	+/(+) –/–	Venier *et al.* 1987
Black pearls L, furnace black n.s.	*Salmonella* microsome test *S. typhimurium* TA98	250 µg/ml (HID) DMSO (5 g/ml)	+	–	Rosenkranz *et al.* 1980
extract of Black pearls L, furnace black Soxhlet extraction of 10 g carbon black with toluene for 48 h		5.0 µg/ml (LED) DMSO (substitution for toluene)	+	+	
extracts of Raven 5750, furnace black Soxhlet extraction of 10 g carbon black with 50 ml toluene for 48 h		500 µg/ml (LED) DMSO (substitution for toluene)	+	(+)	

Table 11. continued

Test substance specific surface area or extraction procedure	Test system	Test concentration solvent[a]	Metabolic activation	Result	References
furnace black N339 100 m^2/g	mouse lymphoma test L5178Y mouse lymphoma cells	40000 µg/ml (HID) culture medium (suspension of carbon black particles)	–/+	–/–	Kirwin *et al.* 1981
	test for sister chromatid exchange CHO cells	1000 µg/ml (HID) DMSO (suspension of carbon black particles)	–/+	–/–	
carbon black (no other details)	micronucleus test M3E3/C3 hamster epithelial cells	1 µg/ml (LED) culture medium (suspension of carbon black particles)	+	+	Riebe-Imre *et al.* 1994

HID: highest ineffective dose
LED: lowest effective dose
[a] with extracts, the extraction solvent was replaced by the solvent given here

Table 12. Genotoxicity of carbon black *in vivo* (from IARC 1996)

Type of carbon black particle size specific surface area	Species, strain, sex	Test system exposure	Result	References
furnace black N339 n.s. 100 m^2/g	*Drosophila melanogaster*	test for somatic mutation, SLRL test, dominant lethal test, test for aneuploidy oral: 1 % in the diet	–	Kirwin *et al.* 1981
Monarch 880, furnace black 16 nm (primary), 0.88 µm (MMAD) 220 m^2/g	Fischer 344/N rat, ♂	test for hprt mutations in type II alveolar cells inhalation: 1.1, 7.1 and 52.8 mg/m^3, 6 h/day, 5 days/week, 13 weeks, recovery period up to 8 months	+	Driscoll *et al.* 1996
Monarch 900, furnace black 15 nm (primary), 230 m^2/g	Fischer 344/N rat, ♀	test for hprt mutations in type II alveolar cells intratracheal instillation: 10 and 100 mg/kg body weight, once, recovery period 15 months	+	Driscoll *et al.* 1997
Elftex 12, furnace black 2 µm (MMAD), 43 m^2/g	Fischer 344/N rat, ♂, ♀	test for DNA adducts in type II alveolar cells inhalation: 6.2 mg/m^3, 16 h/day, 5 days/week, 12 weeks	+	Bond *et al.* 1990
Printex 90, furnace black 0.65 µm (MMAD), 270 m^2/g	Wistar rat, ♀	test for DNA adducts in lung cells inhalation: initially 7.5 mg/m^3 (4 months), then 12 mg/m^3 (20 months); 18 h/day, 5 days/week	–	Gallagher *et al.* 1994
Elftex 12, furnace black 2.0 µm (MMAD), 43 m^2/g	Fischer 344 rat, ♂	test for DNA adducts in lung cells inhalation: 10 mg/m^3, 7 h/day, 5 days/week, 12 weeks	–	Wolf *et al.* 1990

MMAD: mass median aerodynamic diameter
SLRL: sex-linked recessive lethal- mutations

6.2.1 *In vitro*

The *Salmonella* microsome test has yielded not only negative results of studies with carbon black particles (suspended in DMSO) and carbon black pastes but also evidence of mutagenic effects of carbon black extracts even without metabolic activation (Agurell and Löfroth 1993, IARC 1996, Kirwin *et al.* 1981, Rosenkranz *et al.* 1980, Venier *et al.* 1987). The mutagenic effects of carbon black extracts are ascribed to the presence of PAH such as dinitropyrene (Agurell and Löfroth 1993). Printex 90 has been tested both as a suspension in DMSO and in the form of a toluene extract in the *Salmonella* microsome test with and without activation. Both tests yielded negative results (Degussa 1998a, 1998b).

Carbon black (furnace black) from a single production batch was tested as a suspension in a number of genotoxicity tests (Kirwin *et al.* 1981). In both the *Salmonella* microsome test and in a test with L5178Y mouse lymphoma cells (concentrations of 10000–40000 μg/ml) the results were negative with and without a metabolic system. At the highest concentration, only 1 % of the mouse lymphoma cells survived; that is, the carbon black was highly cytotoxic. A suspension of carbon black in DMSO in concentrations of 0.32 to 1000 μg/ml was tested with and without a metabolic system in a test for sister chromatid exchange (SCE) in Chinese hamster ovary cells. There was a slight increase in SCE but because it was not dose-dependent the results were considered to be negative.

Induction of micronuclei by a suspension of carbon black particles (0.1–2.0 μg/ml) was studied *in vitro* in epithelial lung cells from golden Syrian hamsters. There was practically no cytotoxicity. A dose-dependent increase in micronucleus formation was observed. However, the maximum effect was only 150 % of the spontaneous incidence in the controls. Most of the micronucleus formation was ascribed to the clastogenic effects of the fine carbon black particles (detection of a specific anti-centromere-protein antibody) (Riebe-Imre *et al.* 1994).

6.2.2 *In vivo*

The same sample of carbon black which was tested *in vitro* (furnace black, see above) was administered to the larvae of *Drosophila melanogaster* in the diet (concentration 1 %). The tests for somatic mutations, chromosomal aberrations, and aneuploidy, and the dominant lethal test and test for sex-linked recessive lethal mutations all yielded negative results (Kirwin *et al.* 1981).

Rats inhaled carbon black (furnace black) in concentrations of 1.1, 7.1 or 52.8 mg/m^3, 6 h/day, 5 days/week for 13 weeks. Immediately after the end of exposure and 3 and 8 months later, primary cultures of type II alveolar cells were prepared from the experimental animals and tested for induced gene mutations. In the lung cells of the animals killed immediately after the end of exposure to the highest concentration, the hprt mutation frequency was increased by a factor of 4.3 above the control values. Likewise in the animals which had inhaled 7.1 mg/m^3, the hprt mutation frequency was increased significantly. After 3-month and 8-month recovery periods, the hprt mutation frequency in the lung cells was increased only in the animals exposed to the highest concentration, by factors of 3.2 and 2.7, respectively. The genotoxic effects on the alveolar cells were seen only if the experimental animals had been exposed to carbon black concentrations which caused significant inflammatory reactions and epithelial hyperplasia in the lungs (see Section 6.1 and Table 6). The authors consider that these results support the hypothesis that oxidizing agents formed in the cells and increased cell proliferation play a key role in the induction of lung tumours after exposure to carbon black (Driscoll *et al.* 1996). Also after intratracheal instillation of carbon black (furnace black, 10 and 100 mg/kg body weight) an increased hprt mutation frequency was seen in the type II alveolar cells. Epithelial alveolar cells from RLE-6TN rats were co-cultured for 24 hours with the cells from the bronchoalveolar lavage (BAL) of animals exposed to carbon black, to investigate the potential effects of inflammatory cells on *in vivo* mutagenicity. In the RLE-6TN cells the level of gene mutations was increased. This increase could be prevented by pretreating the BAL cells with catalase, which inactivates the oxidizing agents (Driscoll *et al.* 1997).

In two studies (Gallagher *et al.* 1994, Wolff *et al.* 1990) no significant increase in the level of DNA adducts in lung cells of animals exposed to carbon black were detected; another study of DNA adducts yielded positive results (Bond *et al.* 1990). The authors suggested that these adducts are either formed by the PAH present in carbon black or by reactive oxygen species.

6.3 Carcinogenicity

6.3.1 Cell transformation studies

A suspension of carbon black particles (furnace Black N339, 2000–16000 μg/ml) in acetone tested in a cell transformation test with mouse embryo fibroblasts (C3H/10T$^{1}/_{2}$ cells) in the presence of a metabolic system yielded negative results (Kirwin *et al.* 1981). Another cell transformation test was carried out with differentiated and undifferentiated cells from foetal hamster lung (M3E3/C3 cells). In the presence of a metabolic system, carbon black (100–300 μg/ml) produced an increase in the transformation rate. The transforming activity was weaker in undifferentiated cells than in differentiated cells where, at a carbon black concentration of 200 μg/ml, the transformation rate was increased above the control value by a factor of 4 (Riebe-Imre *et al.* 1994).

6.3.2 Long-term studies

Carcinogenicity studies have been carried out with commercially available types of carbon black, with carbon black samples in which the level of organic components has been reduced by extraction with solvents, and with extracts of carbon black. There is no meaningful study in which the carbon black was administered orally. Several epicutaneous tests with mice revealed no tumour-inducing activity of carbon black particles. However, painting mouse skin with extracts of various kinds of carbon black produced skin tumours. Subcutaneous injection of samples of carbon black types with high levels of PAH (300 ppm BaP, suspension in tricaprylin) or of extracts of carbon black induced local sarcomas in the mouse (IARC 1996). The studies in which carbon black was administered by inhalation or intratracheal instillation are shown in detail in Tables 13 and 14.

In a long-term inhalation study with female NMRI mice exposed to carbon black (furnace black, aerodynamic diameter 0.64 µm), no statistically significant increase in the incidence of lung tumours was observed. The mice were exposed for four months, 18 h/day, 5 days/week to a carbon black concentration of 7.4 mg/m^3 and then for 9.5 months to 12.2 mg/m^3 followed by a 9.5-month recovery period. Reduced body weights were recorded for the exposed animals. Under otherwise identical test conditions, female Wistar rats inhaled carbon black in a concentration of 7.4 mg/m^3 for 4 months and then 12.2 mg/m^3 for 20 months. The recovery period was 6 months. Life expectancy was reduced below the control values in the group exposed to carbon black. Body weights were reduced from day 300 to the end of the study; an increase in lung weights was observed. The exposure to carbon black resulted in an increased incidence of benign and malignant tumours (details in Table 13; Heinrich *et al.* 1995).

In another study (Heinrich *et al.* 1994), female Wistar rats were exposed to carbon black (furnace black, aerodynamic diameter 1.1 µm) in a concentration of 6 mg/m^3 for 17 h/day on 5 days/week. The exposure period was either 43 weeks with an 86-week recovery period (group I) or 86 weeks with a 43-week recovery period (group II). Mortality was higher in group II than in group I; the incidence of lung tumours was higher in group I. The lung tumour incidence was 18 % in group I and 8 % in group II (details in Table 13).

Male and female Fischer 344/N rats were exposed to carbon black (furnace black, aerodynamic diameter 2.0 µm) in concentrations of 2.5 (I) or 6.5 mg/m^3 (II), 16 h/day, 5 days/week for up to 24 months, and then observed for the rest of their lives (Mauderly *et al.* 1994, Nikula *et al.* 1995). Only in the female animals was there a significant increase relative to the control values in the incidence of benign and malignant tumours (details in Table 13). None of the animals killed after 3, 6 or 12 months had a tumour; the first lung tumour appeared between 12 and 18 months. In these lung tumours there were no p53 mutations and only in one squamous cell carcinoma were K-ras mutations found (Belinsky *et al.* 1995). In addition to the tumours, in the lungs of the exposed animals squamous cell cysts were found and were considered to be non-neoplastic changes (see Table 13).

Female Wistar rats were given 3 mg doses of carbon black (furnace black) by intratracheal instillation, once weekly for 15 weeks (Pott and Roller 1994, Pott e*t al.* 1994). The animals were observed for 131 weeks after the first instillation. More than 50 % of the exposed and control animals survived 100 weeks; the first rat with a lung tumour died 56 weeks after the first instillation of carbon black. The average life expectancy of the animals with tumours was 110 weeks after the first instillation of the substance. There were no tumours in the control animals. In the group exposed to carbon black, the incidence of lung tumours was 65 % (details in Table 14). The incidence of tumours excluding benign cystic keratinizing squamous cell tumours was 54 %.

Table 13. Carcinogenicity studies with inhaled carbon black

Author:	Heinrich *et al.* 1995
Test substance:	carbon black: Printex 90, furnace black; particle diameter: primary particles 14 nm, aerosol 0.64 µm (MMAD); specific surface area: 227 ± 18.8 m^2/g; EOM 0.04 %, benzo[*a*]pyrene 0.6 pg/mg carbon black, 14 other PAH all < 5 pg/mg carbon black, 1-nitropyrene < 0.5 ng/mg carbon black
Species:	Wistar rat, ♀, 100 exposed to carbon black, 217 clean air controls
Exposure:	4 months to a carbon black concentration of 7.4 mg/m^3, then 20 months to 12.2 mg/m^3, in both cases 18 h/day, 5 days/week average particle concentration 11.6 mg/m^3; cumulative particle dose 102.2 $g/m^3 \times h$ recovery period 6 months, duration of study 30 months
Toxicity:	mortality significantly increased: after 24 and 30 months carbon black 56 % and 92 %, control 42 % and 85 %; body weights significantly decreased from exposure day 300, at end of study carbon black 325 g, control 417 g; lung weights (wet weights) significantly increased after 3 months exposure, maximum after 18 months exposure: carbon black 5 times the control value.
Comments:	incidence of benign and malignant lung tumours increased, first tumour (benign cystic keratinizing squamous cell tumour) after 24 months (carbon black 1/9, control 0/10)

		controls	carbon black
number of animals		217	100
number of lung tumours after 30 months	benign cystic keratinizing squamous cell tumours	1	20
	squamous cell carcinomas	0	4
	adenomas	0	13
	adenocarcinomas	1	13
animals with lung tumours		1	39
animals with lung tumours other than benign cystic keratinizing squamous cell tumours		1	28

Table 13. continued

Author:	Heinrich *et al.* 1995
Test substance:	carbon black: Printex 90, furnace black; particle diameter: primary particles 14 nm, aerosol 0.64 μm (MMAD); specific surface area: 227 ± 18.8 m^2/g; EOM 0.04 %, benzo[*a*]pyrene 0.6 pg/mg carbon black, 14 other PAH all < 5 pg/mg carbon black, 1-nitropyrene < 0.5 ng/mg carbon black
Species:	NMRI mouse, ♀, 80 exposed to carbon black, 80 clean air controls
Exposure:	4 months to a carbon black concentration of 7.4 mg/m^3, then 9.5 months to 12.2 mg/m^3, in both cases 18 h/day, 5 days/week; cumulative particle dose 55.4 $g/m^3 \times h$ recovery period 9.5 months, duration of study 23 months
Toxicity:	mortality increased: after 13.5 months carbon black 20 %, control 10 %; carbon black 50 % after 19 months, control 50 % after 20 months body weights significantly decreased after 11–17 months exposure, after that in the range of the control values lung weights (wet weights) significantly increased after 3 months exposure, maximum after 12 months exposure: carbon black 5 times the control value; slight decrease during recovery period
Comments:	incidence of benign and malignant lung tumours not increased in the group exposed to carbon black

		control	carbon black
incidence of lung tumours after 23 months	adenomas	25 %	11.3 %
	adenocarcinomas	15.4 %	10 %

Table 13. continued

Author:	Heinrich *et al.* 1994
Test substance:	carbon black: Printex 90, furnace black; particle diameter: primary particles 15 nm, aerosol 1.1 µm (MMAD); specific surface area: 230 m^2/g; EOM 0.04 %, benzo[*a*]pyrene 0.6 pg/mg carbon black, 1-nitropyrene < 0.5 ng/mg carbon black
Species:	Wistar rat, ♀, 72 exposed to carbon black, 72 clean air controls
Exposure:	carbon black 6 mg/m^3, (I) exposure for 43 weeks, 5 days/week, 17 h/day, recovery period 86 weeks (II) exposure for 86 weeks, 5 days/week, 17 h/day, recovery period 43 weeks
Toxicity:	mortality increased in group II: 50 % mortality after 728 days in group I, 682 days in II and 740 days in control lung weights (wet weights) increased after 3 months exposure
Comments:	incidence of lung tumours: 18 % in group I, 8 % in II, 0 % in control

		Controls	(I) 43 weeks carbon black	(II) 86 weeks carbon black
number of animals		72	72	72
number of tumours after 129 weeks	benign cystic keratinizing squamous cell tumours	0	7	4
	squamous cell carcinomas	0	1	1
	adenomas	0	2	1
	adenocarcinomas	0	4	0
	hyperplasia and marked proliferation of squamous epithelium, on the borderline between a neoplastic and a non-neoplastic lesion	0	0	6

Table 13. continued

Author:	Mauderly *et al.* 1994, Nikula *et al.* 1995
Test substance:	carbon black: Elftex 12, furnace black; particle diameter bimodal in aerosol: 67 % 2.0 μm (MMAD), 33 % 0.1 μm (MMDD); specific surface area: 43 m^2/g; EOM: 0.04–0.29 %
Species:	Fischer 344 rat, ♂ and ♀, 105–109 animals of each sex in exposure group and clean air control
Exposure:	carbon black 2.5, 6.5 mg/m^3, exposure: 24 months, 5 days/week, 16 h/day; recovery period until spontaneous death
Toxicity:	mortality increased at 2.5 mg/m^3 in ♂, at 6.5 mg/m^3 in ♂ and ♀ body weights decreased in ♂ and ♀ at 2.5 mg/m^3 from day 449, at 6.5 mg/m^3 from day 309 after month 22 at 2.5 mg/m^3 < 10 %, at 6.5 mg/m^3 ♂ 14 %, ♀ 16 % lung weights (wet weights) significantly increased after 3 months exposure to 6.5 mg/m^3 in ♀, after 6 months exposure to 2.5 mg/m^3 in ♀, after 12 months in all groups
Comments:	incidence of benign and malignant lung tumours increased only in ♀, first tumour between 12 and 18 months (carbon black 1/9, control 0/10)

	control		carbon black 2.5 mg/m^3		carbon black 6.5 mg/m^3	
	♂	♀	♂	♀	♂	♀
number of animals	109	105	106	107	106	105
squamous cell carcinomas	1	0	0	0	2	1
squamous cell adenocarcinomas	0	0	0	0	1	1
adenomas	1	0	1	2	0	17
adenocarcinomas	1	0	1	6	1	23
animals with lung tumours	3	0	2	8	4	28
squamous cell cysts (non-neoplastic)	0/86	0/91	1/73	8/90	4/74	13/87

EOM: extractable organic material
MMAD: mass median aerodynamic diameter
MMDD: mass median diffusion equivalent diameter

Table 14. Carcinogenicity studies with carbon black administered by intratracheal instillation

Author:	Pott and Roller 1994, Pott *et al.* 1994
Test substance:	carbon black: Printex 90, furnace black; specific surface area 270 m^2/g
Species:	Wistar rat, ♀, 37 treated with carbon black, 39 vehicle controls
Exposure:	15 doses, once weekly, each dose 3 mg/animal observation period: up to 131 weeks after the first instillation
Toxicity:	mortality of treated and control animals similar; rats with tumours: first died after 56 weeks, average life expectancy 110 weeks
Comments:	incidence of lung tumours increased

		control	carbon black
number of animals		39	37
number of animals with lung tumours after 131 weeks	benign cystic keratinizing squamous cell tumours	0/39	5/39
	squamous cell carcinomas	0/39	9/39
	adenomas	0/39	3/37
	adenocarcinomas	0/39	7/37
percentage of animals with lung tumours		0 %	65 %
percentage of animals with lung tumours other than benign cystic keratinizing squamous cell tumours		0 %	54 %

Table 14. continued

Author:	Dasenbrock *et al.* 1996
Test substance:	(I) carbon black: Printex 90, furnace black, extracted; particle diameter: primary particles 14 nm; specific surface area: 270 m^2/g (II) carbon black: Lampblack 100, extracted; particle diameter: primary particles 95 nm; specific surface area: 22 m^2/g extraction treatment: 3 times with boiling toluene for 30 min
Species:	Wistar rat, ♀, 51–52 per group treated with carbon black, 51 vehicle controls
Exposure:	16–17 doses, once weekly, total dose 15 mg/animal observation period up to 48 weeks after the first instillation
Toxicity:	lung weights significantly increased in groups I and II; lung load of carbon black (1 day after last instillation) in group I: 10.8 mg/g, group II: 11.3 mg/g
Comments:	incidence of lung tumours significantly increased in group I; carcinogenicity dependent on primary particle size and specific surface area

		control	carbon black (I)	carbon black (II)
number of animals		51	52	51
number of lung tumours	benign cystic keratinizing squamous cell tumours	0	9	4
	bronchoalveolar carcinomas	0	4	0
	bronchoalveolar adenomas	0	1	0
percentage of animals with lung tumours		0%	21%	8%

Female Wistar rats were treated by intratracheal instillation of one of two different types of carbon black after solvent extraction, either extracted furnace black (primary particle diameter 14 nm, surface area 270 m^2/g) (group I) or extracted lamp black (primary particle diameter 95 nm, surface area 22 m^2/g) (group II) (Dasenbrock *et al.* 1996). The total dose was 15 mg/animal. In group I, the incidence of animals with lung tumours (10/52, i.e. 21 %) was increased ($p < 0.001$): 9 benign cystic keratinizing squamous cell tumours, 1 bronchoalveolar adenoma and 4 bronchoalveolar carcinomas (several tumours developed in some animals). In group II, benign cystic keratinizing squamous cell tumours developed in 4 animals (8 %). Thus the carcinogenic potency of carbon black under the conditions in this experiment was seen to be dependent on the primary particle size and the specific surface area of the particles.

7 Manifesto (MAK value/classification)

Some of the epidemiological studies of carbon black provide evidence of a possible carcinogenic effect in the lungs, but the evidence is inconclusive. In view of the inconsistency of the results and the methodical problems in some of the studies, it is not possible to derive adequate evidence for a carcinogenic effect of carbon black in man from these epidemiological studies.

In several long-term studies in which carbon black was administered to the rat by inhalation or intratracheal instillation, lung tumours were observed. The lowest carbon black concentration tested in an inhalation study was 2.5 mg/m^3; the animals were exposed for 16 h/day on 5 days/week for two years. On the other hand, when mice were exposed to a carbon black concentration of almost 12 mg/m^3, 18 h/day, 5 days/week for 13.5 months and observed for another 9.5 months, the incidence of lung tumours was not increased. In both the rat and mouse there were not only massive deposits of particles in the lungs but also marked inflammatory, proliferative and even fibrotic changes. The alveolar particle clearance was severely impaired.

With extracts of carbon black applied epicutaneously and subcutaneously to mice, local tumours were induced. A significant increase in the incidence of local tumours was also seen after subcutaneous injection of a carbon black sample containing 300 ppm BaP.

With carbon black, as with other insoluble particles, it is thought that the inflammatory proliferative reaction in the alveolar region and the associated release of active oxygen species is a probable mechanism of the particle-induced tumour development in the rat lung (see Section 3). It is of significance for the assessment of the effects of carbon black on health that the severity of the toxic, inflammatory, proliferative and carcinogenic effects of poorly soluble aerosol particles in the lung increases not so much with the concentration expressed as weight of the material but rather with the specific number of particles and specific surface area, properties which increase with decreasing particle size (see 'Aerosols – Ultrafine aerosol particles, their agglomerates and aggregates', in Volume 16 of the present series).

At present the possibility cannot be excluded that the mechanism of action of carbon black which leads to the development of lung tumours in the rat is also relevant for man. On this basis, carbon black is classified in Carcinogen category 3B. The derivation of a MAK value for carbon black on the basis of the available occupational medical and toxicological data is not possible at present.

8 References

Agurell E, Löfroth G (1993) Impurity variations in a carbon black: characterization by the Ames *Salmonella* mutagenicity assay and polycyclic aromatic hydrocarbon analysis. *Environ Toxicol Chem 12*: 219–223

Beck B, Gohlke R, Sturm W, Bergmann L, Wolf E (1985) Carbon black lung as an occupational disease. *Z Erkr Atmungsorgane 164*: 254–266

Belinsky SA, Mitchell CE, Nikula KJ, Swafford DS (1995) Pulmonary toxicity of inhaled diesel exhaust and carbon black in chronically exposed rats – Part III: Examination of possible target genes. *Health Effects Institute Research Report Number 68, Part III*: 1–26

Blair A, Stewart PA, Hoover RN (1990) Mortality from lung cancer among workers employed in formaldehyde industries. *Am J Ind Med 17*: 683–699

Bond JA, Johnson NF, Snipes MB, Maulderly JL (1990) DNA adduct formation in rat alveolar type II cells: cells potentially at risk for inhaled diesel exhaust. *Environ Mol Mutagen 16*: 64–69

Boorman GA, Brockman M, Carlton WA, Davis JMG, Dungworth DL, Hahn FF, Mohr U, Richter-Reichhelm H-B, Turusov VS, Wagner BM (1996) Classification of cystic keratinizing squamous lesions of the rat lung: Report of a workshop. *Toxicol Pathol 24*: 564–572

Bourguet CC, Checkoway H, Hulka BS (1987) A case-control study of skin cancer in the tire and rubber manufacturing industry. *Am J Ind Med 11*: 461–473

Bowden Dh, Adamson IYR (1984) Pathways of cellular efflux and particulate clearance after carbon instillation to the lung. *J Pathol 143*: 117–125

Cocarla A, Cornea G, Dengel H, Gabor S, Milea M, Papilian VV (1976) Pneumoconiose au noir de fumée (Carbon black pneumoconiosis) (French). *Int Arch Occup Environ Health 36*: 217–228

Creutzenberg O, Bellmann B, Heinrich U, Fuhst R, Koch W, Muhle H (1990) Clearance and retention of inhaled diesel exhaust particles, carbon black, and titanium dioxide in rats at lung overload conditions. *J Aerosol Sci 21, Suppl 1*: S455-S458

Crosbie WA (1986) The respiratory health of carbon black workers. *Arch Environ Health 41*: 346–353

Dasenbrock C, Peters L, Creutzenberg O, Heinrich U (1996) The carcinogenic potency of carbon particles with and without PAH after repeated intratracheal administration in the rat. *Toxicol Lett 88*: 15–21

Degussa (1992) *Was ist Ruß? Informationsschrift der Degussa AG* (What is soot? Information from Degussa) (German)

Degussa (1998a) *Reverse mutation assay using bacteria (Salmonella typhimurium and Escherichia coli) with a DMSO suspension of Printex 90.* BSL Bioservice Project No. 980926, unpublished studies of the Degussa AG

Degussa (1998b) *Reverse mutation assay using bacteria (Salmonella typhimurium and Escherichia coli) with a toluene extract of Printex 90.* BSL Bioservice Project No. 980925, unpublished studies of the Degussa AG

Degussa (1998c) written communication of unpublished data of Degussa AG of 17.12.1998

Driscoll KE, Carter JM, Howard BW, Hassenbein DG, Pepelko W, Baggs RB, Oberdörster G (1996) Pulmonary inflammatory, chemokine and mutagenic responses in rats after subchronic inhalation of carbon black. *Toxicol Appl Pharmacol 136*: 372–380

Driscoll KE, Ceyo LC, Carter JM, Howard BW, Hassenbein DG, Bertram TA (1997) Effects of particle exposure and particle-elicited inflammatory cells on mutation in rat alveolar epithelial cells. *Carcinogenesis 18*: 423–430

Dungworth DL, Mohr U, Heinrich U, Ernst H, Kittel B (1994) Pathologic effects of inhaled particles in rat lungs: associations between inflammatory and neoplastic processes. in: Mohr U, Dungworth DL, Mauderly JL, Oberdörster G (Eds) *Toxic and carcinogenic effects of solid particles in the respiratory tract*, ILSI Monographs, ILSI Press, Washington DC, 75–98

FDA (Food and Drug Administration) (1994) *Determination of PAH content in carbon black. Analytical Test Method No. 63*, FDA, Washington DC

Gärtner H, Brauss FW (1951) Untersuchungen zur Frage der Rußlunge und zur Schädlichkeit des reinen Kohlenstaubanteiles im Staub der Kohlenbergwerke (Carbon black lung and the effects of the pure carbon fraction of coal mine dust) (German). *Medizin Welt 8*: 253–255

Gallagher J, Heinrich U, George M, Hendee L, Phillips DH, Lewtas J (1994) Formation of DNA adducts in rat lung following chronic inhalation of diesel emissions, carbon black and titanium dioxide particles. *Carcinogenesis 15*: 1291–1299

Gardiner K, Trethowan WN, Harrington JM, Calvert IA, Glass DC (1992) Occupational exposure to carbon black in its manufacture. *Ann Occup Hyg 5*: 477–496

Gardiner K, Trethowan NW, Harrington JM, Rossiter CE, Calvert IA (1993) Respiratory health effects of carbon black: a survey of European carbon black workers. *Br J Ind Med 50*: 1082–1096

Heinrich U (1996) Comparative response to long-term particle exposure among rats, mice, and hamsters. *Inhal Toxicol 8 Suppl*: 51–71

Heinrich U, Peters L, Creutzenberg O, Dasenbrock C, Hoymann H-G (1994) Inhalation exposure of rats to tar/pitch condensation aerosol or carbon black alone or in combination with irritant gases. in: Mohr U, Dungworth DL, Mauderly JL, Oberdörster G (Eds) *Toxic and carcinogenic effects of solid particles in the respiratory tract*, ILSI Monographs, ILSI Press, Washington DC, 433–441

Heinrich U, Fuhst R, Rittinghausen S, Creutzenberg O, Bellmann B, Koch W, Levsen K (1995) Chronic inhalation exposure of Wistar rats and two different strains of mice to diesel engine exhaust, carbon black, and titanium dioxide. *Inhal Toxicol 7*: 533–556

Henderson RF, Barr EB, Cheng YS, Griffith WC, Hahn FF (1992) The effect of exposure pattern on the accumulation of particles and the response of the lung to inhaled particles. *Fundam Appl Toxicol 19*: 367–374

Hodgson JT, Jones RD (1985) A mortality study of carbon black workers employed at five United Kingdom factories between 1947 and 1980. *Arch Environ Health 40*: 261–268

IARC (International Agency for Research on Cancer) (1996) Printing processes and printing inks, carbon black and some nitro compounds. *IARC Monographs on the evaluation of carcinogenic risks to humans, Vol 65*: 149–262

Kandt D (1985) Klinische Symptome und Befunde der unteren Atemwege bei rußexponierten Werktätigen (Clinical symptoms and findings in the lower airways of soot-exposed workers) (German). *Z Erkr Atmumgsorgane 165*: 25–41

Kirwin CJ, LeBlanc JV, Thomas WC, Haworth SR, Kirby PE, Thilager A, Bowman JT, Brusick DJ (1981) Evaluation of the genetic activity of industrially produced carbon black. *J Toxicol Environ Health 7*: 973–989

Küpper HU, Breitstadt R, Ulmer WT (1996) Effects on the lung function of exposure to carbon black dusts. *Int Arch Occup Environ Health 68*: 478–483

Lee PS, Gorski RA, Hering WE, Chan TL (1987) Lung clearance of inhaled particles after exposure to carbon black generated from a resuspension system. *Environ Res 43*: 364–373

Mai O (1966) Rußlungen (Soot lung) (German). *Z ges Hyg 12*: 421–425

Mauderly JL, Jones RK, Griffith WC, Henderson RF, McClellan RO (1987) Diesel exhaust is a pulmonary carcinogen in rats exposed chronically by inhalation. *Fund Appl Toxicol 9*: 208–221

Mauderly JL, Snipes MB, Barr EB, Belinsky SA, Bond JA, Brooks AL, Chang I-Y, Cheng YS, Gillet NA, Griffith WC, Henderson RF, Mitchell CE, Nikula KJ, Thomassen DG (1994) Pulmonary toxicity of inhaled diesel exhaust and carbon black in chronically exposed rats. Part I: Neoplastic and nonneoplastic lung lesions. *Health Effects Institute Research Report Number 68*: 1–97

Muhle H, Creutzenberg O, Bellmann B, Heinrich U, Mermelstein R (1990) Dust overloadings of lungs: investigations of various materials, species differences, and irreversibility of effects. *J Aerosol Med 3, Suppl 1*: S111–S128

Muhle H, Bellmann B, Creutzenberg O (1994) Toxicokinetics of solid particles in chronic rat studies using diesel soot, carbon black, toner, titanium dioxide, and quartz. in: Mohr U, Dungworth DL, Mauderly JL, Oberdörster G (Eds) *Toxic and carcinogenic effects of solid particles in the respiratory tract*, ILSI Monographs, ILSI Press, Washington DC, 29–41

Nikula KJ, Snipes MB, Barr EB, Griffith WC, Henderson RF, Maulderly JL (1995) Comparative pulmonary toxicities and carcinogenicities of chronically inhaled diesel exhaust and carbon black in F344 rats. *Fundam Appl Toxicol 25*: 80–94

Oberdörster G, Ferin J, Gelein R, Soderholm SC, Finkelstein J (1992) Role of alveolar macrophage in lung injury: studies with ultrafine particles. *Environ Health Perspect 97*: 193–199

Oberdörster G, Ferin J, Lehnert BE (1994) Correlation between particle size, *in vivo* particle persistence and lung injury. *Environ Health Perspect 102 (Suppl 5)*: 173–179

Oleru UG, Elegbeleye OO, Enu CC, Olumide YM (1983) Pulmonary function and symptoms of Nigerian workers exposed to carbon black in dry cell battery and tire factories. *Environ Res 30*: 161–168

Parent ME, Siemiatycki J, Renaud G (1996) Case-control study of exposure to carbon black in the occupational setting and risk of lung cancer. *Am J Ind Med 30*: 285–292

Pott F, Roller M (1994) Relevance of nonphysiologic exposure routes for carcinogenicity studies of solid particles. in: Mohr U, Dungworth DL, Mauderly JL, Oberdörster G (Eds) *Toxic and carcinogenic effects of solid particles in the respiratory tract.* ILSI Monographs, ILSI Press Washington DC: 109–125

Pott F, Dungworth DL, Heinrich U, Muhle H, Kamino K, Germann P-G, Roller M, Rippe RM, Mohr U (1994) Lung tumours in rats after intratracheal instillation of dusts. *Ann Occup Hyg 38*: 357–363

Pylev LN, Roe F, Vorvik D (1970) The distribution and elimination of ^{3}H-benz[*a*]pyrene in the animal organism after its intratracheal injection with asbestos and carbon black. *Vop Onkol 16*: 61–69

Riebe-Imre M, Aufderheide M, Gärtner-Hübsch S, Peraud A, Straub M (1994) Cytotoxic and genotoxic effects of insoluble particles *in vitro*. in: Mohr U, Dungworth DL, Mauderly JL, Oberdörster G (Eds) *Toxic and carcinogenic effects of solid particles in the respiratory tract, ILSI Monographs*, ILSI Press, Washington DC, 519–523

Robertson JM (1996) Epidemiologic studies in North American carbon black workers. *Inhal Toxicol 8*: 41–50

Robertson JMcD, Ingalls TH (1980) A mortality study of carbon black workers in the United States from 1935–1974. *Arch Environ Health 35*: 181–186

Robertson JMcD, Ingalls TH (1989) A case-control study of circulatory, malignant, and respiratory morbidity in carbon black workers in the United States. *Am Ind Hyg Assoc J 50*: 510–515

Robertson JMcD, Inman KJ (1996) Mortality in carbon black workers in the United States. *J Occup Environ Med 38*: 569–570

Rosenkranz HS, McCoy EC, Sanders DR, Butler M, Kiriazides DK, Mermelstein R (1980) Nitropyrenes: isolation, identification, and reduction of mutagenic impurities in carbon black and toners. *Science 209*: 1039–1043

Rosmanith J, Kandus J, Holuša R (1976) Anthrakofibrose bei der Erzeugung von Kohlenruß (Die Rußlunge) (Anthracofibrosis in carbon black production) (German). *Int Arch Gewerbepath Gewerbehyg 25*: 292–298

Steineck G, Plato N, Gerhardsson M, Norell SE, Hogstedt C (1990) Increased risk of urothelial cancer in Stockholm during 1985–87 after exposure to benzene and exhausts. *Int J Cancer 45*: 1012–1017

Stone V, Shaw J, Brown DM, Macnee W, Faux SP, Donaldson K (1998) The role of oxidative stress in the prolonged inhibitory effect of ultrafine carbon black on epithelial cell function. *Toxicol In Vitro 12*: 649–659

Strom KA, Johnson JT, Chan TL (1989) Retention and clearance of inhaled submicron carbon black particles. *J Toxicol Environ Health 26*: 183–202

Sun JD, Wolff RK, Maio SM, Barr EB (1989) Influence of adsorption to carbon black particles on the retention and metabolic activation of benzo[*a*]pyrene in rat lungs following inhalation exposure or intratracheal instillation. *Inhal Toxicol 1*: 1–19

Taylor GT, Redington TE, Bailey MJ, Buddingh F, Nau CA (1980) Solvent extracts of carbon black – determination of total extractables and analysis for benzo[*a*]pyrene. *Am Ind Hyg Assoc J 41*: 819–825

Valić F, Beritić-Stahuljak D, Mark B (1975) A follow-up study of functional and radiological lung changes in carbon-black exposure. *Int Arch Arbeitsmed 34*: 51–63

Venier P, Tecchio G, Clonfero E, Levis AG (1987) Mutagenic activity of carbon black dyes used in the leather industry. *Mutagenesis 2*: 19–22

Vohler O, von Sturm F, Wege E, von Kienle H, Voll M, Kleinschmit P (1998) Carbon black. in: Elvers B, Hawkins S (Eds) *Ullmann's Encyclopedia of Industrial Chemistry*. Wiley-VCH, Weinheim

Wolff RK, Sun JD, Barr EB, Rothenberg SJ, Yeh HC (1989) Lung retention and binding of [^{14}C]-1-nitropyrene when inhaled by F344 rats as a pure aerosol or adsorbed to carbon black particles. *J Toxicol Environ Health 26*: 309–325

Wolff RK, Bond JA, Henderson RF, Harkema JR, Mauderly JL (1990) Pulmonary inflammation and DNA adducts in rats inhaling diesel exhaust or carbon black. *Inhal Toxicol 2*: 241–254

Zoccolillo L, Liberti A, Coccioli F, Ronchetti M (1984) Routine determination of polycyclic aromatic hydrocarbons in carbon black by chromatographic techniques. *J Chromatogr 288*: 347–355

completed 02.03.1999

Carbon tetrachloride

MAK value (2000)	**0.5 ml/m^3**
Peak limitation (1983)	**Category II,1**
Absorption through the skin (2000)	**H**
Sensitization	**–**
Carcinogenicity (2000)	**Category 4**
Prenatal toxicity (2000)	**Pregnancy risk group D**
Germ cell mutagenicity	**–**
BAT value	**70 µg/l blood**
Synonyms	perchloromethane
Chemical name (CAS)	tetrachloromethane
CAS number	56-23-5
Molecular formula	CCl_4
Molecular weight	153.8
Melting point	–22.92°C
Boiling point	76.7°C
Density at 15°C	1.594 g/cm^3
Vapour pressure at 20°C	122 hPa
$\log P_{ow}$*	2.64
Solubility	785 mg/l water
1 ml/m^3 (ppm) ≙ 6.41 mg/m^3	**1 mg/m^3 ≙ 0.156 ml/m^3 (ppm)**

The MAK value for carbon tetrachloride of 10 ml/m^3 which was valid until the year 2000 was established in 1964 in analogy to the TLV value at the time. In 1982 the substance was classified in Carcinogen category 3 (at that time Section IIIB of the *List of MAK and BAT Values*). The present documentation is based mainly on reviews of the toxicological data in the BUA report from 1990 (BUA 1990) and the IPCS report from 1999 (WHO 1999). More recent information has also been included.

---

* *n*-octanol/water distribution coefficient

1 Toxic Effects and Mode of Action

Carbon tetrachloride is known from experiments with animals to be a classic hepatotoxic substance, which also in man can cause centrilobular fatty degeneration and necrosis of the liver and, at high doses, even hepatic coma. The short-term toxic effects on the skin, eyes, lungs, kidneys, heart, bone marrow and nervous system are less pronounced. In cases of chronic intoxication, which is described less often, micronodular and macronodular liver cirrhosis develops. In rats and mice, carbon tetrachloride causes the development of hepatocellular carcinomas and phaeochromocytomas. Genotoxic effects occur *in vivo* only after the administration of high doses which lead to lipid peroxidation. Carbon tetrachloride has moderately irritative effects on the skin and mucous membranes. In experiments with animals given high maternally toxic doses, carbon tetrachloride has embryotoxic effects, but no teratogenic effects.

2 Mechanism of Action

Biotransformation of carbon tetrachloride yields trichloromethyl radicals and trichloromethylperoxy radicals which can bind to cellular macromolecules such as proteins and lipids (BUA 1990, WHO 1999). DNA adducts have not been investigated in the studies carried out to date; so far only the radioactivity in liver DNA has been determined and the assumption made that it is covalently bound. After intraperitoneal administration of 5 ml/kg body weight (15 % solution) to golden Syrian hamsters, C3H mice and rats, increased radioactivity was detected in liver cells both *in vitro* and *in vivo*. No adducts were determined. The results in the various species were similar (Castro *et al.* 1989, 1990). Male Sprague-Dawley and Osborne-Mendel rats were given intraperitoneal doses of ^{14}C-carbon tetrachloride of 5 ml/kg body weight (5 % solution). The animals were killed 3 hours later. Liver microsomes and cell nuclei were prepared from liver cells. Covalent binding to phospholipids, monoglycerides, diglycerides, cholesterol, free fatty acids, triglycerides and cholesterol esters was detected in both the microsomal and cell nucleus fraction (De Curtis *et al.* 1994, Fanelli and Castro 1995).

Mainly responsible for lipid peroxidation is the trichloromethylperoxy radical, which causes destruction of membrane components (Kappus 1985). There is evidence that "suicide" inactivation of the cytochrome P450 enzyme system is triggered by a radical formed in the metabolism of carbon tetrachloride (Goeptar *et al.* 1995). The reduction of carbon tetrachloride can lead to irreversible inactivation of the cytochrome P450 enzyme and subsequent loss of the prosthetic haem group. Carbon tetrachloride thus inhibits its own metabolism. The observed potentiation of the hepatotoxicity of carbon tetrachloride in rats after short-term and long-term pretreatment with ethanol is attributed to the induction of cytochrome P450-2E1.

The substance has also been shown to have other effects in the liver: it causes not only lipid peroxidation but also an increase in cell proliferation. In rats and mice, a 10 % to

30 % increase in cell proliferation was detected in the liver after gavage administration of carbon tetrachloride doses of 50 to 400 mg/kg body weight. After intraperitoneal injection of doses of 400 mg/kg body weight to F344 rats, proliferation increased by 5 %. An increase in the rate of DNA synthesis was observed 48 hours after administration of single intragastric doses of 2.5 ml/kg body weight to Sprague-Dawley rats (WHO 1999).

Carbon tetrachloride itself and the radicals formed from it are thought to be responsible for the increased cytotoxicity in liver cells, as they inhibit the Ca^{2+}/Mg^{2+}-ATPase in the endoplasmatic reticulum, which leads to a marked decrease in the concentration of calcium in the microsomes and at the same time to a 100-fold increase in calcium in the cytosol (WHO 1999).

After administration of single or three consecutive intraperitoneal doses of carbon tetrachloride of 2 ml/kg body weight to rats, the level of lactic acid in blood increased before the animals could develop liver cirrhosis. Long-term increases in the levels of lactic acid in blood were therefore regarded as evidence of the beginning of liver cirrhosis (WHO 1999).

3 Toxicokinetics and Metabolism

3.1 Absorption, distribution, elimination

3.1.1 Oral administration

Oral absorption of carbon tetrachloride, as of other haloalkanes, depends on the vehicle used. Less of the substance is absorbed after administration in corn oil than after administration in water. 10 % Emulphor solution does not affect the absorption.

Three hours after oral administration of the substance to rats, the radioactivity in the tissues decreased in the following order: liver, kidneys, brain, muscle and blood. Carbon tetrachloride accumulates in the adipose tissue in rats and rabbits (WHO 1999). In female rats given single doses of carbon tetrachloride of 1.5 ml/kg body weight (in olive oil), concentrations of 26 mg/l were determined in the blood after 3 to 6 hours; detectable concentrations were also found in the liver and adipose tissue (WHO 1999). After ingestion of 15 or 3700 mg/kg body weight, rats exhaled 50 % of the administered dose within 3 and 6 hours, respectively (Reynolds *et al.* 1984a, 1984b).

3.1.2 Inhalation

The US Environmental Protection Agency assumes average absorption after inhalation to be 40 % in man (range 30–65 %) (US-EPA 1991). Pharmacokinetic data suggest that carbon tetrachloride accumulates in adipose tissue also in man (Paustenbach *et al.* 1988).

In mice exposed once to high concentrations (256000 mg/m^3) of radioactively labelled carbon tetrachloride for 10 minutes, high levels of radioactivity were found in the adipose tissue, bone marrow, brain, spinal cord, nerves, liver, kidneys, salivary gland and gastrointestinal mucosa (WHO 1999).

In rats exposed to carbon tetrachloride concentrations of 100 or 1000 ml/m^3 (641 or 6410 mg/m^3) for 2 hours, systemic absorption of carbon tetrachloride doses of 17.5 and 179 mg/kg body weight was detected (Sanzgiri *et al.* 1995). Exposure of rats to carbon tetrachloride concentrations of 2, 4 or 8 mg/l for 12 hours produced concentrations in the blood of 12, 20 and 36 mg/l (WHO 1999). In rats exposed to carbon tetrachloride concentrations of 100 ml/m^3 (641 mg/m^3) for 8 to 11.5 hours a day on 1 to 10 days, the highest concentrations were found in adipose tissue, the liver, lungs and adrenal gland (Paustenbach *et al.* 1986a).

After exposure for two weeks (100 ml/m^3, 8 h/day, 5 days/week or 11.5 h/day, 4 days/week), rats exhaled 48 % unchanged and less than 2 % as carbon dioxide. A similar level of radioactivity derived from ^{14}C-carbon tetrachloride was detected in the faeces. The urine contained about 6 % of the total radioactivity. The elimination of carbon tetrachloride via the lungs follows a biphasic course; the half-times for an exposure concentration of 100 ml/m^3 are about 90 and 400 minutes. It was also found that with each further day of exposure, the half-times increased (Paustenbach *et al.* 1986a, 1986b, Veng-Pedersen *et al.* 1987).

In dogs exposed to 15000 ml/m^3 (96150 mg/m^3) for several hours, a steady-state blood concentration of 320 mg/l was reached within 5 hours (WHO 1999).

Three female rhesus monkeys inhaled 46 ml/m^3 (295 mg/m^3) radioactively labelled carbon tetrachloride for 139, 244 or 300 minutes. Within 300 minutes 30 % of the inhaled substance was absorbed; a steady-state concentration in blood had, however, not been reached (WHO 1999).

3.1.3 Dermal application

The level of dermal absorption from the gaseous phase is low. The rate of dermal absorption of liquid carbon tetrachloride by intact mouse skin was found to be 8.3 μg/cm^2 and minute (WHO 1999). The application of 1 ml to the shaved skin of guinea pigs (3.1 cm^2) yielded a blood carbon tetrachloride level of 1 mg/l after one hour. Although exposure was continued, the concentration in blood decreased over the course of the next hour. Local vasoconstriction, rapid transport from the blood to adipose tissue, or metabolism are thought to be responsible (WHO 1999). On the skin of guinea pigs, complete absorption occurred within a few days after application of 0.5 to 2 ml carbon tetrachloride to 3.1 cm^2 skin in a closed glass container (WHO 1999). After 3 volunteers had dipped a thumb into carbon tetrachloride for 30 minutes, concentrations of 3.8 mg/m^3 were detected in the exhaled air. The concentration decreased with a half-time of 2.5 hours (WHO 1999).

3.2 Metabolism

CO_2, chloroform and traces of hexachloroethane have been found as metabolites of carbon tetrachloride (Albano *et al.* 1982, Gregory 1966, Lai *et al.* 1979, Mico *et al.* 1983, Poyer *et al.* 1978). The amount of CO_2 in the exhaled air is about 10 % of the total metabolized carbon tetrachloride. It was deduced from kinetic data that even less carbon tetrachloride is metabolized directly to CO_2, and that CO_2 is also formed indirectly, e.g. from $CHCl_3$ (Paustenbach *et al.* 1988). It was calculated that in the rat exposed to carbon tetrachloride concentrations of 100 ml/m^3, about 60 % of the carbon tetrachloride is metabolized. With repeated exposure to this concentration, accumulation of the substance in adipose tissue was not observed. On the other hand, it was calculated that carbon tetrachloride accumulates in adipose tissue in persons exposed to 5 ml/m^3 (8 hours daily) (Paustenbach *et al.* 1988). In dogs, 12 to 48 hours after single intraperitoneal injections of carbon tetrachloride, the following substances were found in urine: formaldehyde, acetaldehyde, acetone, propyl alcohol, butyl alcohol, pentanal, hexanal and malondialdehyde. The authors state that all these substances provide evidence of lipid peroxidation *in vivo* (WHO 1999, De Zwart *et al.* 1997).

It is known from *in vitro* studies that carbon tetrachloride binds to the cytochrome P450-2E1 of the endoplasmatic reticulum of the liver cell at the oxygen binding site (English and Anders 1985, Wolf *et al.* 1977). Electron transfer results in reductive cleavage of carbon tetrachloride, initially to form the CCl_3 radical (see Figure 1). Hexachloroethane is formed as the condensation product of two CCl_3 radicals. Chloroform, formed by the addition of one hydrogen atom, has also been detected. The CCl_3 radical is held responsible for the observed covalent binding of carbon tetrachloride to proteins and lipids, and reacts also with oxygen to form the trichloromethylperoxy radical (McCay *et al.* 1984). Traces of phosgene, formed either by the oxidation of chloroform or via the trichloromethylperoxy radical, were also detected (Waller and Recknagel 1982) (see Figure 1). CO_2 can also be formed from phosgene. Also carbon monoxide is formed from carbon tetrachloride (Ahr *et al.* 1980). It is probably formed from trichloromethylcarbene, which is formed by the transfer of two electrons on cytochrome P450. Reductive metabolism of carbon tetrachloride to form the CCl_3 radical takes place more rapidly under hypoxic conditions, as are found, for example, in the centrilobular region of the liver (Pohl *et al.* 1984). Under hypoxic conditions also covalent binding of carbon tetrachloride to proteins, lipids and haem is considerably increased (De Groot and Haas 1980, Uehleke and Werner 1975).

Metabolic activation of carbon tetrachloride, which takes place mainly in the endoplasmatic reticulum of the liver cells, has been observed also in the kidneys (Plaa and Larson 1965), the adrenal gland (Colby and Eacho 1985) and the lung (Boyd *et al.* 1979). The metabolism of carbon tetrachloride can be induced, for example, by phenobarbital or DDT (McLean and McLean 1966). It is more pronounced in male rats than in female rats, and prenatally and postnatally is barely detectable. The metabolism of carbon tetrachloride in the liver also depends markedly on the species (very pronounced in rodents and barely detectable in birds).

For short-term inhalation of carbon tetrachloride a maximum metabolic rate of 2.7 mg/kg body weight and hour was calculated (Uemitsu 1986).

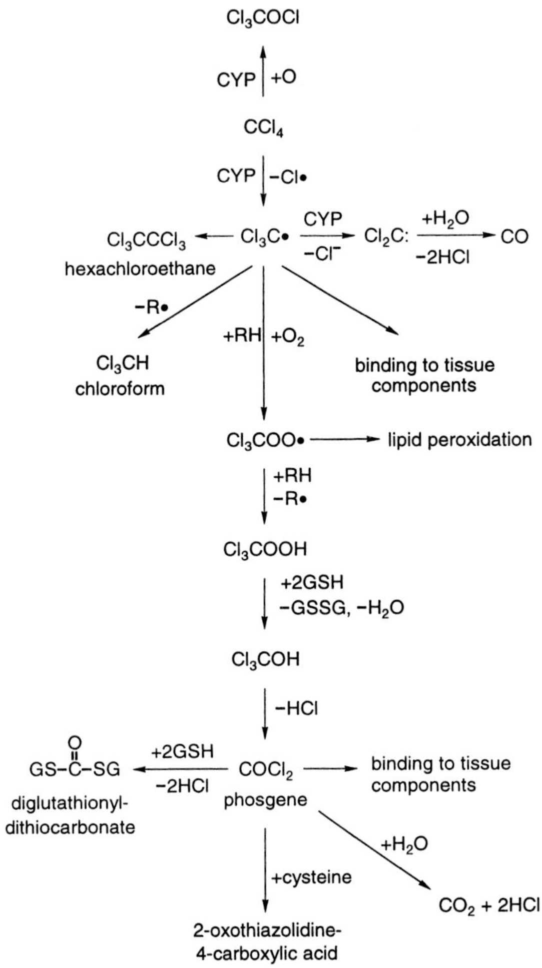

Figure 1. Metabolism of carbon tetrachloride (from WHO 1999)

4 Effects in Man

4.1 Single exposures

Regardless of the route of absorption, the clinical picture of carbon tetrachloride intoxication is dominated in the first 24 hours by gastrointestinal and neurological symptoms. Liver damage occurs at the earliest after 24 hours. In severe cases ascites and hepatic coma develop, and are often accompanied by haemorrhage. Kidney damage is first manifest after 2–3 days, often not until 2–3 weeks after the intoxication (WHO 1999).

The symptoms observed in healthy persons after short-term inhalation of carbon tetrachloride are listed in Table 1 (Lehmann and Flury 1975). The data for the odour threshold differ considerably (Lehmann and Flury 1975, Leonardos *et al.* 1969).

Table 1. Effects of carbon tetrachloride in man (Davies 1934, Lehmann and Flury 1975, Stewart *et al.* 1961)

Concentration [mg/m^3]	Exposure duration [min]	Symptoms
64	180	no effects, no changes in liver enzymes in serum
314	70	decreased level of iron in serum
64–640	not specified	odour threshold
500	150	no effects
1000	30	no effects
1600	20	no effects
2000	30	headache, dizziness, nausea
4000	30	feeling of pressure in the head, dizziness, tiredness, irritation of the mucous membranes
7500	15	fainting
20000	5	no effects
30000	2.5	slight drowsiness
40000	3	feeling of pressure in the head, feeling of inertia, paresthesia of the extremities, tinnitis, slight drowsiness
60000	1.5	paresthesia of the extremities, salivation, weakness, brief spells of fainting
80000	1	paresthesia, unconsciousness
82000	3–7	narcosis

A person who drank 120 to 355 ml carbon tetrachloride in a suicide attempt survived but suffered severe liver function disorders (WHO 1999). However, a woman died after drinking 500 ml carbon tetrachloride. For two days she suffered from vomiting, diarrhoea, oliguria, tachycardia and fever. Histopathological examination revealed centrilobular liver necrosis and fatty degeneration of the liver, and lipid peroxidation was detected by biochemical analysis (Hashimoto *et al.* 1968).

Two workers exposed during cleaning work suffered dizziness, vomiting, stomachache and fever after a latency period of 10 hours. After treatment in hospital, from the third day of illness the increased transaminase and aldolase activities, and the levels of bilirubin in serum and of albumin in urine decreased. The liver biopsy on day 8 of intoxication revealed centrilobular necrosis. The liver biopsy was repeated a month later and showed the liver to be completely healed (Lachnit and Pietschmann 1960).

Abnormal liver function values were reported in 10 of 25 workers accidentally poisoned with carbon tetrachloride after exposure to concentrations of 300 to 500 ml/m^3 (Deng *et al.* 1987).

4.2 Repeated exposures

Workers were exposed to carbon tetrachloride concentrations of 1 ml/m^3 or less (n = 40), 1 to 3 ml/m^3 (n = 54) or 4 ml/m^3 and more (n = 61). The workers were also subdivided into three groups depending on the duration of the activity: < 1 year, 1 to 5 years, > 5 years. The level of alcohol consumption was the same in all exposed workers. The workers were questioned about their age, height, weight, workplace situation, firm, hobbies, state of health and alcohol consumption. Haematological parameters such as haemoglobin level, number of red blood cells and the haematocrit were significantly changed in the middle exposure group, as were the activities of γ-glutamyltransferase and alkaline phosphatase in serum. In the low exposure group only the haematocrit was significantly decreased. No other parameters were investigated. In the high exposure group, however, no changes were observed (Tomenson *et al.* 1995).

17 workers were exposed for about 2 years to carbon tetrachloride concentrations of 290 to 620 mg/m^3. 15 workers complained of nausea, 12 of loss of appetite, 7 of vomiting and flatulence, 10 of feeling unwell and dizzy. Clinical examination did not yield any abnormal findings (Kazantzis and Bomfort 1960).

In 16 workers exposed to carbon tetrachloride concentrations of about 490 to 3900 mg/m^3 (no data for duration of exposure), increased values for serum bilirubin and serum transaminases, and increased amounts of protein in urine were found (Barnes and Jones 1967).

4.3 Local effects on skin and mucous membranes

Three volunteers dipped a thumb into carbon tetrachloride for 30 minutes; this led to slight erythema, which regressed 1 or 2 hours after exposure. The volunteers reported a burning feeling in the thumb within the first 10 minutes of exposure (WHO 1999).

4.4 Allergenic effects

After a woman used an anti-grease lotion for the scalp containing 100 g carbon tetrachloride, dermatitis of the scalp developed. The woman was treated with a test substance containing 10 % carbon tetrachloride, 10 % carbon disulphide and 10 % acetone, corresponding to the amounts in the lotion, and produced a marked reaction after 48 hours (no other details) (Romaguera and Grimalt 1980).

4.5 Reproductive and developmental toxicity

Data for man has been found in only one publication (Dowty *et al.* 1976). The authors analysed 11 blood samples taken from mothers and their offspring after normal births. Carbon tetrachloride was found in both the mothers and offspring; the concentrations determined in the offspring were of the same level as those found in the mothers, or in some cases even higher. Thus it must be assumed that carbon tetrachloride can pass the placental barrier in man. Other data for man are not available (Barlow and Sullivan 1982).

4.6 Carcinogenic effects

There are several epidemiological studies of the carcinogenicity of carbon tetrachloride. In all studies, however, the persons were exposed to a mixture of carbon tetrachloride and other halogenated solvents or other substances. In addition, data for the exposure levels are lacking in these studies. For these reasons the following studies are not included in the present assessment: Blair *et al.* 1990, Bond *et al.* 1986, Chekoway *et al.* 1984, Heinemann *et al.* 1994, Ott *et al.* 1985.

In an epidemiological study carried out in the rubber industry (about 7000 exposed workers studied for 10 years) it was suspected that carbon tetrachloride at the workplace may be responsible for an increased incidence of cancer (Wilkosky *et al.* 1984). It was demonstrated that the incidence of lymphatic leukaemia and lymphosarcomas in workers from these firms correlated with the exposure to carbon tetrachloride. For other solvents there was no correlation. However, other possible carcinogenic substances, e.g. nitrosamines, were not taken into consideration. In a case–control study, the exposure of 342 persons with chronic lymphatic leukaemia to various substances was investigated. A significantly increased risk of developing leukaemia was not observed (Linet *et al.* 1987).

5 Animal Experiments and *in vitro* Studies

5.1 Acute toxicity

5.1.1 Inhalation

After inhalation exposure, LC_{50} values for the mouse of 45000 to 50000 mg/m^3 were determined. The acute inhalation toxicity of carbon tetrachloride is, therefore, low. In the lungs of rats and mice exposed to carbon tetrachloride concentrations between 70000 and 600000 mg/m^3, Clara cell lesions, decreased levels of cytochrome P450, thickening of the septa, alveolar collapse and atypical type II pneumocyte configuration were observed.

In the serum, the levels of aspartate aminotransferase, alanine aminotransferase, sorbitol dehydrogenase and glutamate dehydrogenase were increased; necrosis was observed in the liver (WHO 1999).

5.1.2 Ingestion

The LD_{50} values given for the mouse were 12100 to 14400 mg/kg body weight and for the rat 2821 to 10054 mg/kg body weight; the toxicity of carbon tetrachloride depends markedly on the vehicle used. As after inhalation exposure, mainly affected are the liver and the lungs. Doses of 32 to 2391 mg/kg body weight caused a decrease in the protein, glucose, phospholipid, DNA and RNA concentrations and an increase in the triglyceride, glycogen and cholesterol concentrations in the liver. Decreased cytochrome P450 activity was detected in the lungs and in the liver. In serum, the activity of liver enzymes and the cholesterol concentration were increased in a dose-dependent manner; the glucose level was decreased. In the lung, damage to the Clara cells and their endoplasmatic reticulum was observed; this led to cell necrosis (WHO 1999).

Male and female mice were given single carbon tetrachloride doses of 20 or 100 mg/kg body weight, and male rats doses of 100 or 400 mg/kg body weight by gavage. In the mice of the low dose group, cell proliferation (determined as the number of cells in the S-phase) in the liver was not increased, while in the high dose group it was increased about 80 times. The rats reacted less sensitively to carbon tetrachloride than did the mice. The rate of cell proliferation was increased about 100 times only after doses of 400 mg/kg body weight. Increased unscheduled DNA synthesis was not observed (Mirsalis *et al.* 1985).

5.1.3 Dermal absorption

Dermal LD_{50} values in the range of 8500 to 15000 mg/m^3 were given for rabbits and guinea pigs (WHO 1999). After 16 hours occlusive epicutaneous application of 1 ml carbon tetrachloride to the skin of guinea pigs, centrilobular changes were observed in the liver and first signs of necrosis. Histopathological examination did not reveal any adverse effects on the kidneys (WHO 1999).

5.2 Subacute, subchronic and chronic toxicity

The text which follows describes only those subchronic and chronic toxicity studies which can be used in evaluation of the MAK value. The IPCS report describes all the studies which have been carried out (WHO 1999).

5.2.1 Inhalation

Groups of 10 BDF_1 mice were exposed to carbon tetrachloride concentrations of 10, 30, 90, 270 or 810 ml/m^3 (64, 192, 577, 1731 or 5192 mg/m^3) 6 hours/day, 5 days a week, for 13 weeks. Microscopic examination of the liver revealed cytological changes in the animals of the lowest concentration group only in the males. With increasing concentration the effects increased in severity. The incidence of mitosis was increased; focus formation and increased proliferation were observed. From concentrations of 30 ml/m^3 the body weights of the male animals were reduced. An increase was observed in the liver enzymes in blood from concentrations of 90 ml/m^3. From 270 ml/m^3 haematological changes developed, after concentrations of 810 ml/m^3 the urinary pH decreased (Japan Bioassay Research Centre 1998). From this study, a no observed effect level (NOEL) of 10 ml/m^3 can be derived only for the female animals.

Groups of 50 BDF_1 mice were exposed to carbon tetrachloride concentrations of 5, 25, or 125 ml/m^3 (32.05, 160.25, or 801.25 mg/m^3), 6 hours/day, 5 days a week for 104 weeks. In the lowest concentration group no effects were observed. Survival and body weights were significantly reduced from concentrations of 25 ml/m^3, and haematological and urinary parameters were changed. Microscopic changes developed from 25 ml/m^3 in the liver (thrombus, necrosis, cysts, degeneration), in the kidneys and in the spleen (increasing deposition of haemosiderin and extramedullary haematopoiesis) (Japan Bioassay Research Centre 1998, Nagano *et al.* 1998). From this study, a NOEL of 5 ml/m^3 can be derived for the mouse.

Groups of 50 F344 rats were exposed to carbon tetrachloride for 13 and 104 weeks under the same experimental conditions as were the mice. In the 13-week study body weights were reduced in the highest concentration group; in the low concentration group, granulation was seen in the livers of male and female animals. Haematological changes developed in the female animals from concentrations of 30 ml/m^3 and in the males from 90 ml/m^3. From 90 ml/m^3, the levels of liver enzymes in blood were increased and urinalysis revealed changes. Microscopic examination of the liver revealed fat deposition and cytological changes from concentrations of 30 ml/m^3. There were increases in mitosis, fibrosis, focus formation and proliferation in the female animals from 90 ml/m^3 and in the male animals from 270 ml/m^3. In the high concentration group, vacuolation of the tubules, hyaline degeneration of the glomeruli and accumulation of protein in the kidneys were found (Japan Bioassay Research Centre 1998).

The effects seen in the 2-year study are described below. Survival was reduced in the high concentration group. The animals died of liver tumours or chronic nephropathy. The effects in male and female animals exposed to 5 ml/m^3 were changed nitrate and protein values in the urine. Increasing deposition of haemosiderin in the spleen was observed in the male animals of all concentration groups. In the female animals of all concentration groups, eosinophilia developed in the nasal cavity, and in male animals from 25 ml/m^3. From 25 ml/m^3 body weights were reduced, and the levels of liver enzymes in blood and other haematological parameters were changed. Microscopic changes in the liver (see 13-week study) were observed at the two highest concentrations. The female animals developed chronic nephropathy from concentrations of 25 ml/m^3 and the male animals from 125 ml/m^3 (Japan Bioassay Research Centre 1998, Nagano *et al.* 1998). For the rat no NOEL can be deduced from either the 13-week or the 2-year study.

After inhalation exposure to carbon tetrachloride concentrations of 5 to 400 ml/m^3 a NOEL of 5 ml/m^3 was found for rats, rabbits and male guinea pigs. In the female guinea pigs, increased relative liver weights were observed at this concentration, but no histopathological changes were found (Adams *et al.* 1952). The study design and effects in the animals are summarized in Table 2. Changes in the liver enzymes in blood were not detected; the nose was not investigated.

5.2.2 Ingestion

Groups of 12 male and 12 female CD-1 mice were given carbon tetrachloride doses of 1.2, 12 or 120 mg/kg body weight in corn oil or in 1 % Tween 60 solution 5 times a week for 90 days. This study was carried out to investigate the effects of carbon tetrachloride administered with different vehicles. From 12 mg/kg body weight (vehicle corn oil) and 120 mg/kg body weight (vehicle Tween) the liver enzymes alanine aminotransferase, aspartate aminotransferase and lactate dehydrogenase increased significantly in serum. Hepatomegaly, large fat deposits and necrosis with acute inflammation (only in the male animals) were observed after administration of the substance in corn oil. From this study, a no observed adverse effect level (NOAEL) for hepatotoxic effects of 1.2 mg/kg body weight can be derived for male and female mice (Condie *et al.* 1986).

Groups of 15 to 16 male rats were given carbon tetrachloride doses of 1, 10, or 33 mg/kg body weight on 5 days a week for 12 weeks. In the middle dose group the sorbitol dehydrogenase activity in serum was significantly increased and slight centrilobular vacuolation was observed in the liver. In the high dose group in addition the levels of ornithine carbamyl transferase and alanine aminotransferase were increased. Liver cirrhosis, bile duct proliferation, fibrosis, parenchymal regeneration, hyperplastic nodules and single cell necrosis were found. There were no effects on the kidneys. From this study, a NOEL of 1 mg/kg body weight can be derived for the male rat (Bruckner *et al.* 1986).

Dose–response relationship

After administration of carbon tetrachloride doses of 2 or 10 mg/kg body weight in corn oil for 1, 7 and 14 days to CD-1 mice, there was no increase in the incidence of cell proliferation, in the DNA repair in hepatocytes or in the levels of aspartate aminotransferase and alanine aminotransferase in serum. From doses of 20 mg/kg body weight, on day 7 the levels of aspartate aminotransferase and alanine aminotransferase in serum increased, and from 100 mg/kg body weight also the incidence of cell proliferation was increased. After 14 days the aspartate aminotransferase and alanine aminotransferase levels had returned to the control values, the increase in cell proliferation was not, however, reversible. The study shows that the hepatocyte cytotoxicity increases in parallel with the increase in cell proliferation (Doolittle *et al.* 1987).

Table 2. Effects after long-term inhalation exposure (7 hours/day, 5 days/week, 29 weeks) (Adams *et al.* 1952)

Conc. [mg/m^3]	Rat groups of 15–25 ♂ and ♀	Rabbit groups of 2 ♂ and ♀	Guinea pig groups of 5–9 ♂ and ♀
5	NOEL	NOEL	♂: NOEL; ♀: liver weights* increased, no histopathological changes
10	liver weights increased, liver: slight to moderate fatty degeneration, lipid levels increased, cholesterol increased	NOEL	
25	liver: effects as at 10 ml/m^3, no findings in other organs	liver weights increased, liver: slight to moderate fatty degeneration, beginning cirrhosis	body weights decreased, liver weights increased, liver: slight to moderate fatty degeneration with cirrhosis, lipid levels increased, cholesterol increased, plasma prothrombin time increased
50	liver weights increased, moderate fatty degeneration, slight to moderate cirrhosis, ♀: kidney weights increased, moderate swelling of the tubular epithelium in 4 rats	body weights decreased, liver and kidney weights increased, plasma prothrombin time increased, liver: moderate degeneration and cirrhosis	body weights decreased, liver enlarged, other effects as at 25 ml/m^3
100	plasma prothrombin time increased, phospholipid decreased, neutral fat increased, cholesterol increased	effects as at 50 ml/m^3	body weights decreased, increased mortality in the first 3 weeks, liver: moderate degeneration and cirrhosis, liver weights increased
200	mortality increased (♂ 6/15, ♀ 9/15), body weights and testis weights decreased, liver: fatty degeneration, cirrhosis, necrosis kidney: moderate degeneration of the tubular epithelium, in some cases complete testis atrophy	not specified	mortality increased (♂ 3/8, ♀ 1/8), liver: as at 100 ml/m^3, kidney weights increased, slight degeneration of the tubules
400	mortality increased (♂ 2/15, ♀ 7/15), liver, kidneys and testes as at 200 ml/m^3, congested spleen	not specified	mortality increased (most animals died in the first 4 weeks), liver and kidneys: effects as at 100 ml/m^3, but more pronounced, pathological changes in the adrenal glands, spleen and testes

* all organ weight changes are changes in the relative organ weight

5.3 Local effects on skin and mucous membranes

5.3.1 Skin

Epicutaneous application of 1 ml carbon tetrachloride for periods between 15 minutes and 16 hours caused degenerative changes in the epidermis (WHO 1999). Application of 0.5 ml carbon tetrachloride to the shaved skin of rabbits and guinea pigs caused moderate irritation (WHO 1999). In a Draize skin irritation test, 0.5 ml carbon tetrachloride was applied occlusively to the shaved and scarified skin of rabbits. After 3 days irritative effects were observed (WHO 1999). 10 µl carbon tetrachloride was applied to the skin non-occlusively 3 times a day for 3 days. A skin reaction was observed after 2 days and reddening after 4 days (WHO 1999). Aliquots of 0.1 ml (159 mg) carbon tetrachloride were rubbed into the skin of rabbits and guinea pigs daily for 10 days. Oedema and erythema were observed (WHO 1999).

5.3.2 Eyes

0.1 ml carbon tetrachloride caused mild irritation in the rabbit eye. The eyes were examined 24, 48 and 72 hours after exposure and after a follow-up period of 14 days (WHO 1999).

There are no data available for allergenic effects of carbon tetrachloride.

5.4 Reproductive and developmental toxicity

5.4.1 Fertility

After oral administration of 0.1 to 1.5 mg/kg body weight for 5 days to CD-1 mice, no sperm anomalies were detected (WHO 1999).

Groups of 6 male and 6 female rats were given single intraperitoneal doses of carbon tetrachloride of 3 ml/kg body weight (2378 mg/kg body weight) in coconut oil. After 15 days, the pituitary gland weights in the male animals were significantly increased and the testis weights reduced. Histological examination revealed testis atrophy and changes in spermatogenesis. After 10 days the female cycle stopped, the ovary and uterus weights were reduced and the adrenal gland and pituitary gland weights increased (Chatterjee 1966, 1968).

In an early 3-generation study, groups of 24 albino rats were exposed to carbon tetrachloride concentrations of 50, 100, 200 and 400 ml/m^3 on 5 days a week. The authors reported that fertility in the animals exposed to 200 and 400 ml/m^3 was impaired. Embryotoxic, foetotoxic and teratogenic effects were not, however, mentioned (Smyth *et al.* 1936, Smyth and Smyth 1935).

5.4.2 Developmental toxicity

In Sprague-Dawley rats which had inhaled carbon tetrachloride concentrations of 334 or 1004 ml/m^3 for 7 hours a day on days 6 to 15 of gestation, food consumption and body weights were reduced. In the two dose groups the alanine aminotransferase and *S*-glutamyl transpeptidase activities increased in a manner not dependent on the dose—an indication of liver toxicity. The relative liver weights were significantly increased. Carbon tetrachloride had no influence on the number of implantations, resorptions or pups per litter. The body weights of the foetuses decreased and the body lengths were shorter. Subcutaneous oedema was observed at concentrations of 334 ml/m^3 significantly more often than in the control animals. In the high concentration group, anomalies of the breastbone were significantly more frequent (Schwetz *et al.* 1974).

F344 rats were given gavage doses of carbon tetrachloride of 25, 50 or 75 mg/kg body weight in corn oil or 10 % Emulphor from days 6 to 15 of gestation. Independent of the vehicle, dose-dependent piloerection was seen in the dams from doses of 50 mg/kg body weight. Reduced body weights and kyphosis were observed in the high dose group after administration of the substance in corn oil and from doses of 50 mg/kg body weight in aqueous solution. In the groups given 50 and 75 mg/kg doses in corn oil, 42 and 67 of the litters were resorbed, with the aqueous solution 14 and 8. After administration of 25 mg/kg body weight, the prenatal and postnatal body weights and survival were not affected. No morphological changes were found in the foetuses. The authors give a NOAEL for both the dams and the foetuses of 25 mg/kg body weight after administration in either corn oil or aqueous solution (Narotsky *et al.* 1997).

On 5 consecutive days of gestation, beginning on day 1, 6 or 11, groups of 31 B6D2F$_1$ mice were given carbon tetrachloride doses of 82.6 or 826.3 mg/kg body weight by gavage. No effects were observed in either the dams or the foetuses (Hamlin *et al.* 1993). However, because of the small number of animals investigated and the inadequate documentation of the findings, the study is of only limited use for the evaluation of the developmental toxicity of the substance.

5.5 Genotoxicity

The studies of genotoxicity *in vitro* and *in vivo* are described in detail in tables in the IPCS monograph (WHO 1999). The results *in vitro* and *in vivo* are summarised and evaluated below. More recent studies have also been included.

5.5.1 *In vitro*

Carbon tetrachloride was not found to be mutagenic in the tests for gene mutations or for an SOS response in *Salmonella typhimurium* or *Escherichia coli* either in the absence of a metabolic activation system or in the presence of S9 from the rat, mouse or rabbit. In the yeast *Saccharomyces cerevisiae*, in the absence of an activation system, carbon tetrachloride induced intrachromosomal and mitotic recombination. In V79 cells, a high carbon tetrachloride concentration (2500 µg/ml, about 16 mM) produced aneuploidy, which can probably be attributed to the lipophilia of the substance. Studies of the induction of aneuploidy and micronuclei by carbon tetrachloride in various genetically engineered human lymphoblastoid cell lines which express different cytochrome P450 forms, revealed that lower concentrations (2 mM) are genotoxic only in cells which express CYP2E1. Whether clastogenic effects were involved in the formation of micronuclei was not investigated.

In primary hepatocyte cultures of the rat and mouse, and in human epithelial liver cell lines, carbon tetrachloride led to the induction of DNA strand breaks. In most of the studies strand breaks occurred only at cytotoxic concentrations.

The available *in vitro* studies therefore do not provide any clear evidence of the formation of DNA-reactive metabolites of carbon tetrachloride.

5.5.2 *In vivo*

In *Drosophila melanogaster*, carbon tetrachloride did not lead to sex-specific recessive lethal mutations (SLRL mutations) after feeding or injection.

In rats given oral doses of carbon tetrachloride, no induction of DNA strand breaks or DNA repair synthesis was detected. In rats and mice, no sister chromatid exchange (SCE) or chromosomal aberrations were found. Micronucleus tests with mice also yielded negative results.

Barrows and Shank (1981) described the formation of O^6-methyl guanine and 7-methyl guanine in the liver DNA of rats after oral administration of carbon tetrachloride doses of 1000 mg/kg body weight. The formation of the methylated bases was attributed to the hepatotoxicity caused by this dose, but the mechanism could not be explained or the results confirmed in later studies.

Two research groups showed that the lipid peroxidation occurring *in vivo* after administration of high doses of carbon tetrachloride is accompanied by the increased formation of certain DNA adducts also found in untreated animals and attributed to the reaction of lipid peroxidation products with DNA bases. Wang and Liehr (1995) found an increase in the concentration of lipid peroxides and polar endogenous DNA adducts in the kidneys and liver of hamsters after oral administration of carbon tetrachloride in doses of 0.1 ml/kg body weight (about 159 mg). In the kidneys the incidence of adducts was increased by about 120 %, in the liver by about 50 %. The higher dose of 1.0 ml/kg body weight led in the kidneys to a slight increase in both parameters, in the liver the levels were found to be below those in the untreated animals. This was ascribed to the inactivation of carbon tetrachloride-activating cytochrome P450 forms. Independent of the treatment conditions, there was a linear correlation between the concentration of lipid peroxides and the level of adducts in the organs (Wang and Liehr 1995). The polar DNA adducts were found to be two reaction products of malondialdehyde with deoxyadenosine and one DNA base modification of unknown structure induced by free radicals.

After intraperitoneal administration of carbon tetrachloride doses of 3200 mg/kg body weight to rats, the concentrations of various 1,N^2-propanoguanine adducts in the liver DNA, which may also be detected in untreated animals, were investigated. These exocyclic adducts are derived from the reaction with DNA of α,β-unsaturated aldehydes (enals), formed during the lipid peroxidation of polyunsaturated fatty acids. Formation of these adducts is markedly increased when glutathione levels are reduced. 24 and 72 hours after administration of the substance, significantly increased concentrations of an acrolein–deoxyguanosine adduct were detected, which were about 3.7 and 4.4 times the control value. The concentrations of two crotonaldehyde–deoxyguanosine adducts were increased, but the increase was not statistically significant. 24 hours after administration of the substance, marked zonal necrosis was found, and after 72 hours developed into fibrosis. After 24 hours the binding of 4-hydroxynonenal to liver proteins was increased (Chung *et al.* 1999).

After intragastric administration of carbon tetrachloride doses of 1 ml/kg body weight to rats, the amounts of 4-hydroxynonenal–protein and malondialdehyde–protein adducts (products of lipid peroxidation) in the liver were found by immunochemical analysis to increase with time (Hartley *et al.* 1999).

Twice-weekly administration of carbon tetrachloride doses of 0.75 ml/kg body weight (about 1190 mg) led in mice to a decrease in the frequency of endogenous DNA modifications (so-called I-compounds of type I) in the liver DNA, which persisted for at least 22 weeks (Nath *et al.* 1990, Randerath *et al.* 1999). The structures and biological importance of these modifications are unknown. They are thought to have a physiological function. Their frequency is decreased by numerous hepatocarcinogens and is reduced in tumours (Randerath *et al.* 1999).

5.6 Carcinogenicity

5.6.1 Initiation promotion studies

Five male rats were given intraperitoneal injections of 100 mg diethylnitrosamine. Two weeks later, intragastric doses of carbon tetrachloride of 2 ml/kg body weight were administered 4 or 8 times at intervals of 20 days. Two to three days after the last treatment the animals were killed. Histopathological and histochemical examination of the liver was carried out. Mitogen-induced and compensatory cell proliferation were recorded separately. Body weights were reduced, relative liver weights increased. Cell proliferation (incorporation of [^{3}H-thymidine]) increased to 9 to 16 times that in the control animals without proliferation stimulus. The number of ATPase-negative and GGT-positive foci/cm^2 was significantly increased; this is regarded as evidence of compensatory cell proliferation. After 4 doses of the substance, fat deposits and extensive centrilobular necrosis were observed. After 8 doses, also small nodules were found in the liver (Columbano *et al.* 1990).

5.6.2 Long-term studies

After both oral and inhalation exposure to carbon tetrachloride, the incidence of hepatocellular carcinomas and phaeochromocytomas in mice was increased in a dose-dependent manner. In mice which had inhaled carbon tetrachloride concentrations of 5 ml/m^3, the lowest concentration tested, no tumours were found. After 2 years a complete histopathological examination of all organs was carried out. Therefore from this study a NOEL of 5 ml/m^3 for the formation of hepatocarcinomas in the mouse can be derived (Japan Bioassay Research Centre 1998, Nagano *et al.* 1998). In female rats, oral administration of the substance led to the formation (not dose-dependent) of neoplastic nodules in the liver and hepatocellular carcinomas, while in the male animals only neoplastic nodules were found. After inhalation exposure, hepatocellular adenomas and carcinomas were observed only at the highest concentration tested of 125 ml/m^3. Therefore from this study a NOEL of 25 ml/m^3 for the formation of hepatocarcinomas in the rat can be derived (Japan Bioassay Research Centre 1998, Nagano *et al.* 1998). Data for the incidence of tumours and from other studies are listed in Table 3.

There are several mechanisms which could explain the formation of the tumours. Under discussion for a long time was whether the trichloromethyl radical formed during metabolism can bind covalently to DNA. The studies carried out did not yield conclusive results. Also the mainly negative results in genotoxicity tests, above all *in vivo*, speak against this mechanism. In view of the short lifetime of the trichloromethyl radical, which is mainly found covalently bound to proteins and lipids, binding to the DNA in the cell nucleus is also improbable.

Products of lipid peroxidation, such as malondialdehyde or 4-hydroxynonenal can in principle bind to DNA and cause mutations. They occur, however, also without specific exposure in healthy persons in various organs, including the liver (Marnett 1999).

Table 3. Studies of the carcinogenicity of carbon tetrachloride

Author:	Reuber and Glover 1970				
Substance:	carbon tetrachloride (purity not specified)				
Species:	rat (Japanese, Osborne-Mendel, Wistar, Black Rat, Sprague-Dawley), 12–17 ♂, 12 controls/strain				
Administration route:	subcutaneous, vehicle corn oil				
Concentration:	1000 mg/kg body weight				
Duration:	70–105 weeks, 2/week				
Toxicity:	moderate to severe liver cirrhosis, body weights decreased, changed liver weights, hepatic venous thrombosis, cholangiofibrosis				
	Japanese	Osborne-Mendel	Wistar	Black Rat	Sprague-Dawley
Survivors week 78	12/15	8/13	0	0	0
Average survival	47 weeks	44 weeks	33 weeks	13 weeks	11 weeks

Table 3. continued

Findings:[1]		0 (vehicle controls)	1000 mg/kg body weight
Japanese	liver carcinomas	0/12	12/15
	spleen haemangiomas	not specified	2/15
	thyroid carcinomas	not specified	3/15
	kidney cysts	not specified	3/15
Osborne-Mendel	liver carcinomas	0/13	8/13
	spleen haemangiomas	not specified	1/13
	thyroid carcinomas	not specified	3/13
	kidney cysts	not specified	2/13
	lung metastases	not specified	1/13
Wistar	liver carcinomas	0/12	4/12

[1] no effects in Black rat or Sprague-Dawley rat

Author:	Weisburger 1977
Substance:	carbon tetrachloride (purity tested, no other details)
Species:	rat (Osborne-Mendel), 99 ♂, 99 ♀, controls: 20 ♂, 18 ♀, mouse (B6C3F$_1$), 97 ♂, 87 ♀, controls: 18 ♂, 18 ♀
Administration route:	gavage, vehicle corn oil
Dose:	rat: 50–160 mg/kg body weight (exact details not given) mouse: 1250–2500 mg/kg body weight (exact details not given)
Duration:	78 weeks, 5 days/week
Toxicity:	not specified, carbon tetrachloride served as positive control

		control	low dose	high dose
Osborne-Mendel rats				
thyroid adenomas and carcinomas	♂	1/20	1/49	1/50
	♀	2/20	2/50	4/49
haemangiosarcomas	♂	0/20	4/49	4/50
	♀	0/20	3/50	1/50
neoplastic nodules in the liver	♂	0/20	9/49	3/50
	♀	1/20	11/50	9/49
hepatocellular carcinomas	♂	0/20	2/49	2/50
	♀	1/20	4/50	2/49
B6C3F$_1$ mice				
hepatocellular carcinomas	♂	3/18	49/49	47/47
	♀	1/18	40/40	43/43
adrenal adenomas and chromocytomas	♂	0/18	28/49	28/47
	♀	0/18	15/40	10/43

Table 3. continued

Author:	Nagano *et al.* 1998, Japan Bioassay Research Centre 1998
Substance:	carbon tetrachloride (purity > 99.9 %)
Species:	F344 rat 50 ♂, 50 ♀; BDF_1 mouse 50 ♂, 50 ♀
Administration route:	inhalation
Concentration:	5, 25, 125 ml/m^3
Duration:	6 hours/day, 5 days/week, 104 week
Toxicity:	see Section 5.2.1

		control	5 ml/m^3	25 ml/m^3	125 ml/m^3
F344 rats					
survivors	♂	22/50	29/50	19/50	3/50
	♀	39/50	43/50	39/50	1/50
hepatocellular adenomas	♂	0/50	1/50	1/50	21/50
	♀	0/50	0/50	0/50	40/50
hepatocellular carcinomas	♂	1/50	0/50	0/50	32/50
	♀	0/50	0/50	3/50	15/50
BDF_1 mice					
survivors	♂	35/50	36/50	25/50	1/50
	♀	26/50	24/49	10/50	1/49
hepatocellular adenomas	♂	9/50	9/50	27/50	16/50
	♀	2/50	8/50	17/50	5/50
hepatocellular carcinomas	♂	17/50	12/50	42/50	48/50
	♀	2/50	1/49	33/50	48/49
phaeochromocytomas	♂	0/50	0/50	16/50	31/50
	♀	0/50	0/49	0/50	22/49

The products of lipid peroxidation, also of that caused by carbon tetrachloride, induce the expression of proto-oncogenes such as c-fos and c-jun, which was also detected for carbon tetrachloride (Schiaffonati and Tiberio 1997). The DNA damage (deletion via intrachromosomal recombination) observed *in vitro* in yeast cells (Brennan and Schiestl 1998) is similar to the damage normally observed after oxidative stress (Wang *et al.* 1998). Increased levels of 8-hydroxy-2′-desoxyguanosine, a parameter for DNA changes resulting from oxidative stress, were detected in the liver tissue of rats after administration of carbon tetrachloride (Takahashi *et al.* 1998).

Investigations by Camandola *et al.* (1999) indicate that the lipid peroxidation resulting from exposure to carbon tetrachloride is responsible for the "upregulation" of activator protein-1 (AP-1). It could be completely inhibited by the antioxidant, vitamin E. In addition, it was found that Kupffer's cells play an important role in this gene activation. The authors suspect that products of lipid peroxidation activate the Kupffer's cells, which in turn stimulate gene activation and proliferation in the liver cells.

In all the mechanisms named above, with the exception of the direct alkylation of DNA by trichloromethyl radicals, which is, however, an unlikely mechanism, the effects correlate with carbon tetrachloride-induced lipid peroxidation. The dose–response relationship for lipid peroxidation, however, has a "threshold value", as it takes place in the organism also without specific activation and is inhibited by various antioxidative protective mechanisms. It is also involved to a high degree in the liver toxicity induced by carbon tetrachloride. It can therefore be assumed that in the absence of liver toxicity no liver tumours are formed. The course of the dose–response relationship for the induction of liver tumours in rats and mice confirms this hypothesis.

6 Manifesto (MAK value/classification)

In mice exposed to carbon tetrachloride either by ingestion or inhalation, hepatocellular carcinomas and phaeochromocytomas developed in a dose-dependent manner. After inhalation exposure, the incidence of hepatocellular carcinomas in rats was increased. The studies therefore demonstrate that carbon tetrachloride has carcinogenic effects in animals. The findings listed below suggest that the mechanisms of tumour development are not mainly genotoxic. The genotoxicity studies available for carbon tetrachloride do not indicate that the substance forms DNA-reactive metabolites *in vivo*. The DNA strand breaks induced in hepatocytes occurred only *in vitro* and at cytotoxic concentrations. In general, non-linear concentration–response relationships are assumed also for the aneugenic effects observed in various systems. The increased formation of also endogenously occurring DNA adducts with lipid peroxidation products observed *in vivo* after administration of high carbon tetrachloride doses was closely correlated with increased lipid peroxidation. Cytotoxic effects of carbon tetrachloride are well documented in animal studies and in man. These cytotoxic effects can be put down to the metabolic formation of trichloromethyl and trichloromethylperoxy radicals and the damage to proteins and lipids that this causes. Thus the main effect of carbon tetrachloride is to induce compensatory cell regeneration. Therefore carbon tetrachloride is classified in Carcinogen category 4 of Section III of the *List of MAK and BAT Values*.

A MAK value may be established on the basis of observations in man and the results of animal studies. The NOEL for toxicity in the liver and kidneys of the mouse which has inhaled carbon tetrachloride for 2 years is 5 ml/m^3; cytotoxic effects are seen from concentrations of 25 ml/m^3. For the formation of hepatocellular carcinomas a NOEL of 5 ml/m^3 can also be derived; tumours were found only after concentrations of 25 ml/m^3 and more. For the rat, after inhalation exposure no NOEL could be evaluated, as at concentrations of 5 ml/m^3 haemosiderin deposits were found in the spleen, the nitrate and protein levels in urine were changed and eosinophilia was observed in the nasal cavity. After oral exposure, a NOEL of 1 mg/kg body weight was evaluated for the male rat in a 12-week study.

As a result of marginal effects of exposure to carbon tetrachloride concentrations of about 1 ml/m^3 in man (reduced haematocrit), the MAK value has been lowered to 0.5 ml/m^3. This value is supported by the NOEL of 5 ml/m^3 in the mouse and is not in conflict with the NOEL of 1 mg/kg body weight determined in studies with oral administration to the rat. Exposure to carbon tetrachloride concentrations of 0.5 ml/m^3 for 8 hours a day would correspond to a daily uptake of 0.5 mg/kg body weight for man.

On the basis of the data available at the time, in 1989 carbon tetrachloride was classified in Pregnancy risk group D, as in rats none of the high concentrations tested by inhalation exposure were without embryotoxic effects, and there were no valid studies available with a second species. In the meantime two new studies have become available in which oral administration of 25 mg/kg body weight to the rat and of 82.6 mg/kg body weight to the mouse produced neither embryotoxic nor teratogenic effects. As there is still no study available of inhalation exposure in a second species and the study of oral exposure in the mouse is insufficiently documented, the requirements for classification in Pregnancy risk group C have not been fulfilled. Carbon tetrachloride is therefore still classified in Pregnancy risk group D.

Carbon tetrachloride is still designated with an "H", as absorption through the skin has been demonstrated in both animals and man. After dermal application, degenerative changes were observed in the livers of guinea pigs. Dermal absorption therefore contributes to the systemic toxicity of the substance.

The available data indicate that designation with an "S" (for substances which cause sensitization) is not necessary.

7 References

Adams EM, Spencer HC, Rowe VK, McCollister DD, Irish DD (1952) Vapor toxicity of carbon tetrachloride determined by experiments on laboratory animals. *Arch Ind Hyg Occup Health 6*: 50–66

Ahr HJ, King LJ, Nastainezyk W, Ullrich V (1980) The metabolism of chloroform and carbon monoxide formation from carbon tetrachloride by microsomal cytochrome P-450. *Biochem Pharmacol 29*: 2855–2861

Albano E, Lott KAK, Slater TF, Stier A, Symons MCR, Tomasi A (1982) Spin-trapping studies on the free-radical products formed by metabolic activation of carbon tetrachloride in rat liver microsomal fractions isolated hepatocytes and *in vivo* in the rat. *Biochem J 204*: 593–603

Barlow SM, Sullivan FM (Eds) (1982) *Reproductive Hazards of Industrial Chemicals*, p 200, Academic Press, London,

Barnes R, Jones RC (1967) Carbon tetrachloride poisoning. *Am Ind Hyg Assoc J 28*: 557–560

Barrows LR, Shank RC (1981) Aberrant methylation of liver DNA in rats during hepatotoxicity. *Toxicol Appl Pharmacol 60*: 334–345

Blair A, Stuart PA, Tolbert PE, Grauman D, Moran FX, Vaught J, Reyner J (1990) Cancer and other causes of death among a cohort of dry cleaners. *Br J Ind Med 47*: 162–168

Bond GG, Flores GH, Shellenberger RJ, Cartmille JB, Fishbeck WA, Cook RR (1986) Nested-case control study of lung cancer among chemical workers. *Am J Epidemiol 124*: 53–66

Boyd C, Statham N, Longo NS (1979) The pulmonary Clara cell as a target for toxic chemicals requiring metabolic activation; studies with carbon tetrachloride. *J Pharmacol Exp Ther 212*: 109–114

Brennan RJ, Schiestl RH (1998) Chloroform and carbon tetrachloride induce intrachromosomal recombination and oxidative free radicals in *Saccharomyces cerevisiae. Mutat Res 397*: 271–278

Bruckner JV, MacKenzie WF, Muralidhara S, Luthra R, Kyle GM, Acosta D (1986) Oral toxicity of carbon tetrachloride: Acute, subacute, and subchronic studies in rats. *Fundam Appl Toxicol 6*: 16–34

BUA (GDCh-Advisory Committee on Existing Chemicals of Environmental Relevance) (1990) Tetrachloromethane, BUA Report 45, Hirzel Wissenschaftliche Verlagsgesellschaft Stuttgart

Camandola S, Aragno M, Cutrin JC, Tamagno, E, Danni O, Chiarpotto E, Parola M, Leonarduzzi G, Biasi F, Poli G (1999) Liver AP-1 activation due to carbon tetrachloride is potentiated by 1,2-dibromoethane but is inhibited by α-tocopherol or gadolinium chloride. *Free Rad Biol Med 26*: 1108–1116

Castro GD, Díaz-Gomez MI, Castro JA (1989) Species differences in the interaction between CCl_4 reactive metabolites and liver DNA or nuclear protein fractions. *Carcinogenesis 10*: 289–294

Castro GD, Díaz-Gomez MI, Castro JA (1990) Biotransformation of carbon tetrachloride and lipid peroxidation promotion by liver nuclear preparations from different animal species. *Cancer Lett 53*: 9–15

Chatterjee A (1966) Testicular degeneration in rats by carbon tetrachloride intoxication. *Experientia 22*: 395–396

Chatterjee A (1968) Effect of carbon tetrachloride on gonadal physiology in female rats. *Acta anat 71*: 82–86

Checkoway H, Wilcosky T, Wolf P, Tyroler H (1984) An evaluation of the associations of leukemia and rubber industry solvent exposures. *Am J Ind Med 5*: 239–249

Chung F-L, Nath RG, Nagao M, Nishikawa, Zhou G-D, Randerath K (1999) Endogenous formation and significance of 1, N^2-propanodeoxyguanosine adducts. *Mutat Res 424*: 71–81

Colby HD, Eacho PI (1985) Chemical-induced adrenal injury: Role of metabolic activation. in: Thomas JA (Ed.) *Endocrine Toxicology*, Raven Press, New York, 35–66

Columbano A, Ledda-Columbano GM, Ennas MG, Curto M, Chelo A, Paul P (1990) Cell proliferation and promotion of rat liver carcinogenesis: different effect of hepatic regeneration and mitogen induced hyperplasia on the development of enzyme-altered foci. *Carcinogenesis 11*: 771–776

Condie LW, Laurie RD, Mills T, Robinson M, Bercz JP (1986) Effect of gavage vehicle on hepatotoxicity of carbon tetrachloride in CD-1 mice: Corn oil versus Tween-60 aqueous emulsion. *Fundam Appl Toxicol 7*: 199–206

Davies PA (1934) Carbon tetrachloride as an industrial hazard. *J Am Med Assoc 103*: 962–966

De Curtis V, Gemma S, Sbraccia M, Testai E, Vittozzi L (1994) The contribution of electrophilic and radicalic intermediates to phospholipid adducts formed by halomethanes *in vivo. J Biochem Toxicol 9*: 305–310

De Groot H, Haas W (1980) O_2-independent damage of cytochrome P-450 by CCl_4-metabolites in hepatic microsomes. *FEBS Lett 115*: 253–256

Deng JF, Wang JD, Shih TS, Lan FL (1987) Outbreak of carbon tetrachloride poisoning in a color printing factory related to the use of isopropyl alcohol and an air conditioning system in Taiwan. *Am J Ind Med 12*: 11–19

De Zwart LL, Venhorst J, Groot M, Commandeur JN, Hermanns RC, Meerman JH, Fan Baar BL, Vermeulen NP (1997) Simultaneous determination of eight lipid peroxidation degradation products in urine of rats treated with carbon tetrachloride using gas chromatography with electroncapture detection. *J Chromatogr Biomed Appl B694*: 277–287

Doolittle DJ, Muller G, Scribner HE (1987) Relationship between hepatotoxicity and induction of replicative DNA sysnthesis following single or multiple doses of carbon tetrachloride. *J Toxicol Environ Health 22*: 63–78

Dowty BJ, Laseter JL, Storer JL (1976) The transplacental migration and accumulation in blood of volatile organic constituents. *Pediatr Res 10*: 696–701

English JC, Anders MW (1985) Evidence for the metabolism of *N*-nitrosodimethylamine and carbon tetrachloride by a common isozyme of cytochrome P-450. *Drug Metab Dispos 13*: 449–452

Fanelli SL, Castro JA (1995) Covalent binding of carbon tetrachloride reactive metabolites to liver microsomal and nuclear lipid and phospholipid classes from Sprague-Dawley and Osborne Mendel male rats. *Teratogen Carcinogen Mutagen 15*: 155–166

Goeptar AR, Scheereus H, Vermeuler NPE (1995) Oxygen and xenobiotic reductase activities of cytochrome P_{450}. *Crit Rev Toxicol 25*: 25–65

Gregory NL (1966) Carbon tetrachloride toxicity and electron capture. *Nature (London) 212*: 1460–1461

Hamlin GP, Kholkute SD, Dukelow WR (1993) Toxicology of maternally ingested carbon tetrachloride (CCl_4) on embryonal and fetal development and *in vitro* fertilization in mice. *Zoological Science 10*: 111–116

Hartley DP, Kolaja KL, Reichard J, Petersen D (1999) 4-Hydroxynonenal and malondialdehyde hepatic protein adducts in rats treated with carbon tetrachloride: immunochemical detection and lobular localization. *Toxicol Appl Pharmacol 161*: 23–33

Hashimoto S, Glende EA, Recknagel RO (1968) Hepatic lipid peroxidation in acute fatal human carbon tetrachloride poisoning. *N Engl J Med 279*: 1082–1085

Heinemann EF, Cocco P, Gomez MR, Dosemeci M, Stewart PA, Hayes RB, Zahm SH, Thomas TL, Blair A (1994) Occupational exposure to chlorinated aliphatic hydrocabons and risk of astrocytic brain cancer. *Am J Ind Med 26*: 155–169

Japan Bioassay Research Centre (1998) *Thirteen-week and two-year inhalation studies on F344 rats and BDF_1 mice* (Studies Nos 0020, 0021, 0043 and 0044). Kanagawa, Japan Industrial Safety and Health Association, Japan Bioassay Research Centre (unpublished studies)

Kappus H (1985) Lipid peroxidation: Mechanisms, analysis, enzymology and biological relevance. in: Sies H (Ed.) *Oxidative Stress*, Academic Press, London, 273–309

Kazantzis G, Bomfort RR (1960) Dyspepsia due to inhalation of carbon tetrachloride vapour. *Lancet 1*: 360–362

Lachnit V, Pietschmann H (1960) Activity of serum glutamic-oxaloacetic-transaminase and aldolase in workers exposed to halogenated hydrocarbons. *Ind Med Surg 29*: 523–525

Lai EK, McCay PB, Noguchi T, Fong KL (1979) *In vivo* spin-trapping of trichloromethyl radicals formed from CCl_4. *Biochem Pharmacol 28*: 2231–2235

Lehmann KB, Flury F (1975) *Toxikologie und Hygiene der technischen Lösungsmittel* (Toxicology and occupational hygiene of industrial solvents) (German). Springer Verlag, Stuttgart, 112–118

Leonardos G, Kendall D, Barnard N (1969) Odor threshold determinations of 53 odorant chemicals. *J Air Pollut Control Assoc 19*: 91–95

Linet MS, Stuart WF, Van Natta ML, McCaffrey LD, Szkko M (1987) Comparison of methods for determining occupational exposure in a case-control interview study of chronic lymphocytic leukemia. *J Occup Med 29*: 136–141

Marnett LJ (1999) Lipid peroxidation-DNA damage by malondialdehyde. *Mutat Res 424*: 83–95

McLean AEM, McLean EK (1966) The effect of diet and 1,1,1-trichloro-2,2-bis-(*p*-chlorophenyl)-ethane (DDT) on microsomal hydroxylating enzymes and on sensitivity of rats to carbon tetrachloride poisoning. *Biochem J 100*: 564–571

McCay PB, Lai EK, Poyer JL (1984) Oxygen- and carbon-centered free radical formation during carbon tetrachloride metabolism. Observation of lipid radicals *in vivo* and *in vitro*. *J Biol Chem 259*: 2135–2143

Mico BA, Branchflower RV, Pohl LR (1983) Formation of electrophilic chlorine from carbon tetrachloride-involvement of cytochrome P-450. *Biochem Pharmacol 32*: 2357–2359

Mirsalis JC, Tyson CK, Loh EN, Steinmetz KL, Bakke JP, Hamilton CM, Spak DK, Spalding JW (1985) Induction of hepatic cell proliferation and unscheduled DNA synthesis in mouse hepatocytes following *in vivo* treatment. *Carcinogenesis 6*: 1521–1524

Nagano K, Nishizawa T, Yamamoto S, Matsushima T (1998) Inhalation carcinogenesis studies of six halogenated hydrocarbons in rats and mice. in: Chiyotani K, Hosoda Y, Aizawa Y (Eds) *Advances in the prevention of occupational respiratory diseases*. Amsterdam, Oxford, New York, Elsevier-science Publishers, 741–746

Narotsky MG, Pegram RA, Kavlock RJ (1997) Effect of dosing vehicle on the developmental toxicity of bromodichloromethane and carbon tetrachloride in rats. *Fund Appl Toxicol 40*: 30–36

Nath RG, Li DH, Randerath K (1990) Acute and long term effects of carbon tetrachloride on DNA modifications (I-compounds) in male mouse liver. *Chem Biol Interact 76*: 343–357

Ott MG, Carlo GL, Steinberg S, Bond GG (1985) Mortality among employees engaged in chemical manufacturing and related activities. *Am J Epidemiol 122*: 311–322

Paustenbach DJ, Christian JE, Carlson GP, Born GS (1986a) The effect of an 11.5-hr/day exposure schedule on the distribution and toxicity of inhaled carbon tetrachloride in the rat. *Fundam Appl Toxicol 6*: 472–483

Paustenbach DJ, Carlson GP, Christian JE und Born GS (1986b) A comparative study of the pharmacokinetics of carbon tetrachloride in the rat following repeated inhalation exposure of 8 and 11.5-hr/day. *Fundam Appl Toxicol 6*: 484–497

Paustenbach DJ, Clewell HJ, Gargas ML, Andersen ME (1988) A physiologically based pharmacokinetic model for inhaled carbon tetrachloride. *Toxicol Appl Pharmacol 96*: 191–211

Plaa GL, Larson RE (1965) Relative nephrotoxic properties of chlorinated methane, ethane, and ethylene derivatives in mice. *Toxcol Appl Pharmacol 7*: 37–44

Pohl LR, Schulick RD, Highet RJ, George JW (1984) Reductive-oxygenation mechanism of metabolism of carbon tetrachloride to phosgene by cytochrome P-450. *Mol Pharmacol 25*: 318–321

Poyer JL, Floyd RA, McCay PB, Janzen EG, Davis ER (1978) Spin trapping of the trichloromethyl radical produced during enzymatic NADPH oxidation in the presence of carbon tetrachloride or bromotrichloromethane. *Biochem Biophys Acta 539*: 402–409

Randerath K, Randerath E, Zhou G-D, Li D (1999) Bulky endogenous DNA modifications (I-compounds) – possible structural origins and functional implications. *Mutat Res 424*: 183–194

Reuber MD, Glover EL (1970) Cirrhosis and carcinoma of the liver in male rats given subcutaneous carbon tetrachloride. *J Nat Cancer Inst 44*: 419–427

Reynolds ES, Treinen RJ, Farrish HH, Moslen MT (1984a) Relationship between the pharmacokinetics of carbon tetrachloride conversion to carbon dioxide and chloroform and liver injury. *Arch. Toxicol Suppl 7*: 303–306

Reynolds ES, Treinen RJ, Farrish HH, Treinen Moslen M (1984b) Metabolism of [^{14}C]carbon tetrachloride to exhaled, excreted and bound metabolites. *Biochem Pharmacol 33*: 3363–3374

Romaguera C, Grimalt F (1980) Sensitization to benzoyl peroxide, retinoic acid and carbon tetrachloride. *Contact Dermatitis 6*: 442

Roschlau G, Rodenkirchen H (1969) Histological examination of the diaplacental action of carbon tetrachloride and allyl alcohol in mice embryos. *Exp Pathol 3*: 255

Sanzgiri UY, Kim HJ, Muralidhara S, Edallas E, Bruckner JV (1995) Effect of route and pattern of exposure on the pharmacokinetics and acute hepatotoxicity of carbon tetrachloride. *Tox Appl Pharmacol 134*: 148–154

Schiaffonati L, Tiberio L (1997) Gene expression in liver after toxic injury: Analysis of heat shock response and oxidative stress-inducible genes. *Liver 17*: 183–191

Schwetz BA, Leong BKJ, Gehring PJ (1974) Embryo- and fetotoxicity of inhaled carbon tetrachloride, 1,1-dichloroethane, and methyl ethyl ketone in rats. Toxicol *Appl Pharmacol 28*: 452–464

Smyth HF, Smyth jr HF, Carpenter CP (1936) *J ind Hyg 18*: 277

Smyth H., Smyth jr HF (1935) *Final Report to Producers' Committee*, p 32, cited in National Technical Information Service (NTIS) (1975) *Criteria for a recommended standard ... occupational exposure to carbon tetrachloride*, PB 250424, p 57, Springfield, VA 22161, USA

Stewart RD, Gay HH, Erley DS, Hake CL, Peterson JE (1961) Human exposure to carbon tetrachloride vapor. Relationship of expired air concentration to exposure and toxicity. *J Occup Med 3*: 586–590

Takahashi S, Hirose M, Tamano S, Ozaki M, Orita SI, Ito T, Takeuchi M, Ochi H, Fukada S, Kasai H, Shirai T (1998) Immunohistochemical detection of 8-hydroxy-2′-desoxyguanosine in paraffin-embedded sections of rat liver after carbon tetrachloride treatment. *Toxicol Pathol 26*: 247–252

Tomenson JA, Baron CE, O'Sullivan JJ, Edwards JC, Stonard MD, Walker RJ, Fearnley DM (1995) Hepatic function in workers occupationally exposed to carbon tetrachloride. *Occup Environ Med 52*: 508–514

Tsirel'nikov NI, Tsirel'nikov TG (1976) Morphohistochemical study of the rat placenta after exposure to carbon tetrachloride at different stages of pregnancy (Russian). *Bull exp Biol Med (USSR) 82*: 1262–1265

Tsirel'nikov NI, Dobrovol'skaya SG (1973) Morphohistochemical investigation of the embryonic liver after CCl_4 administration at various stages of ontogeny (Russian). *Bull exp Biol Med (USSR) 76*: 1467–1469

Uehleke H, Werner T (1975) A comparative study on the irreversible binding of labeled halothane trichlorofluoromethane, chloroform, and carbon tetrachloride to hepatic protein and lipids. *Arch Toxicol 34*: 289–308

Uemitsu N (1986) Inhalation pharmacokinetics of carbon tetrachloride in rats based on arterial blood: Inhaled air concentration ratios. *Toxicol Appl Pharmacol 83*: 20–29

US-EPA (1991) *Toxics in the community: National and local perspectives*. Washington, DC, US Environmental Protection Agency, Office of Toxic Substances (EPA 560/4-91-014)

Veng-Pedersen P, Paustenbach DJ, Carlsson GP, Suarez L (1987) A linear systems approach to analyzing the pharmacokinetics of carbon tetrachloride in the rat following repeated exposures of 8 and 11.5h/day. *Arch Toxicol 60*: 355–364

Waller RL, Recknagel RO (1982) Evaluation of a role for phosgene production in the hepatotoxic mechanism of action of carbon tetrachloride and bromotrichloromethane. *Toxicol Appl Pharmacol 66*: 172–181

Wang MY, Liehr JG (1995) Lipid hydroperoxide-induced endogenous DNA adducts in hamsters: possible mechanism of lipid hydroperoxide-mediated carcinogenesis. *Arch Biochem Biophys 316*: 38–46

Wang D, Kreutzer DA, Essigmann JM (1998) Mutagenicity and repair of oxidative DNA damage: insights from studies using defined lesions. *Mutat Res 400*: 99–115

Weisburger EK (1977) Carcinogenicity studies on halogenated hydrocarbons. *Environ Health Perspect 21*: 7–16

WHO (World Health Organization) (1999) *Carbon Tetrachloride IPCS – Environmental health criteria 208*, WHO, Genf

Wilcosky TC, Checkoway H, Marshall EG, Tyroler HA (1984) Cancer mortality and solvent exposure in the rubber industry. *Am Ind Hyg Assoc J 45*: 809–811

Wolf CR, Mansuy D, Nastainczyk W, Deutschmann G, Ullrich V (1977) The reduction of polyhalogenated methanes by liver microsomal cytochrome P-450. *Mol Pharmacol 13*: 698–705

completed 24.2.2000

Coal mine dust

MAK value	**–**
Peak limitation	**–**
Absorption through the skin	**–**
Sensitization	**–**
Carcinogenicity (1998)	**Category 3B**
Prenatal toxicity	**–**
Germ cell mutagenicity	**–**

Abbreviations

CB	chronic bronchitis
CBR	chronic bronchial reaction
CNLD	chronic non-specific lung disease
CURS	chronic unspecific respiratory syndrome
CWP	coal workers' pneumoconiosis
CWSP	coal workers' simple pneumoconiosis
DCO	diffusing capacity
quartz DQ 12	defined species of quartz, < 5 µm, 87 % crystalline quartz, the rest amorphous quartz and low levels of kaolinite
FEV_1	forced expiratory volume: the maximum volume of air which can be expired in one second
FVC	forced vital capacity
IGV	intrathoracic gas volume
MIP	macrophage inflammatory protein
MPG II	sampling device
NSCWP	National Study of Coal Workers' Pneumoconiosis
OR	odds ratio
$PaCO_2$	partial pressure of carbon dioxide in the alveoli
PAH	polycyclic aromatic hydrocarbons
PaO_2	partial pressure of oxygen in the alveoli
PDGF	platelet derived growth factor
PFR	Pneumoconiosis Field Research
PMF	progressive massive fibrosis
ROS	reactive oxygen species
RR	relative risk
R_t	bronchial airway resistance
TBF 50	sampling device with a deposition curve very like that of deposition in the alveoli
TGF	tumour growth factor
TNF	tumour necrosis factor
VC	vital capacity

Unless otherwise specified in this document, the coal miners studied worked in black coal mines, that is, mines for hard coal (bituminous coals and anthracite).

1 Characterization of the Substance

1.1 Occurrence and composition

Coal mine dust is defined, for the purpose of establishing a threshold value, as the respirable fraction[1] of the dust produced during prospecting, extraction and processing of coal. It is equivalent to the dust produced during transport and storage of coal. This fraction is comprised of particles of run-of-mine coal, particles of the minerals present in the coal seam and particles of the associated rock. The mineral fraction can contain quartz in proportions depending on the kind of rock and stratigraphic horizon of the deposits.

The coal deposits in the Lower Rhine/Westphalian and Aachen coal areas are part of the belt of coal along the northern slopes of the American-Variscan mountain range. Stratigraphically they may be classified mainly in the Namurian and Westphalian stages of the Carboniferous. The formation of the deposits was paralic, that is, in a marginal marine environment. The coal seams are autochton, originating close to their present position, and developed from peat deposits. Tectonic activity towards the end of the Carboniferous resulted in the formation of hollows and saddles; the intensity of folding was often greater with increasing depth. Gradual overthrusting and the resulting folding broke up the deposits.

The coal deposits of Saarland are limnic in origin, that is, formed in freshwater lakes. The carboniferous rocks of the Saar take the form of a broad saddle, sinking to the north, broken up by fissures and in places divided into smaller saddles, the geological forms being the result of fissuring and overthrusting processes (Kukuk 1955). The coal seams of the Saar carboniferous rock are generally geologically younger than those in the Ruhr area.

Certain properties of the coal are associated with the degree to which the coal has undergone coalification, the coalification rank. Coalification is an exothermic process which, once started, continues slowly and spontaneously. In this process, the escape of volatile compounds is affected by the stratigraphic depth of the seam. Generally, the loss of volatiles is initially linear but with very lean coal, the process becomes slower with increasing depth. The different parts of a single coal seam which has been split into seams at different depths by tectonic activity can have different coalification ranks (Petrascheck 1961).

Coal may be classified into various types on the basis of the percentage of volatile matter that it contains. High volatile bituminous coal generally has a clearly banded structure with marked clefts. Its content of volatile matter is between 37 % and 45 %. There are also bituminous coals with thinner bands which contain 27 % to 37 % volatile matter and others containing 17 % to 27 % volatile matter which are more fissured and tend to form more dust. Anthracitic low-volatile coal and semi-bituminous coals are harder than are younger types of coal. Their content of volatile matter is given as 10 % to

[1] see the 1997 document on the General Threshold Limit Value for Dust in Volume 12 of the present series

17 %. Anthracite contains between 4 % and 10 % volatile matter. The petrographic classification of coal of different coalification ranks is carried out by reflected light and fluorescence microscopic analysis of the coal constituents (macerals). Three major maceral groups are recognized: vitrinite, liptinite (exinite) and inertinite. Inertinites are oxidized plant material; the inertinite macerals semifusinite and fusinite are comprised of woody cell structures. Vitrinite is derived from humous plant material; the vitrinite maceral telinite contains no visible cell structures. Liptinite consists of fossil resins and waxy plant parts. Fusinite and semifusinite are more common in seams of bituminous coal than, for example, in seams of coking coal or semi-bituminous coal; coking coal and semi-bituminous coal contain more vitrinite than do bituminous coals. Liptinite has quasi-viscous shear properties. Vitrinite and especially inertinite are more brittle because of their higher carbon content and higher levels of aromatic hydrocarbons. Therefore, under mechanical stress such as occurs during coal extraction, they fracture and are found in the respirable dust. The ratio of vitrinite to inertinite is higher in coal dust than in solid coal, how much higher depending on the coalification rank (Schulz *et al.* 1995). This can perhaps be accounted for by the fact that the density of vitrinite decreases with increasing coalification rank. This affects the aerodynamic diameter of the particles. In addition, inertinite particles become longer with increasing coalification whereas vitrinite particles remain relatively constant in form. Liptinite has not been detected in coal mine dust (Wolf and Schulz 1994).

1.2 Chemical and physical properties

Run-of-mine coal contains impurities such as clay, shale and quartz. It can also contain low levels of heavy metals. The sulfur contained in coal is partly in organic and partly in inorganic compounds. All coal macerals can contain pyrite (iron disulfide) (Petrascheck 1961). It occurs as well in hairline cracks, fissures and gaps between layers. Polycyclic aromatic hydrocarbons (PAH) are also present in low levels in coal. The benzo[*a*]pyrene level in coking coal varies between 0.5 and 2 mg/kg. High volatile bituminous coals contain the highest levels, up to 3.7 mg/kg. In anthracite, no benzo[*a*]pyrene was detected by means of a method with a detection limit of 0.04 mg/kg (Bauer and Fricke 1995). The assay determined the benzo[*a*]pyrene level in a toluene:methanol 80:20 mixture after extraction for 36 hours. Longer extraction times did not significantly increase the yield of PAH.

During mining of coal, the coal from the seam is broken up; the amount of dust formed in this process depends on the methods used. It is generally accepted that cutting the coal from the coal face produces more dust than does shearing. The dust formation can be kept to a minimum by

- the choice of cutting tools
- limiting the number of cutters on the drum
- the choice of drum shape
- limiting the speed of drum rotation
- and by optimizing the speed of rotation of the drum for the speed of the shearer.

Dust prevention measures such as wetting the cutters, the mining machines, and the coal during haulage are intended to keep the dust levels down. In addition, a particularly effective way of dust prevention is the wetting of the exposed coalface. Coking coal tends to form more dust than do other kinds of coal. Here, dust formation is favoured by the high vitrinite level. The composition of the dust is determined essentially by the purity of the coal seam being mined, the kind and composition of the impurities in the coal, the kind and composition of the roof and floor material taken along with the seam and the method used for coal extraction.

2 Exposure

2.1 Dust sampling methods

Dust monitoring of all workplaces in coal mines was first introduced in North Rhine-Westphalia in the middle 1950s and became compulsory in 1964 (BVOST 1964). The dust concentrations were measured by tyndallimetry. The conversion values used to calculate concentration values from the tyndallimetric data were obtained from konimetric samples. On the card detailing job description and dust exposure for each individual employee, the dust concentration values were recorded; the tyndallometric determination itself was not carried out for each person separately. Instead, the concentration measured in a return airstream in the mine was assigned to a certain working area upwind of the sampling device. This strategy, the sampling of dust downwind of the miners for determination of individual dust exposure, is still used today. In 1974 the assay procedure was changed. Whereas until 1979, it was merely the time that persons could spend at workplaces with high dust levels which was limited, in 1979 workforce reallocation measures were introduced in order to ensure that personal threshold values were not exceeded over fixed periods of time. To this end, the dust concentrations were classified into exposure grades associated with factors which increased with the concentration, not always linearly, and could be multiplied by the numbers of shifts worked by the individual. Within a period of 5 years a certain threshold value, which depended on the results of occupational medical examinations, was not to be exceeded. As from 01.01.1992, the assessment of individual exposure was again modified in that the period of assessment was reduced to 2 years and the ratio of the measured dust concentration to the threshold value was combined with the number of shifts worked. An exposure value of 220 per year serves as a threshold for persons without dust-related lung changes; that is, within 2 years the individual exposure value may not exceed 440.

Since 1974, the assay has been carried out gravimetrically; the standard and reference device used for respirable dust sampling is the MPG II (Bauer and Bruckmann 1974). Conversion of the old tyndallimetric values to gravimetric values does not necessarily yield reliable values because the tyndallimetric value obtained for a given gravimetric value is a function of the particle size distribution of the particle collective being assayed.

To determine the gravimetric concentration equivalent to a given tyndallimetric value, a universally applicable 'worst-case' method has been recommended, in which the tyndallimetric concentration value is multiplied by a factor of 0.3. The appropriate factor in an individual case, however, can differ enormously from this recommended factor (Bauer 1995a, 1995b).

Thus, in coal mining, unlike industry in general, the concentrations determined at the workplaces serve only as the basis for deciding when reallocation of the workforce is necessary. Reallocation of the workforce is intended to guarantee that the cumulative dust dose for the individual worker, estimated from the number of shifts worked and the respirable dust concentrations during those shifts, is kept below a threshold level.

2.2 Dust and quartz concentrations during coal mining

Comparison of the data for respirable dust concentrations from the years 1958 and 1967 reveals that although the concentration maxima of the frequency distributions were unchanged, the values of the maxima differed. The maximum of the frequency distribution calculated with a conversion factor of 0.3 was about 9.6 mg/m^3. Gravimetric assays carried out in the 'Silikose-Forschungsinstitut' (Institute for Silicosis Research), now called the 'Institut für Gefahrstoff-Forschung' (Institute for Research into Dangerous Substances), revealed respirable dust concentrations in the coal mines of the Ruhr area in the years 1971 and 1972 which were above 4 mg/m^3 in 80 % of samples, above 6 mg/m^3 in 40 % of samples and above 10 mg/m^3 in 20 % of samples (Bauer 1973).

Data from the years 1980 to 1986 reveal time-weighted average dust concentrations between 4.3 and 4.8 mg/m^3, the values being expressed in terms of the total time at the workplace (Woitowitz *et al.* 1989). It is stipulated in the 'Gesundheitsschutz-Bergverordnung' (GesBergV: Health and Safety Act for mines) (Bundesministerium für Wirtschaft 1991) that to obtain a value for the shift these concentration values are to be multiplied by 0.8; this yields values of 3.44 and 3.84 mg/m^3. However, it must be pointed out that for some workers the exposure grades are very different from these values. On 01.01.1992, with the coming into force of the GesBergV (Bundesministerium für Wirtschaft 1991) the actual respirable dust concentration at the workplace was limited to 8 mg/m^3. It is necessary to distinguish clearly between the actual dust concentration at the workplace and the permitted individual dust dose which, as mentioned above, includes not only the concentration value but also the number of shifts worked.

In the coal mines of the Saarland, workplace dust concentrations were determined by tyndallimetric assay until the year 1980. Until 1965 only the workplaces were classified; after 1965, record cards associated the assayed dust concentrations with the miners who worked at the different workplaces. In 1967 a total dust exposure value was established, determined from dust concentration and number of shifts worked, and individual values were not permitted to exceed a threshold during a given exposure period. In the year 1980 the dust assay was first carried out with a gravimetric method, and by the year 1991 the permitted individual average annual dust exposure level had been reduced from 5.2 mg/m^3 to 4 mg/m^3. Since 1992, the GesBergV has ensured harmonization of the assay and assessment procedures in the German coal industry.

Until 1992, the assessment of dust concentrations in the coal mines of Saarland was very different from that in North Rhine-Westphalia. Whereas in Saarland the quartz content of the dust was not determined specifically, in North Rhine-Westphalia quartz concentrations had been measured since the middle 1960s.

Fundamentally, the quartz content (and therefore the resulting respirable quartz dust concentration) is higher in the seams of lower rank coal (lesser degree of coalification) than in the Sprockhövel, Witten, Bochum and lower Essen seams. The quartz content of the respirable dust is also often affected by the coal extraction method. The quartz concentration in respirable dust was generally higher where the coal was extracted by cutting than by shearing. The data of the Silikose-Forschungsinstitut for the Bochum and lower Essen coal seams show quartz concentrations between 2.5 % and 5 % w/w when the coal was extracted by shearing but 5 % to 7 % w/w when extraction was by cutting. When lower rank coal seams were extracted by cutting, the quartz concentrations were generally higher than 7 % w/w. In the Saarland coal mines where extraction was carried out almost entirely by cutting, the quartz levels in the Sulzbach coal seams varied between 11 % and 14 % w/w. For the younger seams, the quartz levels rose between 1981 and 1986 to an average of 17 % w/w. The resulting respirable quartz dust concentrations were more than 0.35 mg/m^3 for the Sulzbach coal seams and between 0.48 mg/m^3 and 0.55 mg/m^3 for the Heiligenwald seams (Woitowitz *et al.* 1989). In the years 1971 and 1972, a study of 78 mines in the Ruhr area revealed for the summed frequency distribution of the respirable quartz dust concentrations a 50 % value of about 0.22 mg/m^3 and an 80 % value of 0.34 mg/m^3 (Bauer 1973). Improved dust prevention measures reduced this value in the respirable dust of the Bochum seam to an average of 0.17 mg/m^3 by the year 1988. In the dust of lower rank coal seams, however, the respirable quartz dust concentrations were still generally markedly higher than 0.25 mg/m^3.

3 Toxic Effects and Mode of Action

Coal mine dust can accumulate as insoluble dust in the lungs of exposed persons or animals and cause not only impaired lung clearance but also inflammatory changes and fibrosis (see Sections 5 and 6).

The pattern of deposition of coal mine dust in the respiratory tract is determined both by the individual pattern of branching and diameters of the airways and also by the size of the dust particles and their physicochemical properties (see the 1997 documentation ‘General Threshold Limit Value for Dust’ in Volume 12 of the present series). Particles of coal mine dust deposited in the upper airways down to the bronchioli can be eliminated by the action of the ciliated epithelium. This epithelium with mucociliary cleansing activity extends from the terminal bronchi to the larynx.

Most deposited dust is transported up to the nasopharyngeal region and then swallowed. Particles of coal mine dust which penetrate to the alveoli are cleared from the lungs more slowly via phagocytosis by alveolar macrophages. Because alveolar clearance is slow, overloading of the lung can result in accumulation of coal mine dust.

Levels of dust in the lungs of miners have been determined in numerous studies. Up to 40 to 60 g respirable dust was found per lung. The retained mass of dust was higher in miners with simple pneumoconiosis or progressive massive pneumoconiosis (PMP) than in miners without pneumoconiosis (IARC 1997).

In man, the retained coal mine dust can cause a number of toxic effects or disorders (see Section 6): impairment of lung function, chronic bronchitis, pulmonary emphysema, coal workers' pneumoconiosis (CWP), progressive massive fibrosis (PMF), rheumatoid arthritis, immune disorders and possibly also bronchial carcinomas and tumours of the digestive organs.

In *in vitro* studies with alveolar macrophages, fibroblasts and erythrocytes, coal mine dust caused haemolysis, cytotoxicity and release of inflammation mediators, growth factors and cytokines.

4 Mechanism of Action

The mechanism of action of coal mine dust is not yet understood. However, a correlation between the exposure to coal mine dust and the development of pathological effects in the lung has been demonstrated. The information available to date on the mechanism of action of coal mine dust is summarized in Figure 1.

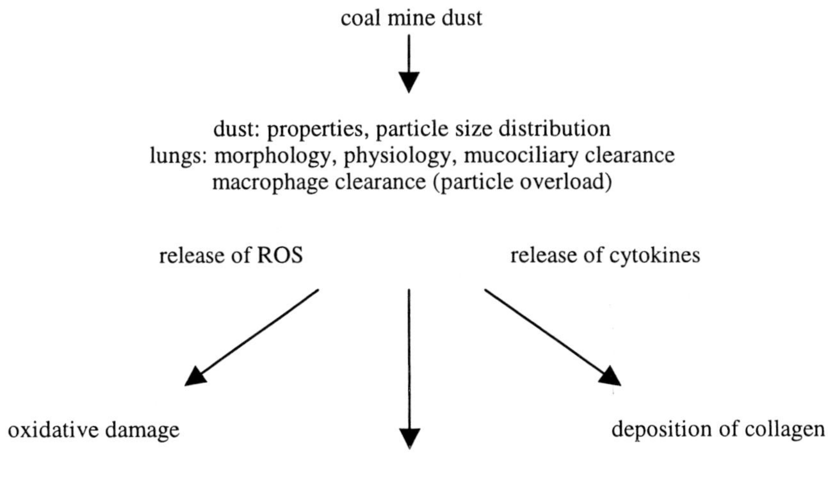

Figure 1. Mechanism of action of coal mine dust (Schins 1996)

With increasing levels of deposited dust, the alveolar macrophages and epithelial cells become activated. This results in increased release of inflammation mediators, reactive oxygen species (ROS), enzymes (elastase, protease, collagenase), cytokines (e.g. TNF, interleukin, MIP) and growth factors (PDGF, TGF). These in turn are responsible for the stimulation of the various kinds of pulmonary fibrosis and emphysema.

Factors such as eating habits, smoking and chewing of tobacco are considered to play a role in the development of stomach cancer (Ames 1982, Ames and Gamble 1983, Ong *et al.* 1985, Wu *et al.* 1990). Nitrosation of coal mine dust in the acid milieu of the stomach is conceivable and could lead to local tumour development. *In vitro* studies have shown that nitrosated extracts of coal mine dust are mutagenic in the *Salmonella* mutagenicity test. They also induce an increase in the incidence of SCE in human lymphocytes and in cell transformation in mouse fibroblasts (NIOSH 1995).

Coal mine dust contains not only quartz and metal compounds but also low levels of polycyclic aromatic hydrocarbons (PAH). The complex PAH mixtures contain substances such as benzo[*a*]pyrene which has been shown to cause bronchial epithelial damage and to have carcinogenic effects on the stomach mucosa (Triolo *et al.* 1977).

The effect of hydroxyl radicals on the development of pneumoconiosis in coal miners has been studied (Dalal *et al.* 1995). It was demonstrated that higher levels of iron in coal mine dust are involved in the generation of hydroxyl radicals, which could play an important role in the development of CWP. Also the findings of Tourmann and Kaufmann (1994a, 1994b) supported the hypothesis that the iron compounds in coal mine dust can have damaging effects. In addition, increased release of iron was demonstrated in persons with lung fibrosis.

5 Toxicokinetics

Man

A correlation has been demonstrated between the development of CWP (nodular pulmonary fibrosis) or PMF and the cumulative deposition of dust in the lungs (Attfield and Seixas 1995, Hurley *et al.* 1982, Rossiter *et al.* 1967, Ruckley *et al.* 1984). No difference was found in the deposition of coal mine dust in the lungs of two groups of miners, one group with CWP, the other without pneumoconiosis (Love *et al.* 1970). Bergmann and Casswell (1972) demonstrated a higher rate of accumulation of coal mine dust in persons working in seams of low rank coal than in persons working in seams of higher rank coal. In addition, a higher rate of accumulation was found in miners with PMF than in those without PMF. Quartz dust was found in the lymph nodes of miners working seams of low rank coal (Chapman and Ruckley 1985). Silicosis with so-called "egg-shell calcification" of the hilar lymph nodes is one morphological form of the highly variable CWP.

Animals

There are no specific studies of the deposition of coal mine dust in animals (IARC 1997). However, there are several inhalation studies which provide information about retained dust mass and elimination. In several studies, the transport of intratracheally instilled mine dust from the lungs to the mediastinal lymph nodes was investigated. In studies of the significance of the mineral and quartz content of coal mine dust, 43 samples differing in their quartz and ash content were instilled intratracheally into Wistar rats in 50 mg doses. Six and twelve months later, the dust levels in the lungs and mediastinal lymph nodes were analysed (Rosmanith *et al.* 1981, 1983). The quartz content of the dusts ranged from 0.3 % to 15 %. The amounts of dust found in the lungs after six months were between 12 g and 29 g and, like the amounts of dust in the lymph nodes, seemed to correlate with the ash, quartz and kaolinite levels in the original dust. A particularly good correlation was found between the quartz level of the dust in the lungs after 6 months and the amount of dust in the lymph nodes after 12 months. This effect, called lymphotropism of quartz, has been described in several publications (Hilscher *et al.* 1983, 1987, Molik and Hilscher 1989). Analysis of the quartz in retained dust demonstrated that the quartz level in the lung dust was the same as that in the original dust but that the level in the lymph node dust was about 5 % higher (Rosmanith *et al.* 1982).

The dust transport through the lymph vessels was shown to depend markedly on the particle size. In other experiments from the same research group, four particle size fractions from each of four dry-filtered dusts with different quartz levels were studied. It was demonstrated that the ability of a dust to enter the lymph nodes increased exponentially with decreasing median particle diameter (Reisner *et al.* 1985a, Rosmanith *et al.* 1985).

The results of these studies yield information about elimination from the lungs including the mean elimination rate or elimination half-time during the period between 6 and 12 months after instillation. The results do not demonstrate dependence on the physicochemical properties of the dust but rather show that for a retained dust weight of about 20 mg, the elimination half-time is very high and in some cases cannot be determined at all because the experimental error is too great. Although an increase in the amount of dust in the mediastinal lymph nodes during the period studied can sometimes be observed and shows that dust is being eliminated from the lungs, the amount of dust involved is exceedingly small in comparison with the amount still in the lungs.

In rats which had inhaled quartz dust in a concentration of 20 mg/m^3, 7 hours daily, 5 days per week for one year, 25 mg of dust was found in the lungs of each animal; this indicates retention of 8 % to 9 % of the inhaled dose (Robertson *et al.* 1988). Rats inhaled genuine mine dust in a concentration of 70.8 mg/m^3, 5 hours daily on 5 days per week for 20 weeks. Six and twelve months after the end of exposure, the average weight of dust per lung was found to be 12.49 and 8.21 mg and in the mediastinal lymph nodes 1.39 and 2.38 mg (Weller and Rosmanith 1983).

6 Effects in Man

6.1 Non-malignant disorders

The disorders which can result from exposure to coal mine dust are described below.

Chronic bronchitis (CB)

Chronic bronchitis involves chronic or recurrent overproduction of bronchial mucous which causes expectoration, coughing and sometimes respiratory distress (Woitowitz 1972). The sequel is a disorder of the bronchopulmonary clearance mechanism, mucociliary insufficiency. The obstructive form of the disorder is also called chronic non-specific lung disease (CNLD). In epidemiological studies, a clinical definition of chronic bronchitis (WHO 1986) based on anamnesis and the results of lung function tests is used. For the induction of chronic bronchitis by coal mine dust, the thoracic fraction of the inhalable dust (previously called total dust) is considered to be the active fraction (ACGIH 1984). However, most epidemiological dose-response studies have been based on analyses of the alveolar fraction (respirable fraction) of the dust (previously called fine dust). But in the development of bronchitis as a result of mucociliary clearance disorders, the alveolar fraction is at most a contributing factor; the weight of dust deposited in the tracheobronchial region plays a more important role in this disease.

Pulmonary emphysema

Pulmonary emphysema is defined morphologically as the irreversible abnormal enlargement of the alveoli distal to the terminal bronchioles with destruction of the alveolar walls (ATS 1962). Generally it is found together with CNLD. A precise diagnosis is generally only possible by pathological-anatomical methods. The pathogenesis involves disorders in the local protease-antiprotease balance in the lungs (Cantin and Chrystal 1985) or an imbalance in the oxidant-antioxidant system (Gadek and Pacht 1990, Laurell and Eriksson 1963). Reactive oxygen species seem to be important in the aetiology of emphysema (Huang *et al.* 1993, 1994).

Coal workers' pneumoconiosis/silicosis

The initial stages of CWP and of emphysema are topographically independent focal inflammatory processes (Leigh *et al.* 1994, Otto and von Hinüber 1972). The various kinds of CWP are caused by inhalation of respirable dust containing quartz. The disease is a chronic progressive fibrosis, initially mainly nodular fibrosis of the lung tissue with involvement of the lymph vessels and with a marked tendency to tissue shrinkage.

An acute form of silicosis (silicotic alveolar proteinosis) develops mostly after exposure to very high concentrations of quartz dust and sometimes after short exposure periods. It has been observed especially in sandblasting operators and tunnel construction workers. The disorder is relatively rapidly fatal, often within a few months. Nowadays, because of the dust prevention measures laid down by law, the disease is practically unknown.

Especially in publications in the English language, the disorder seen in persons exposed to coal mine dust is not called silicosis but coal workers' pneumoconiosis (CWP) and is classified either as coal workers' simple pneumoconiosis (CWSP) or as the less common progressive massive fibrosis (PMF; x-ray opacity at least ≥ A according to ILO (1980)). Especially in the advanced stages of PMF, destruction of pulmonary capillaries can result in enlargement of the right side of the heart (chronic *cor pulmonale*). It has also been observed that many years of exposure to dust are necessary before such severe effects develop. The aetiology of PMF is not yet understood. However, PMF has also been reported in carbon electrode workers (Watson *et al.* 1959) and in so-called coal trimmers who were exposed to quartz-free coal mine dust (Gough 1949).

6.2 Results of epidemiological studies

The epidemiological studies are grouped according to geographic regions because of differences in analytical methods, in the geological situation and the mining methods used in the different countries and on the basis of published data for specific regional cohorts and cases.

6.2.1 British coal mines

Pneumoconiosis Field Research Programme

The Pneumoconiosis Field Research (PFR) Programme is to date the most extensive project for the epidemiological study of collectives of coal miners.

The PFR programme consisted of a series of medical surveys at intervals of about 5 years, beginning in the early 1950s and continuing until the last survey at the beginning of the 1980s. In the first survey, about 30000 miners from 24 collieries took part (Hurley and Maclaren 1987). Of these, 14888 miners from 20 collieries could be examined again in the third survey (Marine *et al.* 1988) and 3600 miners from 10 collieries in the fourth (Collins *et al.* 1988). A total of 53382 coal miners took part in the PFR programme but only 22848 of these were examined in one of the occupational medical examinations (Hurley *et al.* 1987). Participation in the study was voluntary.

The estimated respirable dust exposure levels in the PFR programme (see Tables 1 to 4) were based for the period 1952 to 1964 (in some cases until 1969) on results obtained with the "thermal precipitator" which were converted to gravimetric concentrations on the basis of parallel assays; only for the period after 1965 (or 1969) were gravimetric assay data available. Concentration data for components of the coal mine dust (non-coal fraction = ash, quartz, silicates; see Tables 1 and 2) were also obtained (Hurley *et al.* 1982, Miller *et al.* 1995, Walton *et al.* 1977). For the period before 1952, the exposure levels in the mines were estimated by classifying the miners into job groups on the basis of individual interviews and multiplying the periods spent with the group by the typical respirable dust concentrations to which the groups were exposed, estimated on the basis of assays carried out after 1952 (Marine *et al.* 1988). This probably resulted in a systematic underestimation of the cumulative respirable dust concentration in long-term exposures (Hurley and Maclaren 1987) and thus to an overestimation of the dose-response relationship. However, the size of this effect was not determined.

To record any possible fibrogenic effects such as those seen in CWP, the examinations included the preparation of large format radiographs of the thoracic organs (posterior-anterior). The radiographs were generally evaluated using the ILO classification system. In the course of the PFR programme, various versions of the ILO classification were used, and some of them simultaneously in the individual surveys (Hurley and Maclaren 1987, Hurley *et al.* 1982, Jacobsen *et al.* 1971, Maclaren and Soutar 1985, Miller *et al.* 1995). The results were sometimes presented on a simplified 5 point scale (0, 1, 2, 3, PMF) in the form of current prevalence values (Hurley and Jacobsen 1986, Hurley and Maclaren 1987, Maclaren and Soutar 1985) or, for description of CWSP, on the more detailed 12 point simple pneumoconiosis scale. In the evaluation, the progression from one category of CWSP to the next higher in the course of the interval between examinations was recorded (Jacobsen *et al.* 1971, Walton *et al.* 1977) or the analysis concentrated on the prevalence of a certain minimum category (CWSP ≥ 2/1: Hurley *et al.* 1982, Jacobsen *et al.* 1971; CWSP ≥ 1/0, CWSP ≥ 2/1: Miller *et al.* 1995). The classification of the radiographs into the appropriate ILO class was carried out by one to six readers, some of whom were physicians and some non-medical readers who had been intensively trained in the reading of radiographs (Maclaren and Soutar 1985, Miller *et al.* 1995). A variety of readers were also used in the mines (Hurley *et al.* 1987).

To detect the development of lung function disorders or chronic bronchitis, from the second survey of the PFR programme FEV_1 (forced expiratory volume in one second) and FVC (forced vital capacity) were determined with a spirometer (mean of the values determined in three technically satisfactory forced exhalations). In an interview, the persons were asked about symptoms of respiratory diseases and their smoking habits were recorded (Collins *et al.* 1988, Marine *et al.* 1988).

In association with the PFR programme, autopsy studies (Ruckley *et al.* 1984, 1989) were also carried out and follow-up studies of the mortality of the coal miners (Jacobsen 1976, Miller *et al.* 1981, Miller and Jacobsen 1985). The results of these studies are given in detail by Morfeld and Piekarski (1998a) and are summarized below.

Coal workers' pneumoconiosis/silicosis

The studies have demonstrated the progressive character of CWP from CWSP to PMF and yielded a clear dose-response relationship with exposure to respirable coal mine dust. Even for persons exposed in the low concentration range, the PFR programme revealed that the residual risk associated with all the categories of CWP was above zero. Occasional new cases of PMF were observed even when the cumulative exposure level was as low as 50 to 100 $g \times h/m^3$. In addition, new cases of PMF developed within an observation period of as little as 5 years from a normal initial finding of CWSP = 0; for these persons a dose-response relationship between the development of PMF and the cumulative exposure to respirable coal mine dust could be demonstrated.

The quartz component of respirable coal mine dust has shown to be less important than the total respirable coal mine dust concentration; in fact, the risk relationship for quartz has the opposite sign, there being a higher risk with higher rank coal containing less quartz (Hurley *et al.* 1982, Walton *et al.* 1977). Nonetheless, individual studies have provided evidence that under certain conditions exposure to high levels of respirable quartz dust can promote the development of CWP (Jacobsen and Maclaren 1982, Miller *et al.* 1995).

Tables 1 and 2 summarize the most important studies from the PFR programme in which prevalence of CWSP and of PMF were recorded and exposure to quartz was also taken into consideration. Table 3 shows the evaluation of the main data for the risk of developing CWP as a function of the concentration of respirable coal mine dust.

Table 1. The main studies of CWSP (coal workers' simple pneumoconiosis) from the PFR programme (Hurley *et al.* 1982)

kind of study:	cross-sectional study (prevalence)
size of study:	2600 coal miners
parameter:	CWSP ≥ 2/1
exposure:	on average 33 years working underground, respirable coal mine dust: 183 g × h/m^3 (about 3.2 mg/m^3), respirable quartz dust: 8.8 g × h/m^3 (about 0.15 mg/m^3); 14 % (= 366 miners) exposed to respirable coal mine dust levels above 300 g × h/m^3 (about 5.2 mg/m^3)
results:	a) significant positive association with average respirable coal mine dust exposure levels, time spent underground, cumulative exposure to respirable coal mine dust
	b) no general effect of respirable quartz dust exposure levels
	c) markedly deviating prevalence estimates for individual mines
	d) best average estimate of prevalence (see Jacobsen 1984) for CWSP ≥ 2/1 after 35 years working underground with 1740 working hours per year: concentration of respirable coal mine dust = 3.8 mg/m^3 ⇒ prevalence 3.6 %; 5.0 mg/m^3 ⇒ prevalence 6.5 %
other results:	(Jacobsen and Maclaren 1982, Miller *et al.* 1995, Walton *et al.* 1977): no general effect of exposure to quartz on progression of CWSP, positive association with age of the stratigraphic horizon, evidence of negative association with exposure to silicates (clay minerals); where the quartz content was higher (more than 10 %: Miller *et al.* 1995, 1998; 7.5 %: Walton *et al.* 1977): positive association of CWSP progression and CWSP prevalence with the cumulative exposure to respirable quartz dust independent of the non-quartz components

Table 2. Main studies of PMF (progressive massive fibrosis) from the PFR programme (Hurley *et al.* 1984, 1987)

kind of study:	longitudinal study of prevalence at 5-year intervals
size of study:	52264 observation intervals
parameter:	PMF
exposure:	average cumulative exposure to respirable coal mine dust before the beginning of the observation interval: about 115 g × h/m^3 (for persons with PMF: about 250 g × h/m^3)
results:	a) prevalence of new PMF dependent on CWSP category at the start: CWSP = 0: 0.2 %, CWSP = 1: 4.4 %, CWSP = 2: 12.5 %, CWSP = 3: 13.9 % b) prevalence of new PMF as a function of the cumulative exposure to respirable coal mine dust initial CWSP category = 0: significant positive association initial CWSP category ≥ 1: no additional association
other results:	(Cowie *et al.* 1994, Maclaren and Soutar 1985, Maclaren *et al.* 1989) higher age, lower Quetelet's index, and longer persistence of dust in the lungs are also positively associated with PMF prevalence; no clear evidence of effects of quartz dust; higher prevalence among ex-miners (healthy worker effect)

Table 3. Data from the PFR programme for CWP risk (Hurley and Maclaren 1987)

kind of study:	longitudinal study of change of prevalence at 5-year intervals
size of study:	52264 observation intervals
parameter:	CWSP = 1, CWSP = 2, CWSP = 3, PMF
exposure:	see Hurley *et al.* 1982, 1987 (Tables 1 and 2)
methods:	logistic regression for the estimation of new prevalence values as a function of initial CWSP categories, age, length of interval, cumulative exposure to respirable coal mine dust, coal rank; product calculated by means of matrices of changes in prevalence (Markov process construction)
results:	prevalence (%) of various grades of CWP in persons exposed for an average of 35 years to various respirable dust concentrations assumptions: age = 53 years, exposure period 35 years, 1631 working hours per year, average coal rank 86.2 % carbon

CWP	1	2	3	PMF
concentration [mg/m^3]	prevalence [%]			
1	2.60	0.68	0.05	0.25
2	5.94	1.63	0.15	0.64
3	9.40	2.68	0.27	1.10

higher risk for higher coal rank (older stratigraphic horizon, lower quartz content)

Chronic bronchitis/emphysema

Independent of whether the persons developed CWP and independent of their smoking habits (Jacobsen *et al.* 1971, 1977), the PFR programme revealed a significant adverse effect of exposure to respirable coal mine dust on lung function parameters and on the risk of developing chronic bronchitis. Even for persons exposed to low long-term average concentrations (< 1 mg/m^3), the model calculations revealed a relevant excess risk of developing obstructive chronic bronchitis (WHO definition and FEV_1 < 80 % of the expected value) among British coal miners. Table 4 summarizes the most important studies of adverse effects on lung function and of chronic bronchitis/emphysema from the PFR programme.

Table 4. Studies of adverse effects on lung function, chronic bronchitis and emphysema from the PFR programme (Marine *et al.* 1988)

kind of study:	cross-sectional study (prevalence)
size of study:	3380 coal miners
parameter:	FEV_1 < 80 %, chronic bronchitis (WHO), FEV_1 < 80 % and chronic bronchitis (WHO), FEV_1 < 65 %
exposure:	174 g × h/m^3 (about 3.1 mg/m^3 to 3.7 mg/m^3 for about 30 years working underground) for the middle range cumulative exposure to respirable coal mine dust; 9 % (about 300 coal miners) exposed to respirable coal mine dust levels higher than 348 g × h/m^3 (about 6.1 mg/m^3 to 7.2 mg/m^3 for about 30 years working underground)
results:	all parameters associated positively and significantly with the cumulative exposure to respirable coal mine dust, independently of age, height and smoking habits: all 346 non-smokers with CWSP < 0/1: $\Delta FEV_1 = -1.06$ ml/(g x h/m^3), all 543 non-smokers: $\Delta FEV_1 = -0.94$ ml/(g x h/m^3), all 2837 smokers: $\Delta FEV_1 = -1.02$ ml/(g x h/m^3). Prevalence among non-exposed non-smokers = 1.2 %; relative increase in this prevalence by 10 %, 30 %, 50 %, 100 % (i.e. value doubled) after exposure to average respirable coal mine dust concentrations of 0.3 mg/m^3, 0.8 mg/m^3, 1.2 mg/m^3, 2.0 mg/m^3 at age 55 years (after 35 years exposure to dust with 1700 working hours per year)
other results:	(Collins *et al.* 1988, Love and Miller 1982, Soutar and Hurley 1986): adverse effects on lung function could also be demonstrated longitudinally; in ex-miners in at least the same severity as in active miners; among persons with positive radiograph findings (about 1/1; ILO 1980) the prevalence of chronic bronchitis (WHO) and of respiratory distress were doubled .

Mortality

Studies of the mortality of British coal miners reveal a clear association between the category of CWP demonstrated radiologically and “coal workers’ pneumoconiosis” as the cause of death on the death certificate. A significant trend towards excess mortality (SMR) was found in a group of 5963 coal miners (cause of death “pneumoconiosis”, category established at the beginning of the 30-year observation period, 99.9 % of study population traced) (Atuhaire *et al.* 1985).

Likewise, a clear association was detected between the likelihood of dying from pneumoconiosis and the category of pneumoconiosis established radiologically at the beginning of the 22-year observation period among about 25000 British coal miners (94 % of study population traced). In addition, this study documented a clear relationship between death from pneumoconiosis and the estimated individual cumulative exposure to respirable coal mine dust: of the 164 persons who died of pneumoconiosis, for 5 a cumulative exposure to respirable coal mine dust below 90 $g \times h/m^3$ was estimated whereas for most of the persons the cumulative exposure level was above 180 $g \times h/m^3$ (Miller and Jacobsen 1985).

All of these persons were included in the PFR programme so that the data obtained there for these individuals could be included in the analysis. The evaluation made use of multiple log-linear models. Independent of smoking habits and age at death, a significant association of emphysema prevalence and exposure to respirable dust was demonstrated. This association was only marked for centroacinar emphysema, not for panacinar emphysema. Further subdivision of the data revealed that the association between exposure to respirable dust and the prevalence of centroacinar emphysema could be detected only in the subgroup of miners in whom CWP had been diagnosed. Analysis of the data for 95 non-smokers also revealed a clear association of the prevalence of centroacinar emphysema with the dust exposure even after correction for age at death. Relationships with the lung function parameters determined in field studies (FEV_1, FVC) were studied by means of linear models. After taking into account effects of age, height and smoking habits, it was demonstrated that the FEV_1 and FVC values decreased with increasing exposure to respiratory dust. In addition, in miners in whom emphysema had been diagnosed, the FEV_1 and FVC values were lower. The authors concluded that part of the reduction in the lung function parameters was caused by emphysema (Ruckley *et al.* 1984).

Assessment

It is a weakness of the most extensive analysis of the PFR programme, the assessment of the risk of developing CWP by Hurley and Maclaren (1987), that the radiographs for this analysis were not reassessed according to the ILO classification (ILO 1980) by at least two readers. In addition (also in the opinion of the authors), it must be assumed that the estimated long-term exposure levels tend to be too low because of the unknown levels of early exposures. Even during the observation phases for which exposure concentration values are available, only in some phases were the dust concentrations determined by gravimetric methods. The groups of persons studied in the PFR programme were not complete cohorts of persons starting work in the industry but self-selected cross-sections of the working population which were further reduced before the analysis because the data were incomplete. In addition, healthy worker effects must be assumed which further reduce the applicability of the risk estimates. The proportions of persons in a new diagnostic category were not determined directly but by differences. The results include estimated risks with a negative sign, evidence of validity problems. The effect of calendar year on the dose-response relationship was not examined. Apart from one analysis at the beginning of the 1970s (Jacobsen *et al.* 1971), the analytical models chosen to describe the relationship between exposure to dust and effects were such that zero risk or zero additional risk could be found only at exposure levels of 0 mg/m^3. The selected models are such that if a positive effect coefficient for dust exposure is found in the regression model, then even very small increases in dust exposure levels yield a strictly monotonic increase in risk or excess risk. Thus threshold values may not be determined from such analyses. Similar conclusions with respect to the derivation of a threshold level for respirable coal mine dust also apply to the studies in the PFR programme of the incidence of chronic bronchitis or emphysema and for the mortality studies of British coal miners. Nonetheless, the risk estimates from these model calculations are presented here for orientation purposes because the PFR programme, in spite of its weaknesses, has still provided the most important international database for the assessment of the risk of developing the non-malignant lung diseases of coal miners. However, because of the study limitations mentioned above, it remains unclear to what extent the quantitative estimates of prevalences (Table 3) for persons exposed to average respirable dust concentrations below 2 mg/m^3 for 35 years may be accepted.

6.2.2 American coal mining

General remarks

In analogy to the British PFR programme, in the USA when the Federal Coalmine Health and Safety Act was introduced in 1969, an epidemiological research programme, the National Study of Coal Workers' Pneumoconiosis (NSCWP), was begun and to date has included up to four series of medical examinations between 1969 and 1988. In all, miners from 40 collieries were examined. Participation in the examinations was voluntary, as it was in Pneumoconiosis Field Research in Great Britain: the participation rates were 90 % of the target population in the first examination, 75 % in the second, 52 % in the third and 70 % in the fourth. The numbers of coal miners taking part were 9078 (first examination), 9347 (second), 5275 (third) and 3534 (fourth). During the occupational medical examinations, radiographs of the thoracic organs (posterior-anterior) were prepared for all participants and classified according to various versions of the ILO classification. The radiographs prepared in the various series of examinations were not always assessed by the same readers. Each radiograph was assessed by three to nine readers. However, there are also important assessments (Attfield and Morring 1992b) based on radiograph classifications by single readers.

In addition, lung function tests were carried out; particularly FEV_1 and FVC were determined as the maximum spirometer value obtained in at most 5 and not less than 3 technically satisfactory forced exhalations per examination (Attfield and Hodous 1992). Details of occupational anamnesis, smoking habits and symptoms of airway disorders were obtained by standardized questioning.

The results of the studies of American coal mining are given in detail by Morfeld and Piekarski (1998a) and are summarized below.

Coal workers' pneumoconiosis/silicosis

The tabulated data (Attfield and Morring 1992b) suggest clear dose-response relationships for the estimated cumulative exposure to respirable coal mine dust and stratification of the prevalence of all parameters studied with the extent of coalification such that the highest risk was found for the oldest, highest rank coal. It is conspicuous that severe CWP is found even at low cumulative dust exposure levels. The model calculations for a 40-year exposure to respirable coal mine dust at a concentration of 2.0 mg/m^3 (given an age of 58 years) yield estimated risks of developing CWSP ≥ 1 between 8.2 % and 31.7 %, CWSP ≥ 2 between 1.5 % and 14.2 % and PMF between 1.3 % and 8.9 %, depending in each case on the rank of the coal. The models estimate the background prevalence for exposure to a respirable coal mine dust concentration of 0 mg/m^3 at age 65 years to be 5.6 % for CWSP ≥ 1, 1.8% for CWSP ≥ 2 and 2.5% for PMF (other results in Table 5).

Table 5. CWP prevalence studies from the National Study of Coal workers' Pneumoconiosis

kind of study:	cross-sectional study (prevalence) (Attfield and Morring 1992b)
size of study:	9023 coal miners
parameter:	CWSP ≥ 1, CWSP ≥ 2, PMF
exposure:	on average 21 years working underground; respirable coal mine dust concentration 3.0 mg/m^3 (112 g × h/m^3 cumulative exposure to respirable coal mine dust given 1740 working hours per year)
results:	estimated prevalence (single readings) in persons aged 58 years after exposure to respirable coal mine dust at a concentration of 2.0 mg/m^3 for 40 years CWSP ≥ 1: 8.2 %–31.7 %, CWSP ≥ 2: 1.5 %–14.2 %, PMF: 1.3 %–8.9 %, in each case depending on the coal rank (higher rank means higher prevalence); estimated background prevalence (at 0 mg/m^3) in persons aged 65 years: CWSP ≥ 1: 5.6 %, CWSP ≥ 2: 1.8 %, PMF: 2.5 % after reassessment of a fraction of the radiographs (3 readers) (Attfield *et al.* 1997): CWSP ≥ 1: about 17 % estimated threshold levels of respirable coal mine dust after 40 years exposure (Attfield *et al.* 1997): for CWSP ≥ 1 less than 0.5 mg/m^3, for CWSP ≥ 2 about 1 mg/m^3. Relative prevalence (Attfield *et al.* 1997): increasing the average long-term concentration from 1 mg/m^3 to 2 mg/m^3 resulted in doubling of the prevalence of CWSP ≥ 1, CWSP ≥ 2, PMF
other results:	(Attfield and Seixas 1995, NIOSH 1995) - higher prevalence of CWP after exposure to higher rank coal - excess prevalence of PMF between 0.1 % and 1.7 % after 0.5 mg/m^3, 0.1 % and 4.6 % after 1.0 mg/m^3 and between 0.1 % and 16.7 % after 2 mg/m^3 (long term average concentration of respirable coal mine dust)

Chronic bronchitis/emphysema

The intra-individual reductions in the FEV_1 values of US American coal miners were studied (Attfield 1985). The data were obtained during the first two series of examinations of the NSCWP. 5308 miners took part in the first examination (1969 to 1971) and 1161 miners in the second (1977 to 1981); in both groups, the miners were aged between 20 and 49 years. Analysis of the changes in FEV_1 was carried out for 1072 miners.

Linear models which took into account the effects of age, height, smoking habits and a colliery indicator revealed for a period of 11 years working underground average reductions in FEV_1 between 36 ml and 84 ml, depending on the type of model. The selection phenomena make it difficult to interpret the study.

A cross-sectional analysis of the data of the first series of NSCWP examinations (1969–1971) was published by Attfield and Hodous (1992). Of the total of 9078 miners who were examined, 7139 had white skin and were at least 25 years old. These comprised the study collective. On the basis of the dust concentration analyses carried out between 1970 and 1972, exposure levels for previous periods of underground work were estimated (see Table 6) (Attfield and Morring 1992a). Application of a linear model which took into account height, age, smoking habits and included an indicator for the mining region revealed an average reduction in FEV_1 of 0.69 ml/g × h/m^3, a reduction in FVC of 0.49 ml/g × h/m^3 and for the ratio FEV_1/FVC a reduction of $8 \times 10^{-5}/(g \times h/m^3)$. All effects are significant. In the subgroup of miners with CWSP = 0/0 (n = 913), the reduction in FEV_1 was 0.75 ml/(g × h/m^3) ($p < 0.001$), also a significant effect of the same order of magnitude as the reduction found for the whole group. A significant reduction in lung function parameters (FEV_1, FVC, FEV_1/FVC) was also confirmed in other studies (Seixas *et al.* 1991, 1992, 1993a, 1993b) (see Table 6).

Table 6. The main NSCWP studies of lung function impairment and chronic bronchitis

kind of study:	cross-sectional study (prevalence) (Attfield and Hodous 1992)
size of study:	7139 white-skinned coal miners, at least 25 years old
parameter:	FEV_1, FVC, FEV_1/FVC
exposure:	on average 18 years working underground; 116 g × h/m^3 cumulative exposure to respirable coal mine dust (respirable coal mine dust concentration 3.7 mg/m^3, given 1740 working hours per year)
results:	lung function parameters standardized for height, age, smoking habits and with an indicator for the mining region reduced ($p < 0.05$) with increased cumulative exposure to respirable coal mine dust: FEV_1: –0.69 ml/(g × h/m^3), FVC: –0.49 ml/(g × h/m^3), FEV_1/FVC: $-8 \times 10^{-5}/(g \times h/m^3)$; in subgroup with CWSP = 0/0 (n = 913): FEV_1: –0.75 ml/(g × h/m^3); larger reductions with higher rank coal (adjusted for the cumulative exposure to respirable coal mine dust)
other results:	(Seixas *et al.* 1992, 1993a, 1993b) – significant reduction in the prevalence of each of FEV_1 < 80 %, FEV_1/FVC < 80 %, chronic sputum, chronic bronchitis, respiratory distress with increased cumulative exposure to respirable coal mine dust – high reductions in lung function parameters mainly during the first years of work underground – latency periods for some effects of the exposure to respirable coal mine dust on loss of lung function

Mortality

Autopsies carried out on 266 US American coal miners (including 41 non-smokers) and as controls 75 persons who were not miners (14 non-smokers) were evaluated to establish the presence and severity of pulmonary emphysema as a function of smoking habits, lung dust levels and the pathologically defined pneumoconiosis category (Vallyathan *et al.* 1997). One important result of this study is that both smoking and the total amount of dust in the lungs correlated independently and significantly with the severity of emphysema; a correlation was demonstrated with the amount of coal mine dust but not with the level of the mineral fraction or quartz in the dust in the lungs. The severity of the emphysema was also significantly and positively correlated with the pathologically determined CWP category. The authors conclude that both the smoking of cigarettes and the exposure to coal mine dust have significant effects on the severity of pulmonary emphysema. Suggested pathomechanisms of emphysema development after exposure to coal mine dust are increased protease activities and the oxidative inactivation of α1-antitrypsin.

Attfield *et al.* (1994) and Vallyathan *et al.* (1996) investigated the correlation between radiographic and pathological findings for CWP. For this purpose, autopsy findings for 423 and 430 coal miners from West Virginia were compared with the radiographs available for the same persons. The most important result was that the pathological examination must reveal moderate to severe changes before radiographic diagnosis of the pneumoconiosis is likely. The authors also demonstrated that in about 25 % of cases with radiographically diagnosed PMF, the results of the pathological examination did not confirm the diagnosis but detected, for example, scars resulting from tuberculosis infections or tumours. On the other hand, in 22 % of cases classified as PMF after the pathological examination, no large opacities were seen in the radiograph.

A mortality study of 23232 US coal miners found 188 deaths from pneumoconiosis and a significantly increased SMR of 9.3 (Rockette 1977). The author points out that the expected values, calculated from the data for the total male population of the US, are too high because coal miners were not excluded from the control population; therefore the calculated SMR underestimates the actual risk. Kuempel *et al.* (1995) studied the mortality of 8878 US coal miners and found for the whole collective a significantly increased SMR for pneumoconiosis (3.72; 95 % confidence interval 2.89–4.71). In addition, the study reveals that the relative risk of dying from pneumoconiosis increases significantly for coal miners with increasing exposure to respirable coal mine dust.

Assessment

The results obtained in the National Study of Coal Workers' Pneumoconiosis from US American coal miners reveal a significant positive association of the risk of developing CWP with the cumulative exposure to respirable coal mine dust and indicate that the risk is higher for persons mining higher rank coal. Similar findings were obtained for chronic bronchitis and the development of lung function disorders (FEV_1, FVC, FEV_1/FVC). In individual cases, the appearance of PMF within a 5-year observation period at the beginning of which the radiograph findings were normal (CWSP = 0) was reported. Even for persons exposed to low levels of respirable coal mine dust (since 1970 concentrations of respirable coal mine dust in US coal mines have been limited to a shift maximum value of 2 mg/m^3), the model calculations reveal high risks for developing PMF and marked impairment of lung function. Even if exposure to respirable coal mine dust is assumed to be zero, especially those studies of unexposed cohorts who started work in the industry after 1970 reveal high residual risks for PMF (> 1 %) and for the reduction of FEV_1 to less than 80 % of the expected value (down to 46 %). These unexpectedly high risk estimates for persons exposed to very low levels of dust must be seen in the context of erroneous dust concentration measurements (NIOSH 1995) and the different dust sampling strategy used in US American coal mines. Studies by Prinz and Stolz (1990) reveal that concentration values measured in German coal mines may be expected to be about four times higher than in American mines entirely because of differences in location of the sampling equipment. In all, the results indicate that the dust exposure levels have been systematically underestimated and that the risk estimates are not trustworthy, especially in the low concentration range (< 2 mg/m^3 long-term average concentration). In addition, significant variation in interpretation of the radiographs and selection problems caused especially by the limiting of the analysed collective to those persons for whom the data were complete make the interpretation of the NSCWP results even more difficult. In all, the qualitative results of the NSCWP are considered to be important but the calculated quantitative relationships, especially in the low exposure range, are considered unconvincing and cannot be used in the derivation of a threshold value.

6.2.3 German coal mining

General remarks

Various research programmes have been carried out to study the risk of German coal miners developing CWP, lung function disorders or chronic bronchitis/emphysema. Various research teams have attempted to determine the health risks for the German coal miners by various analyses of the available data.

Under the direction of the medical department of the Silikose-Forschungsinstitutes der Bergbau-Berufsgenossenschaft (Silicosis research institute of the miners' trade association) in Bochum (later called Berufsgenossenschaftliches Forschungsinstitut für Arbeitsmedizin (Research institute for occupational medicine)) in cooperation with works physicians in the Ruhr and Saar regions, the development of CWP, recorded as categories of the 1980 ILO classification, was studied mainly as a function of working time underground (Ulmer *et al.* 1991). The available radiographs were read again independently by at least two physicians. An associated group carried out lung function tests on the coal miners (Reichel 1989).

Under the direction of the Hauptstelle für Staub- und Silikosebekämpfung des Steinkohlenbergbauvereins (Centre for dust and silicosis prevention of the coal mining association) in Essen, extensive studies were carried out of the prevalence of the various categories of CWP in collectives of coal miners of the Ruhr and Saar collieries as a function of parameters such as the estimated cumulative exposure to respirable coal mine dust, the age at entering the occupation and the persistence of the dust in the lungs (Breuer *et al.* 1982, 1983). In these studies, the documentation available from the works physicians was used to find the individual radiographic pneumoconiosis categories; the findings were not recorded according to the ILO classification but according to the Johannesburg classification from 1930 (Russel 1930).

More recent results for the risk of developing CWP and for mortality were obtained in cooperative research by groups from the Institut für Arbeitswissenschaften (Institute for industrial science) of the Ruhrkohle AG and the Institut für Arbeits- und Sozialmedizin (Department of Occupational and Social Medicine) of the University of Cologne (Morfeld *et al.* 1992, Morfeld *et al.* 1997a, 1997b, 1997c). Studies of the incidence of chronic bronchitis among coal miners were also carried out as part of the extensive DFG chronic bronchitis study (DFG 1975, 1981).

The relevant studies are summarized below. They have been reviewed in detail by Morfeld and Piekarski (1998a).

Coal workers' pneumoconiosis/silicosis

Ophoff (1985) studied the effects of the age of the stratigraphic horizon on the development of CWP in the miners working there. For this purpose two groups were established with 109 coal miners in each; the miners had begun work underground between 1949 and 1958, were at least 18 years old in 1958 and had worked in cutting or hauling the coal or in coal preparation. The radiographs available from the occupational medical examinations of these miners in the years 1958 to 1980 were re-examined by an experienced physician and the results recorded using the ILO U/C 1971 classification. For the 218 coal miners, data for dust exposure were taken from the safety files of the Ruhrkohle AG for the period starting in 1978 and a "total exposure level" (calculated from the product of dust concentration classes and shifts worked) for the cumulative exposure to respirable coal mine dust was determined. The main results of this study demonstrated that the group of miners who had worked in coal seams in older stratigraphic horizons were exposed to slightly higher total respirable dust levels, but to levels of respirable quartz dust which were about half those encountered by the coal miners who had mainly worked in younger stratigraphic horizons. CWP of category 2/1 or higher was diagnosed in 18.8 % of the coal miners who had mainly worked in older stratigraphic horizons but in only 1.8 % of those who had mainly worked in younger seams. Therefore, this study provides clear evidence that the prevalence of CWP is positively correlated with the coalification rank, that is, the age of the coal seam (and thus at the same time negatively correlated with the level of exposure to respirable quartz dust). To be criticized in this study is the fact that the radiographs were not read independently by at least two physicians and that multivariate models were not used. Before 1958 analytical values for dust exposure levels were not available and during the whole study only few gravimetric determinations of the respirable dust concentrations.

Another study also provided evidence that miners working in older stratigraphic horizons were classified on average in a higher category than were miners who had worked for the same period in a younger stratigraphic horizon when the small round opacities in their radiographs were classified according to ILO (1980) (Piontek *et al.* 1991).

Schmitt (1991) and Reischig and Schmitt (1994) carried out follow-up studies of miners from the Saar coal mines to assess the further development of CWP when the persons were no longer exposed to dust. The persons included in the study had worked for at least ten years underground. For each person a new radiograph was obtained. The miners who came to the follow-up examination were mainly those already known to have CWP. Among the miners who came to the follow-up examination, the prevalence of confirmed CWP (CWSP $\geq$ 1/1; ILO 1980) for a 30-year period working underground was more than 40 % whereas in the collective of miners who did not come to the follow-up examination the prevalence of confirmed pneumoconiosis for the same period was less than 20 % (Schmitt 1991). The study demonstrated that the prevalence of all categories of CWP tended to increase during the whole period covered by the study, that is, not only during the exposure period (on average about 27 years) but also during the post-exposure observation period (on average about 16 years). Of the 281 miners who had no confirmed dust lung changes at the end of exposure (CWSP = 0/0: n = 137, CWSP = 0/1: n = 96, CWSP= 1/0: n = 48), 43 (15 %) developed CWP in the category 1/1 or higher; five of these persons developed PMF although at their retirement their CWP categories were only 0/1 or 1/0. In three cases (2.2 %), CWSP = 1/1 developed from CWSP = 0/0. The authors concluded that the CWP has a marked tendency to progress even after retirement.

According to Reisner *et al.* (1985b) (see Table 7), until the middle 1950s the average respirable coal mine dust concentrations (for TBF 50) varied, depending on the mine, between 3 and 18 mg/m^3; later reduction of the concentrations down to 2 to 5 mg/m^3 in the middle of the 1970s is described. The highest concentrations of respirable coal mine dust were found in mines with the lowest ash level in respirable dust, that is, in the mines in which coal of higher coalification rank and lower quartz content was mined. From the regression equations in Reisner *et al.* (1985b), the following estimates may be made for the prevalence of CWP of at least grade –I according to the Johannesburg classification (Russel 1930), assuming age 20 years at the start of underground work, a total of 35 years working underground (7000 shifts) and 17.5 years (210 months) persistence of dust in the lungs: assuming an average concentration of respirable coal mine dust of 1 mg/m^3, the prevalence of CWP among persons working younger, lower coalification rank bituminous coals is calculated to be 1.7 %, for those working older, higher rank coking coal 8.6 %; at an average concentration of respirable coal mine dust of 2 mg/m^3, the corresponding values are 6.5 % and 27.7 %. For the prevalence of at least grade 1 CWP, the corresponding calculated values are 0.8 % and 2.4 % for 1 mg/m^3 and 3.2 % and 9.5 % for 2 mg/m^3. Thus these results document the variation in risk with the stratigraphic horizon: the older the coal and the higher its coalification rank (and the less quartz it therefore contains), the higher is the risk for the miners of developing CWP. On the basis of the data from this study, a threshold value was later estimated: for the target value of CWP $\geq$ 1 after 15 years exposure, the threshold value was 6.0 mg/m^3 (Morfeld *et al.* 1997b).

In a new research programme to determine the risk of developing CWP in German coal mines, the first studies investigated prevalence and incidence of CWP in groups of miners from the Walsum colliery (Ruhrbergbau: Zeche N in Reisner *et al.* 1985b) (Morfeld *et al.* 1990, Vautrin 1992, Vautrin *et al.* 1989, 1990); especially the dynamics of disease development was studied (Morfeld 1994, Morfeld *et al.* 1994). The studies aimed to examine complete groups of miners who retired from the mine in a defined calendar year, whereby all data (e.g. time worked underground, calendar year, age) were taken from the records of the routine occupational medical examinations. All the available radiographs of the thorax organs for the miners in the study collective were independently categorized by two physicians according to the 1980 ILO classification. The examination of 478 miners who had worked underground for at least 20 years revealed a significant increase in the incidence of CWP (all categories between 0/1 and 2/2; ILO 1980) with increasing time underground; the risk of developing CWP within 35 years working underground (i.e. for the transition from CWSP = 0/0 to CWSP = 1/1) was given as 31.4 % and the risk of progression of the pneumoconiosis from category 1/1 to 2/2 as 38.5 %, assuming persistence of the initial state (CWSP = 1/1) for 10 years (Morfeld 1994). The study yields some evidence for a risk-reducing effect of workforce reallocation measures in underground coal mining (reduced rate of progression of CWSP = 1/1 to CWSP = 1/2). At the same time, however, it makes it clear that the probability of transition to the next category is mainly a function of the period of time spent in each state, that is, it is dominated by "internal" dynamics which explains why the course of the disease is better described by a stochastic model of a semi-Markov process than of a Markov process. In addition, even after adjustment for age and calendar year, the analysis reveals monotonically increasing incidences of transition from CWSP = 1/1 to CWSP = 2/2. In principle, selection problems and confounding cannot be excluded as explanations for these observations, but the results do support the demand (Woitowitz *et al.* 1989) that in future the development of confirmed CWP (CWSP = 1/1) be prevented as far as possible by appropriate occupational hygiene. In addition, the studies document a significant reduction in the risk of developing CWSP = 1/1 with calendar year (Morfeld *et al.* 1990).

Table 7. The main studies of CWP prevalence in German coal miners

kind of study:	cross-sectional study (prevalence) (Reisner *et al.* 1985b)
size of study:	records for subgroups of 18166 miners (not published in detail)
parameter:	CWP (Johannesburg classification (Russel 1930)) during about 20 years working underground (1954–1974)
exposure:	depending on the mine, 3–18 mg/m^3 (middle 1950s) and 2–5 mg/m^3 (middle 1970s); values converted from tyndallimetric values to TBF 50 equivalents
results:	estimated prevalence after 35 years working underground (7000 shifts) given 17.5 years persistence of dust in the lungs and age 20 years at the start of work in the industry:

concentration [mg/m^3]	CWSP ≥ –I coalification rank low	CWSP ≥ –I coalification rank high	CWSP ≥ I coalification rank low	CWSP ≥ I coalification rank high
1	1.7 %	8.6 %	0.8 %	2.4 %
2	6.5 %	27.7 %	3.2 %	9.5 %

Threshold estimated for CWSP ≥ I (Morfeld *et al.* 1997b): 6.0 mg/m^3 for 15 years exposure

other results: (Ophoff 1985, Piontek *et al.* 1991, Schmitt 1991, Ulmer *et al.* 1987, Wohlberedt 1988a, 1988b):

- marked reduction in risk with calendar year
- progression of the CWP after retirement
- higher prevalence in older stratigraphic horizons (i.e. seams of higher coalification rank with less quartz)

To investigate the effects of the exposure of coal miners to respirable coal mine dust and its components and to various stratigraphic horizons on the current risk of developing CWP, cohorts of persons starting work in the industry in the Ruhr and Saar collieries were selected (to avoid potential selection problems like those in earlier studies). For all persons taking part in the study, complete databases were established for the course of development of CWP and for dust exposure (especially respirable coal mine dust and respirable quartz dust) for the whole period of their work underground (Morfeld *et al.* 1992). Two typical collieries of the Ruhr area (Bergwerk Walsum: bituminous coal, young stratigraphic horizon, coal for power stations; Bergwerk Heinrich Robert: coking coal, older stratigraphic horizon) and all of the Saar mines existing in 1980 (specific stratigraphic horizon, relatively high quartz levels) were studied. The concentrations of respirable dust and respirable quartz dust underground were mainly determined gravimetrically after static and personal sampling with devices designed according to the Johannesburg convention of 1959. To determine the incidence of the various categories of CWP in persons working underground, all the available radiographs were re-evaluated independently by at least two physicians according to the 1980 ILO classification, also making use of the standard film collection of the trade association (Hauptverband der gewerblichen Berufsgenossenschaften e. V.). Follow-up of the morbidity and exposure of the coal miners from the Ruhr area was carried out for the years 1974 to 1991 and for those from the Saar collieries from 1980 to 1991 (Morfeld *et al.* 1995, 1997c). The average period of work underground in the Ruhr cohort (n = 1257) was 11.2 years (maximum 18 years), for the Saar cohort (n = 4521), 8.7 years (maximum 12 years). Exposure levels were estimated from the results of 43842 gravimetric determinations of respirable dust, the individual development of pneumoconiosis was followed by evaluating 47781 radiographs. The most important results are shown in Table 8 (Morfeld and Piekarski 1998a).

Chronic bronchitis/emphysema

The results obtained in lung function studies carried out by the silicosis research institute of the mining trade association are described below. In a sample population of Ruhr coal miners with and without silicosis, aged between 50 and 60 years, and in a similarly aged control collective who were not exposed to dust at work, the average values of the lung function parameters (VC, FEV_1, R_t, PaO_2, $PaCO_2$, DCO, IGV, compliance) were altered only in the miners and independently of whether or not the persons had CWP (Reichel 1989, Reichel *et al.* 1969, Ulmer 1970, Ulmer and Reichel 1972, Ulmer *et al.* 1968). Effects of the method of selection of the collectives were, however, not discussed and multivariate analysis was not carried out, not even for smoking habits. This makes it difficult to interpret the results.

The extensive study of the incidence of chronic bronchitis at work which was carried out with the support of the Deutsche Forschungsgemeinschaft (DFG 1975, 1981) also included collectives of coal miners in coal mines. The definition of the parameter ‘chronic bronchitis’ in this study differs from the otherwise usual definition and is based on the answers to the questions in a complex scheme constructed of combinations of anamnestic, clinical and lung function analytical parameters (Lange and Reiter 1972). In the assessment of obstructive ventilation disorders, the data are used to classify the individual in one of four categories, 0 to 3. In most analyses, classification in one of the two highest categories is considered to indicate the presence of a chronic bronchial reaction. The parameters recorded in the study included age, smoking habits, duration of work at the colliery and estimates of the exposure concentrations of respirable dust and total dust.

Table 8. Incidence of CWP in German coal miners (Morfeld *et al.* 1992, 1995, 1997c)

kind of study:	cohort of persons starting work in the industry, follow-up 1974–1991 (Ruhr), 1980–1991 (Saar)		
size of study:	5778 miners from 2 collieries in the Ruhr area (Heinrich Robert, Walsum) and 6 collieries in the Saar area		
parameter:	CWSP (ILO 1980)		
results:	exposure to respirable dust, incidence of pneumoconiosis; Cox regression: significantly higher risk of CWSP = 0/1 in the older horizons in the Ruhr mines than in the younger Ruhr horizons; weak evidence of a positive association with cumulative level of respirable quartz dust		
	Heinrich Robert (older horizon)	Walsum (younger horizon)	Saar (specific younger horizon)
cohort size	564	693	4521
concentration [mg/m^3]			
coal mine dust			
arithmetic mean	2.9	2.3	1.6
maximum	5.0	5.1	3.7
quartz dust			
arithmetic mean	0.05	0.10	0.21
maximum	0.13	0.21	0.81
incidence during the years working underground (absolute):			
category 0/1	10	4	45
category 1/0	0	0	0

Table 9. Incidence of chronic bronchial reaction in German coal miners (Lange and Ulm 1983)

kind of study:	cross-sectional study (prevalence)
size of study:	1090 miners from coal mines
parameter:	chronic bronchial reaction (CBR)
exposure:	concentration of respirable coal mine dust 8.3 mg/m^3, on average for 25 years working underground
results:	significant positive association of the average concentration of respirable coal mine dust with the CBR prevalence, independent of age, period worked underground and smoking; background prevalence (exposure concentration 0 mg/m^3): 23.9 % (non-smokers), 44.4 % (smokers); average respirable dust concentration required to double the prevalence: 16.9 mg/m^3 (non-smokers), 36.2 mg/m^3 (smokers); estimated threshold for CBR (Ulm *et al.* 1989): 6.75 mg/m^3 given exposure for 26 years

The DFG study (1981) reveals an adverse effect on the bronchopulmonary system of exposure to dust at the workplaces investigated; the effect is independent of age and smoking habits. For the mining collective, independently of the years worked underground (or age) and of smoking habits, the study reveals an effect of exposure to respirable dust on the development of chronic bronchitis, an effect which is significant at the 1 % level. The selection of the collective of coal miners for the DFG study (1981), the method of estimation of the exposure levels and the choice of analytical model are discussed in detail in Morfeld and Piekarski (1994, 1996). A significant limitation of the study is that it was not possible to separate the effects of age from those of time worked underground because of the close correlation of these variables. For an average exposure period of 26 years, a threshold of 6.75 mg/m^3 was calculated (Ulm *et al.* 1989).

Mortality

A historic cohort study of the mortality of 4632 coal miners who had worked underground in the Saar collieries for at least five years and who comprised about 40 % of the underground workers of the Saarbergwerke AG between 1977 and 1979 was carried out on the basis of the causes of death recorded in the death certificates (Lampert *et al.* 1997). The mortality data were complete; 474 miners died in the study period between 01.01.1980 and 31.12.1994. The total mortality ratio (indirectly standardized for age and calendar year on the basis of data for the male population of Saarland) was 0.67. For the cause of death "coal workers' pneumoconiosis" the corresponding value was 3.03 (95 % confidence interval: 1.51 to 5.42). Thus the study documented excess mortality from CWP in the collective of coal miners. However, there is a marked healthy worker effect and the expected values determined from the data for the whole male population of Saarland are too high because the coal miners are included in this population; therefore it may be concluded that the SMR value obtained for the cause of death "coal workers' pneumoconiosis" is too low. For "chronic bronchitis" as cause of death the SMR value obtained was 0.51, for "emphysema" the corresponding value is 0.34. Thus the study in its present form reveals no excess deaths from chronic bronchitis and/or emphysema. An extension of the follow-up period is necessary to increase the number of deaths studied.

Conclusions

The results of the studies carried out by the various research groups document a positive association between the incidence of CWP and the duration of work underground or the level of the cumulative exposure to respirable coal mine dust (see Sections 6.2.1 and 6.2.2). The studies also demonstrate significant progression of the disease even after the end of exposure. At the same time, studies have demonstrated that the pneumoconiosis risk decreases in later calendar years.

All the studies of Ruhr coal mines in which the analysis took the stratigraphic horizon into account demonstrate, whether or not the results were adjusted for other components of the exposure to respirable coal mine dust, a higher risk of developing CWP during extraction of higher rank coal containing less quartz. For the Saar coal mines, there is evidence that the risk of developing CWP is not as low as would be expected from the data for the Ruhr coal mines because of the lower levels of respirable coal mine dust, without considering the higher levels of quartz.

It must be pointed out in this context that the quantitative relationship between exposure levels of respirable coal mine dust and the risks of developing the various grades of CWP derived by Reisner *et al.* (1985b) may only be used for the derivation of a threshold level of respirable coal mine dust after additional corrections have been carried out. And even then, the interpretation is made more difficult by fundamental deficiencies in the data (non-representative selection of a collective comprising about 5 % of the exposed population, radiograph assessment without duplicate readings and without standard films not using the ILO classification, use of non-gravimetric short-term tyndallimetry of konimetric samples for determination of exposure concentrations, stratigraphy of the coal seams not taken into account). More recent studies (Morfeld *et al.* 1995) avoided these problems but because the observation time was limited can only be used with reservations for a cautious estimation of "minimum" threshold levels.

Analyses carried out in the DFG study of chronic bronchitis (DFG 1981) demonstrate that the exposure of coal miners to respirable coal mine dust is significantly negatively correlated with the incidence of chronic bronchial reactions. In this study too, fundamental limitations must be taken into account (selection problems, probable systematic errors in exposure estimates, inability to analyse the effects of time worked underground because of its close correlation with age).

6.2.4 Australian coal mining

There are no large studies of the risk of developing CWP in Australian coal mines. The extensive studies of the risk of developing chronic bronchitis/emphysema (morbidity and autopsy studies) demonstrate an association between exposure to coal mine dust and the development of the disease. Quantitative estimates of risk as a function of underground exposure to respirable coal mine dust were, however, not attempted (Leigh 1990, Leigh *et al.* 1982, 1986, 1993, 1994, 1997).

6.2.5 Coal mining in other countries

The qualitative demonstration that coal mining is associated with a specific risk of developing CWP and chronic bronchitis/emphysema or impairment of lung function has also been carried out in coal mines other than those in Britain, Germany, USA and Australia. Quantitative relationships between the exposure to respirable coal mine dust and the development of CWP or chronic bronchitis/emphysema or impaired lung function have not been published in a form suitable for use in the derivation of a threshold value. The one exception is the study of Sardinian coal mining by Carta *et al.* (1996) who emphasize the importance of the parameter "impaired lung function" for the establishment of a threshold value for coal mine dust. Significant adverse effects on the measured lung function parameters were recorded in persons who had been exposed underground to concentrations of respirable coal mine dust of 1.7 and 3.0 mg/m^3 for 10 years.

6.3 Specific adverse effects of quartz in coal mine dust

It is generally agreed that pure respirable quartz dust has cytotoxic and silicogenic effects and also that exposure to respirable coal mine dust is the cause of CWP (WHO 1986). In German coal mines, the level of quartz in respirable coal mine dust varies widely. In younger stratigraphic horizons, the level of quartz in the respirable coal mine dust is generally much higher than in older seams (Morfeld *et al.* 1995): quartz makes up about 12 % of the dust in Saar mines (younger stratigraphic horizon) but, for example, only about 2 % in the Ruhr mine Heinrich Robert (older stratigraphic horizon) and the higher level of quartz would be expected to result in higher levels of quartz-specific damage in persons exposed to respirable coal mine dust from younger horizons. In other words, we would expect the risk to be negatively associated with the age of the stratigraphic horizon (i.e. with the coalification rank of the coal) for persons exposed to a fixed concentration of respirable coal mine dust. However, epidemiological studies have demonstrated that for a given level of exposure to respirable coal mine dust, the risk of developing CWP (silicosis) is positively associated with the coalification rank: the older the coal seam and the higher the coalification rank of the coal being mined, the more likely is it that the radiograph will reveal an abnormal classification according to ILO 1980 (see Section 6.2.2, especially Attfield and Morring 1992a, 1992b, Attfield and Seixas 1995, Hurley and Maclaren 1987, Hurley *et al.* 1982, Morfeld *et al.* 1997c, Reisner *et al.* 1985b).

Respirable coal mine dust is considered to be the main agent which causes CWP ("silicosis"). But the quartz in the coal mine dust cannot be considered to be the substance causing the damage, independent of the coalification rank of the mined coal and independent of the silicate content of the respirable coal mine dust.

6.4 Carcinogenic effects

There are numerous epidemiological studies of the carcinogenicity of coal mine dust. The results of these studies have been reviewed and evaluated by the IARC (IARC 1997, see especially their Table 26). The review is summarized below.

Most epidemiological studies have yielded data for the cancer risk for the occupation coal mining but not for the exposure levels involved. The lung cancer and stomach cancer risks have been studied most intensively. In addition, a few studies have investigated the risk of cancer in other organs such as the urinary bladder. Because there are generally no precise data for the composition of the coal mine dust (e.g. proportions of coal, metals, quartz), the epidemiological studies are difficult to evaluate.

The cohort studies have revealed no consistent causal relationship between exposure to coal mine dust and the development of lung cancer. There is no consistent evidence for a relationship between lung cancer risk and exposure duration, cumulative exposure level or radiographic evidence of pneumoconiosis.

On the other hand, the data from both cohort studies and case-control studies are consistently indicative of an increased risk of stomach cancer in coal miners. However, there is no consistent evidence of a dose-response relationship between the level of exposure to coal mine dust and the risk of developing stomach cancer.

A meta-analysis (Morfeld and Piekarski 1998b) of the results of 13 SMR follow-up studies (Armstrong *et al.* 1979, Atuhaire *et al.* 1985, Christie *et al.* 1995, Costello *et al.* 1974, Kuempel *et al.* 1995, Lampert *et al.* 1997, Miller *et al.* 1997, Ortmeyer *et al.* 1973, 1974, Rockette 1977, Starzynski *et al.* 1996, Swaen *et al.* 1995, Une *et al.* 1995) has been carried out. In 9 of these studies the parameter "cancer" was studied, in 11 "lung cancer", in 10 "stomach cancer", in 6 "bladder cancer" and in 5 "leukaemia". In addition, the analysis included 15 OR case-control studies (Ames 1983a, 1983b, Ames *et al.* 1983, Bartsch and Rondia 1995, Bolm-Audorff *et al.* 1997, Coggon *et al.* 1990, Cordier *et al.* 1993, Gilman *et al.* 1985, Golka *et al.* 1995, González *et al.* 1991, Meijers *et al.* 1990, Morabia *et al.* 1992, Schifflers *et al.* 1987, Swaen *et al.* 1987, Weinberg *et al.* 1985): 6 studies of lung cancer, 4 stomach cancer, 3 bladder cancer and one leukaemia.

Some of the studies reviewed by the IARC (1997) were not and other studies were included in the meta-analysis. A more exact account of the studies used is given by Morfeld and Piekarski (1998b). All studies with external controls or control populations were pooled using the method of precision weighting recommended by Greenland (1987).

The results of the meta-analysis are given below.

1. Taken as a whole, the studies do not yield evidence of an increased total cancer risk (RR (relative risk) = 0.95) for coal miners.
2. Nor do they demonstrate an increased incidence of lung cancer (RR: 0.96). However, there are indications that the risk may have been underestimated especially in follow-up studies because of survivor effects and it is noteworthy that three case-control studies (Bartsch and Rondia 1995, Bolm-Audorff *et al.* 1997, Morabia *et al.* 1992) and the extensive US American follow-up study by Rockette (1977) found significantly increased risks of lung cancer in coal miners, even though smoking habits were also accounted for in the three case control studies. It must be pointed out here that, of the studies with negative results, neither the American studies (Ames 1983a, Ames *et al.* 1983, Kuempel *et al.* 1995) nor the British studies (Miller *et al.* 1981, 1997, Miller and Jacobsen 1985) exclude an increasing lung cancer risk with increasing cumulative exposure to respirable coal mine dust with sufficient statistical certainty.

   More detailed analysis of the German follow-up study (Morfeld *et al.* 1997b) suggests that excess lung cancer risks were masked by healthy worker selection processes, especially in the mortality studies with internal controls. When the survivor-selection processes are explicitly taken into account in appropriate models, the sign of the effect of duration of work underground on lung cancer mortality changes from negative to positive and there is evidence that the positive effect of the time since the start of exposure on lung cancer mortality was underestimated by about 150 %. In addition, when the follow-up period is extended to 15 years and the pneumoconiosis category at the beginning of follow-up was taken into account in this study (without correction for smoking habits), the risk of lung cancer for coal miners with radiographic signs of pneumoconiosis (> –I Johannesburg classification (Russel 1930), SMR = 1.28) was significantly increased (about 2.5 times) above that for miners without radiographic signs of pneumoconiosis (O, X in the Johannesburg classification, SMR = 0.51) (Lampert *et al.* 1997, Morfeld *et al.* 1997c). This excess risk is, however, not associated with the development of PMF. Thus, although the initial estimates of the risk of developing lung cancer for coal miners were modest, it is not possible to refute the suspicion that exposure to respirable coal mine dust results in an excess risk.

3. The overall results for stomach cancer (RR = 1.30) reveal a significantly increased risk for coal miners. However, it is not generally accepted that these results indicate a carcinogenic effect of swallowed coal mine dust and there are several hypotheses to account for the increased risk of stomach cancer among coal miners. The potential confounders which have been suggested include chewing of tobacco and use of snuff, consumption of spirits, working at night and in shifts, exposure to *Helicobacter pylori*, eating habits (high levels of salt and fat in the diet) and socio-economic status (Ames 1982, Atuhaire *et al.* 1985, Boeing 1991, Boeing and Frentzel-Beyme 1991, Boeing *et al.* 1991, Finkelstein *et al.* 1987, Harrison and Morris 1989, Hirayama 1986, 1992). Because of the complexity of the potential confounders, it is not possible to make deductions as to causality from the epidemiological studies of Miller *et al.* (1981), Miller and Jacobsen (1985) or Swaen *et al.* (1995). In addition, the evidence of a causal effect of the cumulative exposure to respirable coal mine dust in the development of stomach carcinoma conceived by Miller *et al.* (1981) was not confirmed in the more recent analysis by Miller *et al.* (1997). The inverse relationship between risk and the pneumoconiosis category found in the Dutch study (Swaen *et al.* 1995) was not confirmed in the British study (Miller *et al.* 1981) and so an interpretation of the results of Swaen *et al.* (1995) in the sense of the causality model of Meyer *et al.* (1980) is not justified at present.
4. The available studies of the risk of developing leukaemia, or cancer of the bladder, prostate or nose do not yield evidence of a causal relationship between exposure to coal mine dust and development of these tumours (see Morfeld and Piekarski 1998b).

6.5 Genotoxicity studies in man

Chromosomal aberrations were studied in peripheral lymphocytes of groups of 23 to 31 persons who were exposed to coal mine dust (Sram *et al.* 1985). The first group of exposed persons was employed in open-cast mining, the second in underground mines, the third in processing plants; the control group was not exposed to any known mutagens. An increased incidence of chromosomal aberrations was found only in the lymphocytes of the persons exposed in open-cast mining. The authors suggested that this group were exposed to higher levels of polycyclic aromatic hydrocarbons formed in combustion processes.

As a marker for oxidative DNA damage in lymphocytes, the ratio of 7-hydro-8-oxo-2′-deoxyguanosine to deoxyguanosine was determined (Schins *et al.* 1995). Lymphocytes from 38 retired miners (30 healthy and 8 with CWSP) and from 24 control persons were examined. The ratio was higher in the miners than in the control persons. There was, however, no difference between the healthy miners and those with CWSP. No relationship was found between oxidative DNA damage and the cumulative exposure to dust or the total period of exposure.

There are no data available for allergenic effects of coal mine dust.

6.6 Other effects

In studies of coal miners with and without silicosis, it was demonstrated that macrophages from the bronchoalveolar lavage fluid released cytokines including TNF-α and interleukin-1 (Lassalle *et al.* 1990, Vanhee *et al.* 1994) and increased concentrations of immunoglobulin A, G and M were detected (Calhoun *et al.* 1986). Studies of peripheral lymphocytes of miners have not yielded consistent results. In 82 miners with silicosis, the number of circulating T lymphocytes was found to be significantly reduced and that of circulating B lymphocytes increased; in the hilar lymph nodes the T helper cell count was increased (Watanabe *et al.* 1987). These results indicate that, in persons with silicosis, T cells migrate into the lymph nodes; this may be expected to cause local B cell stimulation and thus to promote immunoglobulin production. The release of tumour necrosis factor (TNF) from the blood monocytes of miners was higher than that from monocytes of control persons (Schins 1996). Alveolar macrophages from miners with PMF released less transforming growth factor (TGF-β) than did those from miners with simple pneumoconiosis (IARC 1997).

In coal mine dust recovered from the lungs and lymph nodes of coal miners, long-lived radicals were detected (exposed persons: $5.13 \pm 1.3 \times 10^{17}$, controls: $< 10^{15}$ spin/g) (Dalal *et al.* 1991). The concentration of radicals in the coal mine dust was shown to correlate with the pneumoconiosis category.

In the bronchial lavage fluid of persons exposed to coal mine dust, the concentration of reactive oxygen species (ROS) was higher than in control persons (Hansen and Mossman 1987).

7 Animal Experiments and *in vitro* Studies

There are no data available for acute toxic effects of coal mine dust nor for local effects on skin and mucous membranes.

7.1 Subchronic and chronic non-neoplastic effects

Histological examination of the lungs of rats which had inhaled coal mine dust without added quartz for long periods or been treated intratracheally with coal mine dust containing little quartz revealed aggregations of macrophages containing dust, signs of inflammation (alveolitis, alveolar lipoproteinosis), mild fibrosis and hyperplasia (Busch *et al.* 1981, Karagianes *et al.* 1981, Le Bouffant *et al.* 1982, Lewis *et al.* 1986, 1989, Ray *et al.* 1951b, Robertson *et al.* 1988). Many of these changes were said to be like the symptoms of human silicosis caused by exposure to dust mixtures but the hyperplasia and bronchiolization of the alveoli (transformation of the squamous alveolar epithelium to columnar epithelium) were exceptions (Busch *et al.* 1981, Karagianes *et al.* 1981).

Many studies with animals and *in vitro* systems have investigated the effect of quartz in the induction of fibrogenesis by coal mine dust (Kysilka *et al.* 1985, Le Bouffant *et al.* 1982, Martin *et al.* 1972, Ray *et al.* 1951a, 1951b, Reisner *et al.* 1985a, Rosmanith *et al.* 1981, 1982, 1983, 1985, Ross *et al.* 1962, Woitowitz *et al.* 1989). The results are summarized below.

- The fibrogenicity of mine dusts is dependent on their quartz content. However, the correlation is only weak.
- The addition of quartz to coal mine dusts increases their fibrogenicity in the rat lung.
- The fibrogenicity of mine dusts is considerably lower (that is, by more than one order of magnitude) than would be expected from their quartz content.

The last result and also the observation that the effects of pure quartz are reduced by coal combustion residues (Martin *et al.* 1972) can be explained in terms of the inhibitory effects of clay minerals on fibrogenesis (Bruch *et al.* 1983, Le Bouffant *et al.* 1977, Walton *et al.* 1977). Such phenomena have been studied by means of Auger spectroscopy and thermoluminescence procedures (Kriegseis and Scharmann 1982). The authors demonstrated that the surface of quartz particles in mine dusts are generally covered with quartz-free inorganic material. They considered it likely, however, that the electronic structure of the SiO_2 could be exposed at the surface at a small number of submicroscopic points. *In vitro* studies of alveolar macrophages have demonstrated that the quartz in coal mine dust has much weaker effects than the same amount of pure quartz (mainly between 5 % and 30 % of the effects of DQ 12) (Rehn *et al.* 1995). The specific effects of real respirable coal mine dust are perhaps caused by components other than quartz, namely by particles containing iron and manganese. This is suggested by the results of *in vitro* studies (Tourmann and Kaufmann 1994a, 1994b). The *in vitro* findings suggest that the fibrogenic effects of respirable coal mine dusts are a result of the permeability coefficients and surface tension of the dust suspensions (Brammertz *et al.* 1995).

7.2 Carcinogenicity

There are three carcinogenicity studies of inhaled coal mine dust in rats. Karagianes *et al.* (1981) and Lewis *et al.* (1989) investigated especially the situation of coal miners exposed to a combination of coal mine dust and diesel motor emissions. Martin *et al.* (1977), on the other hand, aimed to clarify the significance of the quartz content of mine dust. From the dust concentrations and exposure times chosen in the last-mentioned study, the theoretical dust concentration which would yield the same cumulative exposure level after exposure during 40 hours weekly for two years may be calculated to be 47 mg/m^3 (equivalent concentration or time-weighted mean value). After two years almost 100 mg dust per rat lung was found. In one group of rats exposed to practically quartz-free coal mine dust, tumours were diagnosed histologically in 4 of 36 female Sprague-Dawley rats and, in the gross pathological examination, 3 tumours were detected in 45 lungs. Because apparently only 6 control animals were examined histologically, the confirmed lung tumour incidence is not statistically significant although an incidence as high as 11 % is clearly above the expected value in this strain of rat (Sprague-Dawley). A second group was exposed to coal mine dust to which so much quartz had been added that about 10% quartz was recovered from the lung dust. Lung tumours (epidermoid tumours and adenocarcinomas) developed in 32/72 of the rats exposed to such mixtures. The methods used make it difficult to interpret the results of this study (Martin *et al.* 1977).

In the studies of Karagianes *et al.* (1981) and Lewis *et al.* (1989), the tumour incidence was not clearly increased in any of the groups; nonetheless, the studies cannot be considered to have yielded negative results. The levels of coal mine dust to which the animals were exposed were many times lower than in the study described above. In the experiments of Karagianes *et al.* (1981) the longest observation period was only 20 months, at which time only 6 animals were still alive. Even the 24-month observation period of Lewis *et al.* (1989) must be seen as a limitation of the study because lung tumours often do not develop in rats until the animals are old; 30 months is the necessary observation period for such studies. The limitations of these studies are also revealed by the fact that an increase in tumour incidence could not be induced even by exposure to diesel motor emissions which have been shown to have such effects in other studies.

Interim results are available from a carcinogenicity study in progress in which rats are being exposed to one of five dusts by intratracheal instillation. Two of the dusts were prepared by milling coal, two appear to be coal mine dusts and one is designated as rock dust. After two years, primary lung tumours were diagnosed in the histological examination in animals from all of the exposure types (all five dusts). The doses administered were 60 and 120 mg per rat; there were 48 animals per group. Quartz DQ 12, administered in doses of 5, 10 and 20 mg, also caused a significant increase in the incidence of lung tumours. Primary lung tumours were not found in the 48 control animals. However, the results of this study have not yet been finally evaluated (Roller *et al.* 1998).

Because previous studies by Wagner and Wagner (1972) had revealed an increase in the incidence of histiocytic lymphomas in rats treated by intrapleural injection of quartz, Wagner (1976) investigated whether such tumours developed after intrapleural application of coal mine dust or synthetic soot (probably single doses of 20 mg in 0.4 ml saline). No increases in tumour incidence were found (coal: 1/16 rats, mean survival period 690 days; soot: 2/16 rats, 618 days; control 1/15 rats, 720 days).

7.3 Effects on the immune system

In several studies, an increase in the number of macrophages and other inflammatory cells was seen in the lung lavage fluid of rats exposed to coal mine dusts and a change in the proportions of the various cell types, especially an increase in the proportion of neutrophilic granulocytes (Bowden and Adamson 1978, Brain 1971, Brown *et al.* 1992, Castranova *et al.* 1985, Chen *et al.* 1988, Donaldson *et al.* 1990, Kusaka *et al.* 1990). The studies have been reviewed in detail in the IARC monograph (IARC 1997).

7.4 Genotoxicity

In vitro

Nitrosation of coal mine dust can occur *in vivo* at the low pH found in the stomach and can result in the development of stomach tumours (IARC 1997, NIOSH 1995; see Section 6.4). Therefore various genotoxicity tests have been carried out with nitrosated coal mine dust extracts.

With extracts of dust from coal mines, no evidence of mutagenic effects was found in the gene mutation test with *Salmonella typhimurium* (TA98) either in the presence or absence of a metabolic activation system. In the presence of S9 mix at the highest concentration (about 312 mg/plate), a borderline positive result was obtained. In contrast, after nitrosation of the dust samples from coal mines, positive results were obtained in the *S. typhimurium* strains TA98 and YG1024, but not in TA1535 (Green *et al.* 1983, Hahon *et al.* 1988, Krishna *et al.* 1987, Stamm *et al.* 1994, Whong *et al.* 1983) (cited from Table 28 of IARC 1997).

In the mouse lymphoma test, an increased incidence of mutations was found with concentrations of at least 1 µl/ml of nitrosated coal mine dust extract (Tucker *et al.* 1984).

Nitrosated coal mine dust extracts induced sister chromatid exchange in Chinese hamster cells and human lymphocytes without metabolic activation. Samples which were not nitrosated were tested only with human lymphocytes and yielded negative results (Tucker *et al.* 1984). Nitrosated coal mine dust extract induced the formation of micronuclei in BALB/c-3T3 cells. Non-nitrosated samples were not tested (Gu *et al.* 1992).

In a cell transformation test with BALB/c-3T3 cells, a concentration-dependent increase in the numbers of transformed cells was obtained both with an extract of coal mine dust and with nitrosated samples at concentrations between 2.05 and 8.32 mg/ml. The effect was more pronounced with the nitrosated samples (Wu *et al.* 1990).

In vivo

In rats which had inhaled coal dust in a concentration of 2 mg/m^3, 7 hours daily, 5 days per week for 24 months, no micronuclei were found in the bone marrow and no mutagenic metabolites in the urine (Green *et al.* 1983, Ong *et al.* 1985).

7.5 Other effects

In vitro, coal mine dust was shown to induce haemolysis of sheep erythrocytes (Gormley *et al.* 1979). A number of samples of coal mine dust in concentrations between 20 and 40 mg/ml were shown to be cytotoxic for macrophages isolated from various species (rat, guinea pig, pig, rabbit) and for cells of a permanent macrophage-like cell line (P388D1) (Gormley *et al.* 1979). The concentration of quartz in the coal mine dust did not correlate with the cytotoxicity. Similar results have been obtained in other studies (Bruch and Rehn 1994, Massé *et al.* 1994, Reisner and Robock 1977, Robock and Reisner 1982). That dusts containing quartz have macrophage-activating effects may be shown in the rat model with intrapulmonary alveolar macrophages which release increased levels of preinflammatory cytokines such as TNF-α, interleukin-6 or insulin-like growth factor after exposure to quartz. Extrapulmonary effects indicative of systemic stimulation may also be demonstrated with peritoneal macrophages (Mohr *et al.* 1991). The cytokines have macrophage-activating, pro-inflammatory and fibroblast-stimulating activity and they promote the development of granulomas and formation of new connective tissue. Because of these findings, continuous macrophage stimulation is considered to be significant in the development of silicosis (Davis 1986). Coal mine dust in a concentration of 1 mg/ml stimulated the release of pro-inflammatory and fibrogenic cytokines from human alveolar macrophages *in vitro* more than did quartz (0.03 mg/ml) or titanium dioxide (1 mg/ml) (Gosset 1991).

Tourmann and Kaufmann (1994a, 1994b) suggested that the iron in coal mine dust could play a role in its toxic effects, namely by the formation of hydroxyl radicals in the Fenton reaction (Dalal *et al.* 1995).

The effects of radicals including reactive oxygen species in the lung could include damage to cell membranes by lipid peroxidation, oxidation of proteins and DNA damage (Fubini *et al.* 1995). The oxidative DNA damage is probably the result of reactions with hydroxyl radicals (Arumoa *et al.* 1989, Schraufstätter and Cochrane 1991) and can result in cell death or in tissue proliferation. It plays a role in carcinogenesis (Janssen *et al.* 1993), in the pathogenesis of emphysema (Huang *et al.* 1993) and in other toxic effects.

It is assumed that mechanical processes such as the milling of coal mine dust generate radicals on newly formed surfaces (Dalal *et al.* 1989, Vallyathan *et al.* 1988).

Numerous *in vitro* and some *in vivo* studies have demonstrated that coal mine dust can induce the release of inflammation mediators, growth factors (PDGF, TGF) and cytokines (TNF, interleukin, etc.) (Borm *et al.* 1988, Gosset 1991, Lassalle *et al.* 1990).

8 Derivation of a threshold value for respirable coal mine dust

Morfeld *et al.* (1997b) and Morfeld and Piekarski (1998a) estimated a threshold value for the concentration of respirable coal mine dust on the basis of the results of German studies of the risk of developing chronic bronchitis (chronic bronchial reaction, DFG study 1975, 1981) or CWP (category = 1/1, ILO 1980). CWP proved to be the more sensitive parameter and a model with a threshold proved to provide a better fit for the dose-response relationship than did the normal kind of regression model. The threshold value was found to lie between 1.5 and 6 mg/m^3 depending on whether or not the threshold of effect is assumed to decrease with time. In extensive analyses, Morfeld *et al.* (1998a) found the 97.5th percentile of the respirable dust concentration to be 3.2 times the arithmetic mean value of 2.6 mg/m^3. At these concentrations there was no occupational medical evidence of adverse effects on health. From this result, therefore, an excursion factor of 3 was derived.

From the results of British studies, Morfeld and Piekarski (1998a) determined the risk of developing CWP to be 2.5 % after exposure for 35 years to an average concentration of 1 mg/m^3 and 6 % at 2 mg/m^3. These risk estimates were, however, carried out without assuming the existence of a threshold.

9 Manifesto (MAK value/classification)

There are no data available for exposed persons from which a concentration of coal mine dust without effects in the sense of a no observed adverse effect level (NOAEL) can be derived. A threshold of 1.5 mg/m^3 for the respirable fraction of coal mine dust was derived from the results of the epidemiological studies of miners in German coal mines by means of a threshold model (to avoid the development of coal workers' pneumoconiosis category 1/1 within life expectancy, even when the decrease in the threshold of effect until age 75 is taken into account). Because of the wide scatter of the data, however, considerable uncertainty is associated with this value.

In vitro genotoxicity studies have yielded no evidence of genotoxic effects. However, after nitrosation, coal mine dust is mutagenic and clastogenic. It is conceivable that nitrosation of coal mine dust could take place in the acid milieu of the human stomach. However, this has been demonstrated to date only in *in vitro* studies. In rats which have inhaled coal mine dust, no increase in tumour incidence is observed. In still incomplete studies in which animals were exposed to coal mine dust by intratracheal instillation, no evidence of carcinogenic effects was found; final assessment of the results has, however, not yet been carried out.

Epidemiological studies suggest a carcinogenic potential with the target organs lung and stomach. An increase in the incidence of lung cancer could not be demonstrated. However, there are indications that the risk has been underestimated, especially in follow-up studies because of "survivor effects". More detailed analysis of the German follow-up study suggests that excess lung cancer risks were masked by healthy worker selection processes, especially in the mortality studies with internal controls. Thus, although the initial estimates of the risk of developing lung cancer for coal miners were modest, it is not possible to refute the suspicion that exposure to respirable coal mine dust results in an excess risk.

Most epidemiological studies reveal an increased risk of stomach cancer among coal miners but find no correlation with exposure level or exposure time.

Therefore coal mine dust has been classified in Section III, Category 3B of the *List of MAK and BAT Values*. A scientifically based MAK value cannot be established.

The epidemiological findings and results of animal studies indicate that the fibrotic effects of coal mine dust are largely independent of the quartz content up to a quartz level of 7 %. More research is necessary to establish the effects of higher proportions of quartz.

7 References

ACGIH (American Conference of Governmental Industrial Hygienists) (1984) *Particle size-selective sampling in the workplace.* Report of the ACGIH Technical Committee on air sampling procedures, ACGIH, Cincinnati, OH, USA

Ames RG (1982) Gastric cancer in coal miners: some hypotheses for investigation. *J Soc Occup Med 32*: 73–81

Ames RG (1983a) Does coal workers pneumoconiosis predict for lung cancer? Some evidence from a case-control study. *J Soc Occup Med 33*: 141–144

Ames RG (1983b) Gastric cancer and coal mine dust exposure: a case-control study. *Cancer 52*: 1346–1350

Ames RG, Gamble JF (1983) Lung cancer, stomach cancer, and smoking status among coal miners. *Scand J Work Environ Health 9*: 443–448

Ames RG, Amandus H, Attfield M, Green FY, Vallyathan V (1983) Does coal mine dust present a risk for lung cancer? A case-control study of US coal miners. *Arch Environ Health 38*: 331–333

Armstrong BK, McNulty JC, Levitt LJ (1979) Mortality in gold and coal miners in Western Australia with special reference to lung cancer. *Br J Ind Med 36*: 199–205

Arumoa OI, Halliwell B, Dizdaroglu M (1989) Iron ion-dependent modification of bases in DNA by the superoxide radical-generating system hypoxanthine/xanthine oxidase. *J Biol Chem 264*: 13024–13028

ATS (American Thoracic Society) (1962) Chronic bronchitis, asthma, and pulmonary emphysema: a statement by the committee on diagnostic standards for non tuberculous respiratory diseases. *Am Rev Respir Dis 85*: 762

Attfield MD (1985) Longitudinal decline in FEV_1 in United States coal miners. *Thorax 40*: 132–137

Attfield MD, Hodous TK (1992) Pulmonary function of US coal miners related to dust exposure estimates. *Am Rev Respir Dis 145*: 605–609

Attfield MD, Morring K (1992a) The derivation of estimated dust exposure for US coal miners working before 1970. *Am Ind Hyg Assoc J 53*: 248–254

Attfield MD, Morring K (1992b) An investigation into the relationship between coal workers pneumoconiosis and dust exposure in US coal miners. *Am Ind Hyg Assoc J 53*: 486–492

Attfield MD, Seixas NS (1995) Prevalence of pneumoconiosis and its relationship to dust exposure in a cohort of US bituminous coal miners and ex-miners. *Am J Ind Med 27*: 137–151

Attfield MD, Vallyathan V, Green FHY (1994) Radiographic appearances of small opacities and their correlation with pathology grading of macules, nodules and dust burden in the lungs. *Ann Occup Hyg 38*: 782–789

Attfield MD, Kuempel E, Wagner GR (1997) Exposure-response for coal workers pneumoconiosis in underground coal miners. A discussion of issues and findings. *Ann Occup Hyg 41, Suppl 1*: 341–345

Atuhaire LK, Campbell MJ, Cochrane AL (1985) Mortality of men in the Rhondda Fach 1950–80. *Br J Ind Med 42*: 741–745

Bartsch P, Rondia D (1995) *Recherche épidémiologique sur l'origine professionelle des cancers primaires des bronches* (Epidemiological study of the occupational origins of primary bronchial cancer) (French). "Health Hazards" Impulse Program Final Research Report. Prime Minister's Office – Federal Office for Scientific, Technical and Cultural affairs, Brussels

Bauer H-D (1973) Umstellung des Meß- und Beurteilungsverfahrens zur Einstufung von Arbeitsplätzen im Steinkohlebergbau (Changes in the assay and assessment procedures for the classification of workplaces in coal mines) (German). *Kompaß 83*: 1–10

Bauer H-D (1995a) Staubjahre – Ein Modell zur Ermittlung der kumulativen Staubdosis (Dust years – a model for determining the cumulative dust dose) (German). *Kompaß 105*: 317–323

Bauer H-D (1995b) *Staubjahre* (Dust years) (German). BIA-Report 7/95, Hauptverband der gewerblichen Berufsgenossenschaften, St. Augustin

Bauer H-D, Bruckmann E (1974) Neues gravimetrisches Staubsammelgerät zur Messung des atembaren Feinstaubes nach MAK-Werten (New dust sampler for gravimetric analysis of respirable fine dust in line with MAK values) (German). *Kompaß 84*: 96–102

Bauer H-D, Fricke H (1995) *Feststellung und Differenzierung von polyzyklischen aromatischen Kohlenwasserstoffen im Steinkohlestaub* (Detection and differentiation of polycyclic aromatic hydrocarbons in coal dust) (German). Wissenschaftlicher Abschlußbericht zum EGKS-Forschungsvorhaben 7263-02/098/01, Forschungszeitraum 01.10.1993–31.03.1995, Kommission der Europäischen Gemeinschaften, Generaldirektion Beschäftigung, Soziale Angelegenheiten und Bildung

Bergmann I, Casswell C (1972) Lung dust and lung iron contents of coal workers in different coalfields in Great Britain. *Br J Ind Med 29*: 160–168

BVOST (1964) *Bergverordnung für die Steinkohlenbergwerke im Verwaltungsbezirk des Oberbergamtes in Dortmund vom 18. Dezember 1964* (Mining ordinance for the coal mines in the administrative district of the Dortmund head office) (German), H Bellmann, Buchdruckerei und Verlag, Dortmund

Bundesministerium für Wirtschaft (1991) Bergverordnung zum gesundheitlichen Schutz der Beschäftigten (Gesundheitsschutz-Bergverordnung-GesBergV) vom 31. Juli 1991 (Mining ordinance for protection of workers' health) (German), *Bundesgesetzblatt I*: 1751–1767

Boeing H (1991) Epidemiological research in stomach cancer: progress over the last ten years. *J Cancer Res Clin Oncol 117*: 133–143

Boeing H, Frentzel-Beyme R (1991) Regional risk factors for stomach cancer in the FRG. *Environ Health Perspect 94*: 83–89

Boeing H, Frentzel-Beyme R, Berger M, Berndt V, Göres W, Körner M, Lohmeier R, Menarcher A, Männl HFK, Meinhardt M, Müller R, Ostermeier H, Paul F, Schwemmle K, Wagner KH, Wahrendorf J (1991) Case-control study on stomach cancer in Germany. *Int J Cancer 47*: 858–864

Bolm-Audorff U, Möhner, Abrens W, Brüske-Hohlfeld I, Jöckel K-H, Morfeld P, Pohlabeln H, Wichmann HE (1997) Das Lungenkrebsrisiko von quarzstaubexponierten Beschäftigten (Lung cancer risk for workers exposed to quartz dust) (German). *Verhandlungen der Deut Gesell für Arbeits- und Umweltmedizin 37*: 139–142

Borm PJA, Palmen N, Engelen JJM, Buurman WA (1988) Spontaneous and stimulated release of tumor necrosis factor-alpha (TNF) from blood monocytes of miners with coal workers' pneumoconiosis. *Am Rev Respir Dis 138*: 1589–1594

Bowden DH, Adamson IYR (1978) Adaptive responses of the pulmonary macrophagic system to carbon. *Lab Invest 38*: 422–429

Brain JD (1971) The effects of increased particles on the number of alveolar macrophages. in: Walton WH (Ed.) *Inhaled particles III, Vol 1*, The Gresham Press, Old Wokey, Surrey, England, 209–223

Brammertz A, Rosmanith J, Weller W, Schyma SB (1995) Abhängigkeit der biologischen Wirkung von physikalischen Kenngrößen der in Intratrachealversuchen getesteten Stäube (Dependence of the biological effects on the physical properties of dusts used in intratracheal tests) (German). *Silikosebericht Nordrhein-Westfalen, Vol 19*, 241–246

Breuer H, Reisner MTR, Kotitschke G, Niesert E (1982) *Epidemiologische Untersuchungen zu den Auswirkungen der Feinstaubbelastung am Arbeitsplatz auf Lunge und Atemwege bei Ruhrbergleuten* (Epidemiological studies of the effects of occupational exposure to fine dust on the lungs and respiratory passages of Ruhr miners) (German). Final report No. 7256-31/023/01, Commission der European Communities

Breuer H, Reisner MTR, Kotitschke G, Niesert E (1983) *Epidemiologische Untersuchungen zu den Auswirkungen der Feinstaubbelastung am Arbeitsplatz auf Lunge und Atemwege bei Saarbergleuten, 2. Phase* (Epidemiological studies of the effects of occupational exposure to fine dust on the lungs and respiratory passages of Saar miners) (German). Final Report No. 7256-31/022/01, Commission der European Communities

Brown GM, Brown DM, Donaldson K (1992) Persistent inflammation and impaired chemotaxis of alveolar macrophages on cessation of dust exposure. *Environ Health Perspect 97*: 91–94

Bruch J, Rehn B (1994) Correlation of *in vitro* and *in vivo* studies on the bioeffects of mineral particles. in: Davis JMG, Jaurand M-C (Eds) *Cellular and molecular effects of mineral and synthetic dusts and fibres, NATO ASI series, H 85*, 263–272

Bruch J, Joos M-R, Reisner MTR, Rosmanith J (1983) Untersuchungen zur spezifischen Schädlichkeit von Feinstäuben des Steinkohlebergbaus. XVII: Vergleich der Ergebnisse von Tier- und Zellversuchen (Studies of specific damage caused by fine coal mine dusts. XVII: Comparison of the results of animal and cell studies) German). in: Ergebnisse von Untersuchungen auf dem Gebiet der Staub- und Silikosebekämpfung im Steinkohlebergbau. *Silikosebericht Nordrhein-Westfalen 14*: 241–255

Busch RH, Filipy RE, Karagianes MT, Palmer RF (1981) Pathologic changes associated with experimental exposure of rats to coal dust. *Environ Res 24*: 53–60

Calhoun WJ, Christman JW, Ershler WB (1986) Raised immunoglobulin concentrations in bronchoalveolar lavage fluid of healthy granite workers. *Thorax 41*: 266–273

Cantin AM, Chrystal RG (1985) Oxidants, antioxidants and the pathogenesis of emphysema. *Eur J Respir Dis 66*: 1–17

Carta P, Aru G, Barbieri MT, Avataneo G, Casula D (1996) Dust exposure, respiratory symptoms, and longitudinal decline of lung function in young coal miners. *Occup Environ Med 53*: 312–319

Castranova V, Bowman L, Reasor MJ, Lewis T, Tucker J, Miles PR (1985) The response of rat alveolar macrophages to chronic inhalation of coal dust and/or diesel exhaust. *Environ Res 36*: 405–419

Chapman JS, Ruckley VA (1985) Microanalyses of lesions and lymph nodes from coal miners' lungs. *Br J Ind Med 42*: 551–555

Chen BH, Meng ZH, Huang Y, Mao HQ, Wang SY (1988) Pulmonary injury in laboratory animals induced by Huai-Nan coal mine respirable dust. *Biomed Environ Sci 1*: 247–252

Christie DGS, Brown AM, Taylor RJ, Secombe MA, Coates MS (1995) Mortality in the New South Wales coal industry, 1973–1992. *Med J Aust 163*: 19–21

Coggon D, Barker DJP, Cole RB (1990) Stomach cancer and work in dusty industries. *Br J Ind Med 47*: 298–301

Collins HPR, Dick JA, Bennett JG, Pern PO, Richards MA, Thomas DJ, Washington JS, Jacobsen M (1988) Irregularly shaped small shadows on chest radiographs, dust exposure, and lung function in coal workers pneumoconiosis. *Br J Ind Med 45*: 43–55

Cordier S, Clavel J, Limasset JC, Boccon-Gibod L, Moual N, Mandereau L (1993) Occupational risks of bladder cancer in France: a multicentre case-control study. *Int J Epidemiol 22*: 403–411

Costello J, Ortmeyer CE, Morgan WKC (1974) Mortality from lung cancer in US coal miners. *Am J Public Health 64*: 222–224

Cowie HA, Hurley JF, Soutar CA, Pern P, Hutchison P (1994) Further studies of progressive massive fibrosis (PMF) among miners and ex-miners. *Ann Occup Hyg 38, Suppl 1*: 791–797

Dalal NS, Suryan MM, Vallyathan V, Green FHY, Jafari B, Wheeler R (1989) Detection of reactive free radicals in fresh coal mine dust and their implication for pulmonary injury. *Ann Occup Hyg 33*: 79–84

Dalal NS, Jafari B, Petersen M, Green FHY, Vallyathan V (1991) Presence of stable coal radicals in autopsied coal miners' lungs and its possible correlation to coal workers' pneumoconiosis. *Arch Environ Health 46*: 366–372

Dalal NS, Newman J, Pack D, Leonard S, Vallyathan V (1995) Hydroxyl radical generation by coal mine dust: possible implication to coal workers' pneumoconiosis (CWP). *Free Radical Biol Med 18*: 11–20

Davis G (1986) Pathogenesis of silicosis: current concepts and hypotheses. *Lung 164*: 139–154

DFG (Deutsche Forschungsgemeinschaft) (1975) *Forschungsbericht "Chronische Bronchitis und Staubbelastung am Arbeitsplatz"—Arbeitsmedizinische Querschnittsuntersuchungen zur Bedeutung chronisch-inhalativer Belastungen für das bronchopulmonale System, Teil 1* (Research report "Chronic bronchitis and occupational dust exposure"—Occupational medical cross-sectional studies of the effects of long-term inhalation exposure on the bronchopulmonary system. Part 1) (German), Harald Boldt Verlag, Boppard

DFG (Deutsche Forschungsgemeinschaft) (1981) *Forschungsbericht "Chronische Bronchitis und Staubbelastung am Arbeitsplatz"—Arbeitsmedizinische Längsschnittuntersuchungen zu den Auswirkungen inhalativer Noxen am Arbeitsplatz, Teil 2* (Research report "Chronic bronchitis and occupational dust exposure"—Occupational medical longitudinal studies of the effects of long-term inhalation exposure on the bronchopulmonary system. Part 2) (German), Harald Boldt Verlag, Boppard

Donaldson K, Brown GM, Brown DM, Robertson MD, Slight J, Cowie H, Jones AD, Bolton RE, Davis JMG (1990) Contrasting bronchoalveolar leukocyte responses in rats inhaling coal mine dust, quartz, or titanium dioxide: effects of coal rank, airborne mass concentration, and cessation of exposure. *Environ Res 52*: 62–76

Finkelstein M, Liss GM, Kramer F, Kusiak RA (1987) Mortality among workers receiving compensation awards for silicosis in Ontario 1940–85. *Br J Ind Med 44*: 588–594

Fubini B, Bolis V, Cavenago A, Volante M (1995) Physicochemical properties of crystalline silica dusts and their possible implication in various biological responses. *Scand J Work Environ Health 21*: 9–14

Gadek JE, Pacht ER (1990) The protease-antiprotease balance within the human lung: implications for the pathogenesis of emphysema. *Lung 168*: 552–564

Gilman PA, Ames RG, McCawley MA (1985) Leukemia risk among US white male coal miners. *J Occup Med 27*: 669–671

Golka K, Bandel T, Urfer W, Bolt HM (1995) *Berufliche Risikofaktoren für Tumoren der Harnblase und der Prostata* (Occupational risk factors for tumours of the urinary bladder and the prostate) German). Schriftenreihe der Bundesanstalt für Arbeitsschutz, Fb714, Wirtschaftsverlag NW, Verlag für neue Wirtschaft

González CA, Sanz M, Marcos G, Pita S (1991) Occupation and gastric cancer in Spain. *Scand J Work Environ Health 17*: 240–247

Gormley IP, Collings P, Davis JMG, Ottery J (1979) An investigation into the cytotoxicity of respirable dusts from British collieries. *Br J Exp Pathol 60*: 526–536

Gosset P (1991) Production of tumor necrosis factor alpha and interleukin 6 by human alveolar macrophages exposed *in vitro* to coal mine dust. *Am J Respir Cell Mol Biol 5*: 431–436

Gough J (1949) Pneumoconiosis in coal trimmers. *J Pathol Bacteriol 51*: 277–285

Green FHY, Boyd RL, Danner-Rabovsky J, Fisher MJ, Moorman WJ, Ong T-M, Tucker J, Vallyathan V, Whong W-Z, Zoldak J, Lewis T (1983) Inhalation studies of diesel exhaust and coal dust in rats. *Scand J Work Environ Health 9*: 181–188

Greenland S (1987) Quantitative methods in the review of epidemiologic literature. *Epidemiol Rev 9*: 1–30

Gu Z-W, Whong W-Z, Wallace WE, Ong TM (1992) Induction of micronuclei in BALB/c-3T3 cells by selected chemicals and complex mixtures. *Mutat Res 279*: 217–222

Hahon N, Booth JA, Stewart JD (1988) Interferon induction inhibition and mutagenic activity of nitrosated coal dust extract. *Environ Res 45*: 213–223

Hansen K, Mossman BT (1987) Generation of superoxide (O_2^-) from alveolar macrophages exposed to asbestiform and nonfibrous particles. *Cancer Res 47*: 1681–1686

Harrison JD, Morris DL (1989) Dyspepsia in coal miners and the general population: a comparative study. *Br J Ind Med 46*: 428–429

Hilscher W, Kemper M, Rosmanith J, Molik B, Grover R (1983) Untersuchungen zur spezifischen Schädlichkeit von Feinstäuben des Steinkohlebergbaus. XV: Morphometrische Untersuchungen zur fibrogenen und lymphocytären Wirkung von Grubenstäuben in den Mediastinallymphknoten nach intratrachealer Verabreichung (Studies of the specific damage caused by fine dust from coal mines. XV: Morphometric studies of fibrogenic and lymphocytic effects of mine dusts in the mediastinal lymph nodes after intratracheal instillation) (German). in: *Ergebnisse von Untersuchungen auf dem Gebiet der Staub- und Silikosebekämpfung im Steinkohlebergbau. Silikosebericht Nordrhein-Westfalen 14*: 213–225

Hilscher W, Kah R, Rosmanith J, Grover R, Hilscher B (1987) Morphometrische Untersuchungen an grubenstaubinduzierten Granulomen in den Lungen und Lymphknoten von Ratten nach intratrachealer Instillation (Morphometric studies of granulomas induced by mine dust in the lungs and lymph nodes of rats after intratracheal instillation) (German). in: *Ergebnisse von Untersuchungen auf dem Gebiet der Staub- und Silikosebekämpfung im Steinkohlebergbau. Silikosebericht Nordrhein-Westfalen 16*: 287–303

Hirayama T (1986) A large scale cohort study of dietary habits and cancer mortality (Japanese). *Gan No Rinsho 32*: 610–622

Hirayama T (1992) Life-style and cancer: from epidemiological evidence to public behavior change to mortality reduction of target cancers. *J Natl Cancer Inst Monogr 12*: 65–74

Huang X, Laurent PA, Zalma R, Pezerat H (1993) Inactivation of alpha 1-antitrypsin by aqueous coal solutions: possible relation to the emphysema of coal workers. *Chem Res Toxicol 6*: 452–458

Huang X, Zalma R, Pezerat H (1994) Factors that influence the formation and stability of hydrated ferrous sulfate in coal dusts: possible relation to the emphysema of coal miners. *Chem Res Toxicol 7*: 451–457

Hurley JF, Jacobsen M (1986) Occupational hygiene implications of new results on progressive massive fibrosis in working coal miners. *Ann Am Conf Gov Ind Hyg 14*: 85–89

Hurley JF, Maclaren WM (1987) *Dust-related risks of radiological changes in coal miners over a 40-year working life*: report on work commissioned by NIOSH. Report No. TM/87/09, Institute of Occupational Medicine, Edinburgh, UK

Hurley JF, Burns J, Copland L, Dodgson J, Jacobsen M (1982) Coal workers' simple pneumoconiosis and exposure to dust at 10 British coal mines. *Br J Ind Med 39*: 120–127

Hurley JF, Maclaren WM, Alexander WP, Cowie AJ, Collins HP, Ewing A, Hazledine DJ, Jacobsen M, Munro L, Soutar CA (1984) *Factors influencing the occurrence of progressive massive fibrosis in British coal miners.* Report No. TM/84/2, Institute of Occupational Medicine, Edinburgh, UK

Hurley JF, Alexander WP, Hazledine DJ, Jacobsen M, Maclaren WM (1987) Exposure to respirable coal mine dust and incidence of progressive fibrosis. *Br J Ind Med 44*: 661–672

IARC (International Agency for Research on Cancer) (1997) *Silica, some silicates, coal dust and para-aramid fibrils. IARC monographs on the evaluation of carcinogenic risks to humans, Vol 68*, IARC, Lyon, 337–406

ILO (International Labour Organization) (1980) *Guidelines for the use of ILO International Classification of Radiographs of Pneumoconioses.* ILO Publications; International Labour Office, 1211 Geneve 22, Switzerland, rev. ed.

Jacobsen M (1976) *Dust exposure, lung diseases and coal miners mortality.* Doctoral thesis, University of Edinburgh

Jacobsen M. (1984): Coalworkers' Pneumoconiosis: results from epidemiological studies in Britain. *Proceedings of the VIth International Pneumoconiosis Conference*, Bochum, FRG, 20th September 1983, 92–102

Jacobsen M, Maclaren WM (1982) Unusual pulmonary observations and exposure to coal mine dust: a case-control study. *Ann Occup Hyg 26*: 753–765

Jacobsen M, Rae S, Walton WH, Rogan JH (1971) The relation between pneumoconiosis and dust-exposure in British coal mines. in: Walton WH (Ed.) *Inhaled particles III, Vol 2*, Unwin Brothers, London, 903–917

Jacobsen M, Burns J, Attfield MD (1977) Smoking and coal workers simple pneumoconiosis. in: Walton WH, McGovern B (Eds) *Inhaled Particles IV*, Pergamon Press, Oxford, 759–772

Janssen YMW, Van Houten B, Borm PJA, Mossman BT (1993) Biology of disease: cell and tissue responses to oxidative damage. *Lab Invest 69*: 261–274

Karagianes MT, Palmer RF, Busch RH (1981) Effects of inhaled diesel emissions and coal dust in rats. *Am Ind Hyg Assoc J 42*: 382–391

Kriegseis W, Scharmann A (1982) Specific harmfulness of respirable dusts from West German coal mines. V: influence of mineral surface properties. *Ann Occup Hyg 26*: 511–525

Krishna G, Nath J, Soler L, Ong T (1987) *In vivo* induction of sister chromatid exchanges in mice by nitrosated coal dust extract. *Environ Res 42*: 106–113

Kukuk P (1955) *Geologie, Mineralogie und Lagerstättenlehre*, Springer Verlag, Berlin

Kuempel ED, Stayner LT, Attfield MD, Buncher CR (1995) Exposure response analysis of mortality among coal miners in the United States. *Am J Ind Med 28*: 167–184

Kusaka Y, Brown GM, Donaldson K (1990) Alveolitis caused by exposure to coal mine dusts: production of interleukin-1 and immunomodulation by bronchoalveolar leukocytes. *Environ Res 53*: 76–89

Kysilka Z, Knie B, Rosmanith J (1985) Untersuchungen zum Einfluß der Korngröße und des Mineralgehaltes auf die spezifische Schädlichkeit von Feinstäuben des Steinkohlebergbaus. IV. Änderungen im Hydroxyprolin- und Lipidgehalt in Rattenlungen beim Intratrachealtest (The effect of particle size and mineral content on the specific damage caused by fine dust from coal mines. IV. Changes in the hydroxyproline and lipid levels in rat lungs in the intratracheal test) (German). in: *Ergebnisse von Untersuchungen auf dem Gebiet der Staub- und Silikosebekämpfung im Steinkohlebergbau. Silikosebericht Nordrhein-Westfalen 15*: 217–231

Lampert K, Morfeld P, Piekarski C, Bauer H-D, Stegmaier C, Ziegler H, Dohm G (1997) *Längsschnittstudie zur Krebsmortalität saarländischer Steinkohlebergleute (Phase I und II)* (Longitudinal study of cancer mortality of Saarland miners) (German). Studie der Arbeitsgemeinschaft des Saarlandes zur Erforschung und Förderung des Gesundheitsschutzes im Bergbau eV (AGiB)

Lange HJ, Reiter R (1972) Datenverdichtung bei der Auswertung multivariater epidemiologischer Studien (Data reduction in the evaluation of multivariate epidemiological studies) (German). *Methods Inf Med 11*: 253–257

Lange HJ, Ulm K (1983) *Mathematische Modelle zur Frage eines Allgemeinen Staubgrenzwertes; epidemiologische Auswertungen mit Daten des DFG-Schwerpunktprogramms "chronische Bronchitis (CB)"* (Mathematical models for a general threshold limit for dust; epidemiological analysis of data from the DFG chronic bronchitis study) (German), Verlag Chemie, Weinheim

Lassalle P, Gosset P, Aerts C (1990) Abnormal secretion of interleukin-1 and tumor necrosis factor-α by alveolar macrophages in coal workers's pneumoconiosis: comparison between simple pneumoconiosis and progressive massive fibrosis. *Exp Lung Res 16*: 73–80

Laurell CB, Eriksson S (1963) The electrophoretic alpha-1-globulin pattern of serum in alpha-1-antitrypsin deficiency. *Scand J Clin Lab Invest 15*: 132–140

Le Bouffant L, Daniel H, Martin JC, Bruyère S (1977) Variation in the toxicity of various silicas in relation to the duration of presence in the body. *Acad Sci Hebd Seances Acad Sci D 285 (5)*: 599–602

Le Bouffant L, Daniel H, Martin JC, Bruyère S (1982) Effect of impurities and associated minerals on quartz toxicity. *Ann Occup Hyg 26*: 625–634

Leigh J (1990) 15 year longitudinal studies of FEV_1 loss and mucus hypersecretion development in coal workers in New South Wales, Australia. *Proceedings of the VIIth International Pneumoconiosis Conference 1988*, Pittsburgh, 112–121

Leigh J, Outhred KG, McKenzie HI, Wiles AN (1982) Multiple regression analysis of quantified aetiological, clinical and postmortem pathological variables related to respiratory disease in coal workers. *Ann Occup Hyg 26*: 383–400

Leigh J, Wiles AN, Glick M (1986) Total population study of factors affecting chronic bronchitis prevalence in coalmining industry of New South Wales, Australia. *Br J Ind Med 43*: 263–271

Leigh J, Beck R, Cole B (1993) Relationship between emphysema severity, pneumoconiosis histopathology and lung coal and silica content in coal workers. *Proceedings of the 8th International Conference on Occupational Diseases 1992*, Prague, 613–618

Leigh J, Driscoll TR, Cole BD, Beck RW, Hull BP, Yang J (1994) Quantitative relation between emphysema and lung mineral content in coal workers. *Occup Environ Med 51*: 400–407

Leigh J, Todorovic M, Driscoll T (1997) Histological changes in hilar lymph glands in relation to coal workers' progressive massive fibrosis and lung coal and quartz contents. Evidence in support of the "central" hypothesis. *Ann Occup Hyg 41*: 374–378

Lewis TR, Green FHY, Moorman WJ, Burg JAR, Lynch DW (1986) A chronic inhalation toxicity study of diesel engine emissions and coal dust, alone and combined. *Dev Toxicol Environ Sci 13*: 361–380

Lewis TR, Green FHY, Moorman WJ, Burg JR, Lynch DW (1989) A chronic inhalation toxicity study of diesel engine emissions and coal dust, alone and combined. *J Am Coll Toxicol 8*: 345–375

Love RG, Miller BG (1982) Longitudinal study of lung function in coal miners. *Thorax 37*: 193–197

Love RG, Muir DCF, Sweetland KF (1970) Aerosol deposition in the lungs of coal workers. in: Walton WH (Ed.) *Inhaled particles III*, Unwin Brothers, Old Woking, UK, 131–137

Maclaren WM, Soutar CA (1985) Progressive massive fibrosis and simple pneumoconiosis in ex-miners. *Br J Ind Med 42*: 734–740

Maclaren WM, Hurley JF, Collins HPR, Cowie AJ (1989) Factors associated with the development of progressive massive fibrosis in British coal miners: a case control study. *Br J Ind Med 46*: 597–607

Marine WM, Gurr D, Jacobsen M (1988) Clinically important respiratory effects of dust exposure and smoking in British coal miners. *Am Rev Respir Dis 137*: 106–112

Martin JC, Daniel-Moussard H, Le Bouffant L, Policard A (1972) The role of quartz in the development of coal workers pneumoconiosis. *Ann N Y Acad Sci 200*: 127–141

Martin JC, Daniel H, Le Bouffant L (1977) Short- and long-term experimental study of the toxicity of coal-mine dust and of some of its constituents. in: Walton WH (Ed.) *Inhaled Particles IV, Vol 1*, Pergamon Press, Oxford, 361–371

Massé J, Larivée P, Sébastien P, Bégin R (1994) The cytotoxicity of respirable coal dusts. in: Davis JMG, Jaurand MC (Eds) *Cellular and molecular effects of mineral and synthetic dusts and fibres*, Nato ASI Series H 85: 388–396

Meijers JMM, Swaen GMH, van Vliet K, Borm PJA (1990) Epidemiologic studies on inorganic dust-related lung diseases in the Netherlands. *Exp Lung Res 16*: 15–23

Meyer MB, Luk GD, Sotelo JM, Cohen BH, Menkes HA (1980) Hypothesis: the role of the lung in stomach carcinogenesis. *Am Rev Respir Dis 121*: 887–892

Miller BG, Jacobsen M (1985) Dust exposure, pneumoconiosis, and mortality of coal miners. *Br J Ind Med 42*: 723–733

Miller BG, Jacobsen M, Steele RC (1981) *Coal miners' mortality in relation to radiological category, lung function and exposure to airborne dust. IOM Report TM/81/10*, Institute of Occupational Medicine, Edinburgh, UK

Miller BG, Hagen S, Love RG, Cowie HA, Kidd MW, Lorenzo S, Tielemans ELJP, Robertson A, Sourtar CA (1995) *A follow-up study of coal miners exposed to unusual concentrations of quartz. IOM Report TM/95/03*, Institute of Occupational Medicine, Edinburgh, UK

Miller BG, Buchanan D, Hurley JF, Robertson A, Huchison PA, Kidd MW, Pilkington A, Soutar CA (1997) *The effects of exposure to diesel fumes, low-level radiation, and respirable dust and quartz, on cancer mortality in coal miners. IOM Report TM/97/04*, Institute of Occupational Medicine, Edinburgh, UK

Mohr C, Gemsa D, Graebner C (1991) Systemic macrophage stimulation in rats with silicosis: enhanced release of tumor necrosis factor-α from alveolar and peritoneal macrophages. *Am J Respir Cell Mol Biol 5*: 395–402

Molik B, Hilscher B (1989) Comparative presentation of the results of animal experiments on the specific harmfulness of dry filter dusts. in: *Ergebnisse von Untersuchungen auf dem Gebiet der Staub- und Silikosebekämpfung im Steinkohlebergbau. Silikosebericht Nordrhein-Westfalen 17:* 273–283

Morabia A, Markowitz S, Garibaldi K, Wynder EL (1992) Lung cancer and occupation: results of a multicentre case-control study. *Br J Ind Med 49*: 721–727

Morfeld P (1994) *Konstruktion stochastischer Modelle zur epidemiologischen Analyse des Krankheitsverlaufsprozesses der Bergarbeiter-Pneumokoniose auf einem Steinkohlebergwerk des Ruhrreviers* (Construction of stochastic models for epidemiological analysis of the course of

CWP in a mine in the Ruhr area) (German). *Schriftenreihe der Bundesanstalt für Arbeitsmedizin, Sonderschrift 6*

Morfeld P, Piekarski C (1994) Zur Frage der Anerkennung von chronischer Bronchitis und/oder Emphysem als Listen-Berufskrankheit der Deutschen Steinkohlebergarbeiter (On the question of recognition of chronic bronchitis and/or emphysema as a listed occupational disease of German coal miners) (German). in: *Berichtsband zur Tagung der Arbeitsgemeinschaft des Saarlandes zur Erforschung und Förderung des Gesundheitsschutzes im Bergbau eV (AGiB), 27. bis 29. April 1994* in Saarbrücken, 112–120

Morfeld P, Piekarski C (1996) *Chronische Bronchitis und Emphysem als Berufskrankheit der Steinkohlebergleute* (Chronic bronchitis and/or emphysema as an occupational disease of coal miners) (German). *Schriftenreihe Zentralblatt für Arbeitsmedizin, Band 15*, Dr. Curt Hafner Verlag, Heidelberg

Morfeld P, Piekarski C (1998a) *Epidemiologie der Pneumokoniose und der chronischen Bronchitis im Steinkohlebergbau. Ableitung von Grenzwerten für Steinkohlegrubenstaub* (Epidemiology of pneumoconiosis and chronic bronchitis in coal mining. Derivation of threshold values for coal mine dust) (German). *Schriftenreihe der Bundesanstalt für Arbeitsschutz und Arbeitsmedizin, Sonderschrift 45*

Morfeld P, Piekarski C (1998b) Epidemiologische Erkenntnisse zur kanzerogenen Wirkung von Steinkohlegrubenstäuben (Epidemiology of the carcinogenic effects of coal mine dusts) (German). *Silikosebericht Nordrhein-Westfalen 20*: 281–324

Morfeld P, Kampmann B, Vautrin H-J, Piekarski C (1990) The development of coal workers' pneumoconiosis in relation to time underground, calendar time and age in a German coal mine. in: Sakurai H, Okazaki I, Omae K (Eds) *Occupational epidemiology, Proceedings of the Seventh International Symposium on Epidemiology in Occupational Health, Tokyo, Japan, 11–13 October 1989*, Elsevier, Amsterdam, 147–150

Morfeld P, Vautrin H-J, Rohleder F, Kösters A, Lampert K, Frese M, Kampmann B, Piekarski C (1992) Epidemiologische Untersuchungen zur Feinstaubexposition und zur Inzidenz der Bergarbeiter-Pneumokoniose (Streuungskategorie 1/1, ILO 1980) in untertägigen Betrieben des Ruhr- und Saar-Kohlenbergbaus zwischen 1974 und 1990 (Epidemiological study of the exposure to fine dust and the incidence of CWP (1/1, ILO 1980) in underground workings of the Ruhr and Saar coal mines 1974–1990) (German). *Zentralbl Arbeitsmed 42*: 92–111

Morfeld P, Vautrin H, Kampmann B, Piekarski C (1994) Zur Effektivität der Begrenzung des Pneumokonioserisikos durch spezielle Arbeitsschutzmaßnahmen im Deutschen Steinkohlebergbau—Ansätze einer epidemiologischen Überprüfung (The effectiveness of attempts to limit the pneumoconiosis risk by industrial hygiene in German coal mines–an epidemiological analysis) (German). *Informatik Biom Epidemiol Med Biol 25*: 39–47

Morfeld P, Vautrin H-J, Rohleder F, Kösters A, Lampert K, Frese M, Kampmann B, Piekarski C (1995) *Epidemiologische Kohortenstudie zur Inzidenz von Streuungskategorien kleiner rundlicher Schatten im Röntgenbild (ILO 1980) bei Steinkohlebergleuten und zur individuellen Exposition gegenüber Feinstäuben sowie Quarzfeinstäuben unter Berücksichtigung verschiedener Lagerstätten des deutschen Steinkohlebergbau* (Epidemiological cohort study of the incidence of the various categories (ILO 1980) of small round opacities in the X-rays of coal miners and individual exposure to fine dusts and fine quartz dusts in various deposits of German coal) (German). Abschlußbericht im Rahmen des KEG-Vorhabens "Charakterisierung von Grubenstäuben hinsichtlich ihrer spezifischen Schädlichkeit (Pneumokoniosen)" FK 7280/03/027

Morfeld P, Lampert K, Ziegler H, Stegmaier C, Dohm G, Piekarski C (1997a) Coal mine dust exposure and cancer mortality in German coal miners. *J Appl Occup Environ Hyg 12*: 909–914

Morfeld P, Vautrin HJ, Kösters A, Lampert K, Piekarski C (1997b) Limit value assessment for respirable coal mine dust in Germany. *J Appl Occup Environ Hyg 12*: 840–848

Morfeld P, Vautrin HJ, Kösters A, Lampert K, Piekarski C (1997c) Components of coal mine dust exposure and the occurrence of pre-stages of pneumoconiosis. *J Appl Occup Environ Hyg 12*: 973–979

NIOSH (National Institute for Occupational Safety and Health) (1995) *Occupational exposure to respirable coal mine dust. Criteria for a recommended standard.* DHHS (NIOSH) Publication No. 95–106

Ophoff B (1985) *Pneumokoniosehäufigkeit und -ausprägungsgrad in Abhängigkeit von der Staubbelastung in unterschiedlichen Flözhorizonten des Steinkohlebergbaus* (Incidence of pneumoconiosis and severity as a function of the dust exposure in various stratigraphic horizons in coal mining) (German). Doctoral thesis, University of Düsseldorf

Ong T, Whong W-Z, Xu J, Burchell B, Green FHY, Lewis T (1985) Genotoxicity studies of rodents exposed to coal dust and diesel emission particulates. *Environ Res 37*: 399–409

Ortmeyer CE, Baier EJ, Crawford GM (1973) Life expectancy of Pennsylvania coal miners compensated for disability. *Arch Environ Health 27*: 227–230

Ortmeyer CE, Costello J, Morgan WKC, Swecker S, Perterson M (1974) The mortality of Apalachian coal miners, 1963 to 1971. *Arch Environ Health 29*: 67–72

Otto H, von Hinüber G (1972) Zur Häufigkeit des Emphysems, der Tuberkulose und des Bronchuskarzinoms bei Staublungenerkrankungen (The incidence of emphysema, tuberculosis and bronchial carcinoma in persons with dust lung disease) (German). *Prax Klin Pneumol 26*: 145–158

Petrascheck WE (1961) *Lagerstättenlehre* (Mineral deposits) (German), Springer Verlag, Wien

Piontek K, Hoffarth HP, Robock K, Baur X, Ulmer WT (1991) Einfluß der Schichthorizonte auf die Entwicklungsgeschwindigkeit der Bergarbeiter-Pneumokoniose (Silikose) (Effect of the stratigraphic horizon on the rate of development of CWP (silicosis)) (German). *Arbeitsmed Sozialmed Praeventivmed 26*: 89–93

Prinz B, Stolz R (1990) Effect of the measuring strategy on the respirable dust concentration in the breathable air at underground workplaces. *Proceedings of the VIIth International Pneumoconioses Conference, Pittsburgh 1988*, 1148–1151

Ray SC, King EJ, Harrison CV (1951a) The action of anthracite and bituminous coal dusts mixed with quartz on the lungs of rats. *Br J Ind Med 8*: 74–76

Ray SC, King EJ, Harrison CV (1951b) The action of small amounts of quartz and larger amounts of coal and graphite on the lungs of rats. *Br J Ind Med 8*: 68–73

Rehn B, Zawatzki S, Bruch J (1995) In vitro Untersuchungen zur Hemmung der Quarzstimulation von Alveolarmakrophagen durch Polyvinyl-N-Oxyd (PVNO). Eine zellbiologische Methode zur Bestimmung der biologischen Quarzwirkung von Grubenmischstäuben (*In vitro* studies of the inhibition of the quartz stimulation of alveolar macrophages with PVNO. A cell-biological method to determine the biological activity of quartz in mixed mine dusts) (German). *Silikosebericht Nordrhein-Westfalen 19*: 353–358

Reichel G (1989) Die Bedeutung der beruflichen Staubbelastung unter Tage für die Entwicklung der obstruktiven Atemwegserkrankungen (Occupational underground dust exposure and the development of obstructive airway diseases) (German). *Atemwegs Lungenkr 15*: 471–475

Reichel G, Ulmer WT, Buckup H, Stempel G, Werner U (1969) Die chronisch obstruktiven Atemwegserkrankungen des Bergmannes (Chronic obstructive airway diseases in miners) (German). *Dtsch Med Wochenschr 94*: 2375–2380

Reischig HL, Schmitt R (1994) Entwicklung der Mischstaubpneumokoniose nach Ende der Staubexposition—Nachuntersuchung ausgeschiedener Bergleute im saarländischen Steinkohlebergbau (Further development of mixed dust pneumoconiosis after the end of exposure–follow-up study of retired miners from Saarland coal mines) (German). *Berichtsband zur Tagung der Arbeitsgemeinschaft des Saarlandes zur Erforschung und Förderung des Gesundheitsschutzes im Bergbau eV (AGiB), 27. bis 29. April 1994* in Saarbrücken, 30–35

Reisner MTR, Robock K (1977) Results of epidemiological, mineralogical and cytotoxicological studies on the pathogenicity of coal-mine dusts. in: Walton WH, McGovern B (Eds) *Inhaled particles IV*, Pergamon Press, Oxford, 793–716

Reisner MTR, Armbruster L, Kühn L (1985a) Untersuchungen zum Einfluß der Korngröße und des Mineralgehaltes auf die spezifische Schädlichkeit von Feinstäuben des Steinkohlebergbaus. I. Herstellung und Analyse der Staubfraktionen (The effect of particle size and mineral content on the specific damage caused by fine dust from coal mines. I. Production and analysis of the dust fractions) (German). in: *Ergebnisse von Untersuchungen auf dem Gebiet der Staub- und Silikosebekämpfung im Steinkohlebergbau. Silikosebericht Nordrhein-Westfalen 15*: 169–184

Reisner MTR, Kotitschke G, Niesert E (1985b) Pneumokoniose und Staubexposition—epidemiologische Untersuchungen im Steinkohlenbergbau an der Ruhr über einen Zeitraum von 20 Jahren (Pneumoconiosis and dust exposure—epidemiological studies in Ruhr coal mining during 20 years) (German). *Silikosebericht Nordrhein-Westfalen, Band 15*, 445–492

Robertson A, Bolton RE, Miller BG, Chapman JS, Dodgson J, Jones AD, Niven KJ, Davis JMG (1988) The effect of quartz content on the pathogenicity of coal mine dusts. *Ann Occup Hyg 32, Suppl 1*: 621–633

Robock K, Reisner MTR (1982) Specific harmfulness of respirable dusts from West German coal mines. I: Results of cell tests. *Ann Occup Hyg 26*: 473–479

Rockette HE (1977) Cause specific mortality of coal miners. *J Occup Med 19*: 795–801

Roller M, Pott F, Althoff G-H, Friemann J, Mogilevski G (1998) Zwischenergebnisse nach zwei Jahren aus einer Lebenszeitstudie zur Kanzerogenität von Steinkohlestäuben in der Rattenlunge (Interim results after 2 years of a life-time study of the carcinogenicity of coal dusts in the rat lung) (German). *Silikosebericht Nordrhein-Westfalen 20*: 267–279

Rosmanith J, Reisner MTR, Prajsnar D, Breining H, Ehm W (1981) Untersuchungen zur spezifischen Schädlichkeit von Feinstäuben des Steinkohlebergbaus. VIII. Die Beziehungen zwischen der Zusammensetzung des Feinstaubs, der retinierten Staubmasse und der fibrotischen Reaktion in Lungen und Lymphknoten der Ratte beim Intratrachealtest (Studies of the specific damage caused by fine dust from coal mines. VIII: The relationship between the composition of the fine dust, the weight of retained dust and the fibrotic reaction in the rat lung and lymph nodes in the intratracheal test) (German). in: *Ergebnisse von Untersuchungen auf dem Gebiet der Staub- und Silikosebekämpfung im Steinkohlebergbau. Silikosebericht Nordrhein-Westfalen 13:* 241–249

Rosmanith J, Reisner MTR, Prajsnar D, Breining H, Ehm W (1982) Specific harmfulness of respirable dusts from West German coal mines. II: Results of intratracheal tests on rats. *Ann Occup Hyg 26*: 481–490

Rosmanith J, Reisner MTR, Prajsnar D, Breining H, Ehm W (1983) Untersuchungen zur spezifischen Schädlichkeit von Feinstäuben des Steinkohlebergbaus. XIII. Die Beziehungen zwischen der Zusammensetzung des Feinstaubs, der retinierten Staubmasse und der fibrotischen Reaktion in Lungen und Lymphknoten der Ratte beim Intratrachealtest (2. Mitteilung) (Studies of the specific damage caused by fine dust from coal mines. XIII: The relationship between the composition of the fine dust, the weight of retained dust and the fibrotic reaction in the rat lung and lymph nodes in the intratracheal test (Part 2)) (German). in: *Ergebnisse von Untersuchungen auf dem Gebiet der Staub- und Silikosebekämpfung im Steinkohlebergbau. Silikosebericht Nordrhein-Westfalen 14*: 177–194

Rosmanith J, Reisner MTR, Hilscher W, Prajsnar D, Ehm W (1985) Untersuchungen zum Einfluß der Korngröße und des Mineralgehaltes auf die spezifische Schädlichkeit von Feinstäuben des Steinkohlebergbaus. II. Retention des Staubes in der Rattenlunge und dessen Penetration in die Lymphknoten beim Intratrachealversuch (The effect of particle size and mineral content on the specific damage caused by fine dust from coal mines. II. Retention of dust in the rat lung and its penetration into the lymph nodes in the intratracheal test) (German). in: *Ergebnisse von Untersuchungen auf dem Gebiet der Staub- und Silikosebekämpfung im Steinkohlebergbau. Silikosebericht Nordrhein-Westfalen 15*: 185–204

Ross HF, King EJ, Yoganathan M, Nagelschmidt G (1962) Inhalation experiments with coal dust containing 5 per cent, 10 per cent, 20 per cent and 40 per cent quartz: tissue reactions in the lungs of rats. *Ann Occup Hyg 5*: 149–161

Rossiter CE, Rivers D, Bergmann I, Casswell C, Nagelschmidt G (1967) Dust content, radiology and pathology in simple pneumoconiosis of coal workers (further report). in: Davies CN (Ed.) *Inhaled Particles and Vapours II, Proceedings of an International Symposium organized by the British Occupational Hygiene Society, Cambridge, 28 Sept–1 Oct 1965*, Pergamon Press, Oxford, 419–437

Ruckley VA, Gauld SJ, Chapman JS, Davis JMG (1984) Emphysema and dust exposure in a group of coal workers. *Am Rev Respir Dis 129*: 528–532

Ruckley VA, Fernie JM, Cambell SJ, Cowie HA (1989) *Causes of disability in coal-miners: a clinico-pathological study of emphysema, airways obstruction and massive fibrosis.* Report No. TM/89/05, Institute of Occupational Medicine, Edinburgh, UK

Russel AE (1930) in: International Labour Office: silicosis records of the international conference held at Johannesburg 13–27 August 1930, Studies and Report Series F. *Ind Hyg 13*: 68–72

Schifflers E, Jamart J, Renard V (1987) Tobacco and occupation as risk factors in bladder cancer: a case-control study in southern Belgium. *Int J Cancer 39*: 287–292

Schins RPF (1996) *Biomarkers of mineral dust induced lung disorders. Molecular epidemiologic studies in coal workers.* Doctoral thesis, University of Maastricht

Schins RPF, Schilderman PAEL, Borm PJA (1995) Oxidative DNA damage in peripheral blood lymphocytes of coal workers. *Int Arch Occup Environ Health 67*: 153–157

Schmitt R (1991) *Entwicklung der Mischstaubpneumokoniose nach Ende der Staubexposition – Nachuntersuchung ausgeschiedener Bergleute im saarländischen Steinkohlebergbau* (Further development of mixed dust pneumoconiosis after the end of exposure—Follow-up study of retired miners from the Saarland coal mines) (German). Doctoral thesis, Ruhr-Universität Bochum

Schraufstätter IU, Cochrane CG (1991) Oxidants: types, sources and mechanisms of injury. in: Crystal RG, West JB, Weibel ER (Eds) *The lung: scientific foundations*, Raven Press, New York, 1803–1810

Schulz HM, Wolf M, Hagemann HW (1995) Organisch-petrochemische Untersuchungen an Kohle und Grubenstäuben des H1-Programms sowie an rückgewonnenen Grubenstäuben (Petrochemical analysis of coal and mine dusts from the H1 programme and of recovered mine dusts) (German). *Silikosebericht Nordrhein-Westfalen 19*: 147–159

Seixas NS, Hewett P, Robins TG, Haney R (1991) Estimation of cumulative exposures for the national study of coal workers' pneumoconiosis. *Appl Occup Environ Hyg 6*: 1032–1041

Seixas NS, Robins TG, Attfield MD, Moulton LH (1992) Exposure-response relationship for coal mine dust and obstructive lung disease following enactment of the federal coal mine health and safety act of 1969. *Am J Ind Med 21*: 715–734

Seixas NS, Robins TG, Attfield MD, Moulton LH (1993a) Longitudinal and cross sectional analyses of exposure to coal mine dust and pulmonary function in new miners. *Br J Ind Med 50*: 929–937

Seixas NS, Robins TG, Becker M (1993b) A novel approach to the characterization of cumulative exposure for the study of chronic occupational disease. *Am J Epidemiol 137*: 463–471

Soutar CA, Hurley JF (1986) Relation between dust exposure and lung function in miners and ex-miners. *Br J Ind Med 43*: 307–320

Sram RJ, Hola N, Kotesovec F, Vavra R (1985) Chromosomal abnormalities in soft coal open-cast mining workers. *Mutat Res 144*: 271–275

Stamm SC, Zhong BZ, Whong WZ, Ong T (1994) Mutagenicity of coal dust and smokeless-tobacco extracts in *Salmonella typhimurium* strains with differing levels of *O*-acetyltransferase activities. *Mutat Res 321*: 253–264

Starzynski Z, Marek K, Kujawska A, Szymczak W (1996) Mortality among different occupational groups of workers with pneumoconiosis: results of the register-based cohort study. *Am J Ind Med 30*: 718–725

Swaen GM, Aerdts CWHM, Slangen JJM (1987) Gastric cancer in coal miners: final report. *Br J Ind Med 44*: 777–779

Swaen GM, Meijers JMM, Slangen JJM (1995) Risk of gastric cancer in pneumoconiotic coal miners and the effect of respiratory impairment. *Occup Environ Med 52*: 606–610

Tourmann JL, Kaufmann R (1994a) Laser microprobe mass spectrometric (LAMMS) study of quartz-related and non-quarz-related factors of the specific harmfulness of coal mine dusts. *Ann Occup Hyg 38*: 455–467

Tourmann JL, Kaufmann R (1994b) Biopersistence of the mineral matter of coal mine dusts in silicotic human lungs: is there a preferential release of iron? *Environ Health Perspect 102*: 265–268

Triolo AJ, Aponte GE, Herr DL (1977) Induction of aryl hydrocarbon hydroxylase and forestomach tumors by benzo[*a*]pyrene. *Cancer Res 37*: 3018–3021

Tucker JD, Whong W-Z, Xu J, Ong T-M (1984) Genotoxic activity of nitrosated coal dust extract in mammalian systems. *Environ Res 35*: 171–179

Ulm K, Lange H-J, Pache L (1989) Statistisches Verfahren zur Begründung von Schwellenwerten (Statistical procedure for the derivation of threshold values) (German). Proceedings of the 34th annual meeting of the GMDS, Aachen, *Medizinische Informatik und Statistik 71*: 50–56

Ulmer WT (1970) The relationship between dust exposure and chronic bronchitis as emphysema. Pneumoconiosis—*Proceedings of the International Conference Johannesburgh 1969*: 328–336

Ulmer WT, Reichel G (1972) Epidemiological problems of coal workers' bronchitis in comparison with the general population. *Ann N Y Acad Sci 200*: 211–219

Ulmer WT, Reichel G, Werner U (1968) Die chronisch obstruktive Bronchitis des Bergmannes (Chronic obstructive bronchitis in miners) (German). *Int Arch Gewerbepathol Gewerbehyg 25*: 75–98

Ulmer WT, Bengtsson U, Reischig HL, Speckmann B, Vautrin HJ, Zimmermann I (1987) Lungenfunktionsuntersuchungen bei Bergleuten (Lung function tests in miners) (German). *Arbeitsmed Sozialmed Praeventivmed 22*: 169–174

Ulmer WT, Hoffarth HP, Bengtsson U, Reischig HL (1991) Die Bergarbeiter-Pneumokoniose: Grundlagen, neue Aspekte für Prävention, Diagnose und Therapie (CWP: fundamentals, new aspects for prevention, diagnosis and therapy) (German). *Pneumologie 45*: 410–413

Une H, Esaki H, Osajima K, Ikui H, Kodama K, Hatada K (1995) A prospective study on mortality among Japanese coal miners. *Ind Health 33*: 67–76

Vallyathan V, Shi X, Dalal NS, Irr W, Castranova V (1988) Generation of free radicals from freshly fractured silica dust: potential role in acute silica-induced lung injury. *Am Rev Respir Dis 138*: 1213–1218

Vallyathan V, Green FHY, Brower P, Attfield MD (1996) The role of coal mine dust exposure in the development of pulmonary emphysema. *Inhaled Particles VIII*, 26th to 30th August 1996, Cambridge, UK, Book of Abstracts: 44

Vallyathan V, Green FHY, Brower P, Attfield MD (1997) The role of coal mine dust exposure in the development of pulmonary emphysema. *Ann Occup Hyg 41*: 352–357

Vanhee D, Gosset P, Wallaert B, Tonnel AB (1994) Role of macrophage-derived cytokines in coal workers pneumoconiosis. *Ann N Y Acad Sci 88*: 183–192

Vautrin HJ (1992) *Pneumokonioseepidemiologie für ein Bergwerk am Niederrhein* (Epidemiology of pneumoconiosis in a Lower Rhine mine) (German). Doctoral thesis, University of Cologne

Vautrin HJ, Morfeld P, Kampmann B (1989) Vergleichende Untersuchungen zur Pneumokonioseentwicklung auf einem Bergwerk der Ruhrkohle AG im Hinblick auf eine Festsetzung des Grenzwertes für quarzhaltigen Kohlegrubenstaub (Comparative studies of the development of pneumoconiosis in a Ruhrkohle AG mine for the establishment of a threshold level of quartz-containing coal mine dust) (German). *Silikosebericht Nordrhein-Westfalen 17*: 297–333

Vautrin HJ, Morfeld P, Kampmann B (1990) Prevalences, incidence densities and cumulative incidences of pneumoconiotic changes for two groups of miners of a mine in Western German coal mining. *Proceedings of the VIIth International Pneumoconiosis Conference 1988*, Pittsburgh, 132–152

Wagner MMF (1976) Pathogenesis of malignant histiocytic lymphoma induced by silica in a colony of specific-pathogen-free Wistar rats. *J Nat Cancer Inst 57*: 509–518

Wagner MMF, Wagner JC (1972) Lymphomas in the Wistar rat after intrapleural inoculation of silica. *J Nat Cancer Inst 49*: 81–91

Walton WH, Dodgson J, Hadden GG, Jacobsen M (1977) The effect of quartz and other non-coal dusts in coal workers pneumoconiosis. Part I: Epidemiological studies. in: Walton WH (Ed.) *Inhaled particles IV*, Pergamon Press, Oxford, 669–690

Watanabe S, Shirakami A, Takeichi T (1987) Alterations in lymphocyte subsets and serum immunoglobulin levels in patients with silicosis. *J Clin Lab Immunol 23*: 45–51

Watson AJ, Black J, Doig AT, Nagelschmidt G (1959) Pneumoconiosis in carbon electrode workers. *Br J Ind Med 16*: 274–385

Weinberg GB, Kuller LH, Stehr PA (1985) A case-control study of stomach cancer in a coal mining region of Pennsylvania. *Cancer 56*: 703–713

Weller W, Rosmanith J (1983) Untersuchungen zur spezifischen Schädlichkeit von Feinstäuben des Steinkohlebergbaus. XVIII. Vergleich der Wirkung eines grubenechten Staubes im Intraperitonealtest, im Intratrachealtest und im Inhalationstest (Studies of the specific damage caused by fine dust from coal mines. XVIII: Comparison of the effects of a genuine mine dust in the intraperitoneal test, in the intratracheal test and in the inhalation test) (German). in: *Ergebnisse von Untersuchungen auf dem Gebiet der Staub- und Silikosebekämpfung im Steinkohlebergbau. Silikosebericht Nordrhein-Westfalen 14*: 257–264

WHO (World Health Organization) (1986) *Recommended health-based limits in occupational exposure to selected mineral dusts (silica, coal).* Technical Report Series No. 734, WHO, Genf

Whong W-Z, Long R, Ames RG, Ong T-M (1983) Role of nitrosation in the mutagenic activity of coal dust: a postulation for gastric carcinogenesis in coal miners. *Environ Res 32*: 298–304

Wohlberedt F (1988a) Silikoseerkrankungen im Bergbau der Bundesrepublik Deutschland, Teil 1. (Silicosos in German mines, Part 1) (German) *Kompaß 98*: 274–281

Wohlberedt F (1988b) Silikoseerkrankungen im Bergbau der Bundesrepublik Deutschland, Teil 2. (Silicosos in German mines, Part 2) (German) *Kompaß 98*: 342–351

Woitowitz H-J (1972) Das chronische, unspezifische respiratorische Syndrom (CURS) (The chronic, unspecific respiratory syndrome (CURS)) (German). *Arbeitsmed Sozialmed Arbeitshyg 7*: 22–26

Woitowitz H-J, Armbruster L, Bauer H-D, Breuer H, Bruch J, Lange H-J, Rödelsperger K, Sieben J, Stalder K (1989) Überprüfung des Grenzwertes für quarzhaltigen Kohlegrubenstaub (Examination of the threshold value for coal mine dust containing quartz) (German). *Zentralbl Arbeitsmed 39*: 132–146

Wolf M, Schulz HM (1994) *Charakterisierung von Grubenstäuben hinsichtlich ihrer spezifischen Schädlichkeit (Pneumokoniosen)* (Characterization of the specific harmful effects (pneumoconiosis) of mine dusts) (German). Wissenschaftlicher Abschlußbericht des EGKS, Forschungsvorhaben Nr 7263-02/079/01

Wu Z-L, Chen J-K, Ong T, Matthews EJ, Whong W-Z (1990) Induction of morphological transformation by coal-dust extract in BALB/3T3 A31-1-13 cell line. *Mutat Res 242*: 225–230

completed 1.4.1998

Diisobutyl ketone

MAK value	**not yet established, see Section IIb of the *List of MAK and BAT Values***
Peak limitation	–
Absorption through the skin	–
Sensitization	–
Carcinogenicity	–
Prenatal toxicity	–
Germ cell mutagenicity	–
BAT value	–
Synonyms	–
Chemical name (CAS)	2,6-dimethyl-4-heptanone
CAS number	108-83-8
Structural formula	$((CH_3)_2CH–CH_2)_2CO$
Molecular formula	$C_9H_{18}O$
Molecular weight	142.2
Melting point	–41.5°C
Boiling point	168°C
Density at 20°C	0.81 g/cm^3
Vapour pressure at 20°C	2.2–11 hPa
$\log P_{ow}$*	3.1 (calculated)
1 ml/m^3 (ppm) ≙ 5.8 mg/m^3	**1 mg/m^3 ≙ 0.17 ml/m^3 (ppm)**

The MAK value for diisobutyl ketone of 50 ml/m^3 was established in 1958 in analogy to the TLV value at the time. The present documentation is based on more recent reviews of the toxicological data for diisobutyl ketone (ACGIH 1996, BIBRA 1995).

---

* *n*-octanol/water distribution coefficient

1 Toxic Effects and Mode of Action

Diisobutyl ketone is a clear, colourless liquid with a sweet smell, and is miscible with alcohol, diethyl ether and chloroform. The acute toxicity of the substance is low for all routes of absorption. Diisobutyl ketone is moderately irritative to the skin and mucous membranes. A threshold value of 50 ml/m^3 has been reported for the irritative effects on the eyes and respiratory passages in man. Higher concentrations can impair central nervous functions. There is no information available for the effects on man of repeated exposure to diisobutyl ketone. After short-term exposure the target organs in rats are regarded as being the kidneys and liver. Histopathological damage occurs in female rats after exposure to concentrations from 900 ml/m^3, in male rats in the kidneys after as little as 300 ml/m^3. Short-term exposure to concentrations of 100 ml/m^3 had no effects in rats or guinea pigs. Medium-term ingestion of diisobutyl ketone doses of 2000 mg/kg body weight and day led in rats to increased liver, kidney and adrenal gland weights and to unspecific changes in the liver, kidneys and stomach. Double this dose is lethal within 3 days.

Diisobutyl ketone was not found to have mutagenic effects in prokaryotes in various test systems and did not induce chromosomal aberrations in cultured rat hepatocytes. There is no information available for the genotoxic potential of the substance *in vivo*, nor for the carcinogenic or sensitizing potential, nor the toxic effects on reproduction.

2 Mechanism of Action

The kidney damage observed in male rats in an inhalation study (Dodd *et al.* 1987) is regarded as species-specific and sex-specific α_2-microglobulin nephropathy because of the accumulation of hyaline droplets in the cells of the proximal tubules. The mechanism of effects on the kidneys found also in female animals after higher doses is not known.

There are no data available for the toxicokinetics or metabolism of diisobutyl ketone.

3 Effects in Man

The odour threshold for diisobutyl ketone is below 0.11 ml/m^3 (BIBRA 1995).

During exposure for 15 minutes to diisobutyl ketone, most of the 12 volunteers of both sexes found 50 ml/m^3 to be irritative to the eyes, and concentrations above 50 ml/m^3 (no other details) also irritative to the nose and throat. 25 ml/m^3 was regarded by the test persons as an acceptable concentration for 8-hour exposure (Silverman *et al.* 1946). Two and three persons who were exposed for 3 hours to diisobutyl ketone concentrations of 50 ml/m^3 and 100 ml/m^3, respectively, felt slight irritation of the eyes and nose only at the beginning of exposure. They could smell and taste the substance during the whole exposure period. Other effects occurred only at the high concentration. One person reported slight irritation of the throat, another watering of the eyes. Two persons complained after 2 hours of slight headaches and 2 persons became slightly dizzy when after 3 hours they returned to fresh air. The exposure had no effects on the pulse, blood pressure, urine status or the results of a co-ordination test (Carpenter *et al.* 1953).

In a 60-year-old laboratory assistant who was repeatedly exposed at the workplace to unknown concentrations of diisobutyl ketone, severe headaches and a 20-minute loss of sight occurred after exposure for about a month (White *et al.* 1993). To what extent the complaints are associated with the exposure to diisobutyl ketone cannot be assessed, in particular because of the lack of documentation of other possible confounders.

In an American tannery the incidence of testicular cancer (3 cases among 80 employees in the period from 1975 to 1987) was markedly increased. In the subsequent evaluation of the occupational hygiene at the workplace in 1989, a mixture of 4 solvents was detected, including diisobutyl ketone with an average concentration of 2.6–4.0 mg/m^3. This, however, was not held responsible for the observed effects, but earlier exposure to dimethylformamide (Calvert *et al.* 1990).

4 Animal Experiments and *in vitro* Studies

There are no data available for allergenic effects, toxic effects on reproduction or carcinogenicity of diisobutyl ketone.

4.1 Acute toxicity

4.1.1 Inhalation

Rats and guinea pigs survived exposure to an atmosphere saturated with diisobutyl ketone vapour for 7.5 to 16 hours. The estimated concentration was 3200 ml/m^3 (ACGIH 1996, BIBRA 1995). On the other hand, in another study rats tolerated exposure to an atmosphere saturated with diisobutyl ketone vapour for 4 hours at the longest without deaths (Smyth *et al.* 1949). Exposure of Sherman rats for 8 hours to concentrations of 2000 ml/m^3 was lethal for 5 of 6 (Smyth *et al.* 1949) and 7 of 12 animals (Carpenter *et al.* 1953); in the latter study the female animals were found to be the more sensitive. Wistar rats of both sexes survived exposure to 2000 ml/m^3 for 8 hours. Exposure for 6 hours to 1500 ml/m^3 led in rats to motor co-ordination disturbances and apathy (BIBRA 1995).

In mice exposed for one hour to an atmosphere saturated with diisobutyl ketone vapour (about 3200 ml/m^3), somnolence, motor co-ordination disturbances and paralysis in the hind limbs were seen. At the same concentration, narcosis of the animals occurred after 3 hours and death after 7 hours. During exposure for 7.5 hours, the substance was narcotic and lethal from concentrations in the air of 10200 mg/m^3 (about 1700 ml/m^3). Histopathological examination of the mice killed revealed accumulation of fluid in the lungs, liver and kidneys, haemorrhage in the lungs, and pneumonia (BIBRA 1995).

In male Swiss-OF1 mice, an RD_{50} value of 287 ml/m^3 was determined after exposure for 15 minutes to diisobutyl ketone. Even the lowest tested concentration of 184 ml/m^3 significantly reduced the respiration rate by 26 % (De Ceaurriz *et al.* 1984). The reduction in the respiration rate is interpreted as the result of sensory irritation of the mucous membranes of the upper respiratory tract caused by the substance. After exposure of mice for 4 hours, a value of 310 ml/m^3 was determined in a swimming test as the EC_{50} for behavioural changes (De Ceaurriz *et al.* 1984). These behavioural changes indicate central nervous effects.

4.1.2 Ingestion, dermal absorption and intraperitoneal administration

For the rat after oral administration of diisobutyl ketone and a follow-up period of 14 days, an LD_{50} value of 5750 mg/kg body weight was determined (Smyth *et al.* 1949). For the mouse, oral LD_{50} values in the range from 1400 to 2820 mg/kg body weight were given. In the mice that died, accumulated blood was observed in the stomach wall and duodenum (BIBRA 1995).

In rabbits the dermal LD_{50} after occlusive application for 24 hours was found to be 16000 mg/kg body weight (Smyth *et al.* 1949). Undiluted diisobutyl ketone was applied to the skin of 2 rabbits for 8.5 hours. No evidence of systemic toxic effects was found, and gross pathological examination did not reveal changes in the most important internal organs (no other details) (BIBRA 1995).

Intraperitoneal administration of 1600 mg/kg body weight was survived by all rats (BIBRA 1995).

4.2 Subacute, subchronic and chronic toxicity

4.2.1 Inhalation

Groups of 10 F344 rats per concentration and sex were exposed for 6 hours daily, 5 days per week for 9 weeks to concentrations of diisobutyl ketone of 98, 300 or 905 ml/m^3. The high concentration caused lacrimation, indicating mild eye irritation. In the female animals of this group, body weights were slightly but significantly reduced. At the end of the experiment, in the male animals of all exposure groups the drinking water consumption and urine excretion were increased and thus the osmolarity of the urine decreased. Only in the two high concentration groups were the absolute and relative liver weights increased. The same was true for the kidney weights in the male animals, although in the female animals only the relative kidney weights were significantly increased. The only unusual finding in the extensive histopathological examination was the damage in the proximal tubules of the kidneys in the male animals of the 300 and 905 ml/m^3 groups which was more severe than that in the control animals and was attributed to the increased accumulation of hyaline droplets. In another group of 10 male and 10 female animals exposed in the same way to 905 ml/m^3 but observed for 14 days afterwards, the findings corresponded qualitatively to those described above, but in some cases were considerably less marked (Dodd *et al.* 1987). A NOAEC (no observed adverse effect concentration) for diisobutyl ketone of 98 ml/m^3 can be deduced from this study.

Groups of 15 male and 15 female Sherman rats were exposed to diisobutyl ketone concentrations of 125, 250, 530, 920, or 1650 ml/m^3 for 7 hours a day, on 5 days a week for 6 weeks. In the high concentration group all the female animals and 2 of the males died on the first day. In the surviving male animals co-ordination disturbances were observed on day 2 of exposure but not on the following days. At the end of the experiment the body weights of the 12 surviving animals were reduced and the liver and kidney weights increased. The histopathological examination revealed foamy swelling in the liver and pulmonary congestion. In the 920 and 530 ml/m^3 groups, in the female animals also in the 250 ml/m^3 group, the only unusual finding was increased liver and kidney weights. For both sexes, 125 ml/m^3 was found to be the concentration that had no effect. This was also true for guinea pigs, which were exposed in the same way. Liver weights were reduced only in the male guinea pigs after exposure to 250 ml/m^3. Higher concentrations were not tested with this species. Microscopic examination did not reveal pathological changes at either exposure concentration (Carpenter *et al.* 1953).

Twelve mice survived twelve 3-hour exposures to diisobutyl ketone concentrations of 14400 to 22200 mg/m^3 (about 2500–3800 ml/m^3) (no other details) (BIBRA 1995).

4.2.2 Ingestion and dermal absorption

Gavage of diisobutyl ketone doses of up to 2000 mg/kg body weight and day for 3 weeks was not found to be lethal for rats (number of animals not specified). At the highest dose, transient central nervous disturbances, increased liver and kidney weights, and slight unspecific tissue changes in the liver and kidneys were observed. Administration for 90 days of doses of 2000 mg/kg body weight and day caused an increase in the adrenal gland weights and a decrease in brain weights. The blood sugar levels of the animals were reduced. Unspecific tissue changes were found in the stomach, liver and kidneys. Evidence of neurotoxic effects was not observed. All the animals given a dose of 4000 mg/kg body weight and day died after 3 days. The deaths were attributed to severe central nervous disturbances, liver damage and dehydration. Damage to the lungs (fluid accumulation and swelling) and kidneys was also reported (BIBRA 1995). Rats given 100 mg diisobutyl ketone with the diet daily for 8 days (about 250–1000 mg/kg body weight and day), excreted increased amounts of vitamin C with the urine (Longenecker *et al.* 1939). Other parameters were not investigated.

Undiluted diisobutyl ketone was applied to the skin of 2 rabbits twelve times for 5 hours within 21 days. No evidence was found of systemic toxic effects and gross pathological examination did not reveal changes in the most important internal organs. In the upper layers of skin, local inflammatory changes were observed (no other details) (BIBRA 1995).

4.3 Local effects on skin and mucous membranes

4.3.1 Skin

The non-occlusive application of pure diisobutyl ketone to rabbit skin produced slight irritation after 24 hours (Smyth *et al.* 1949). Under semi-occlusive conditions no skin irritation was visible in 6 male and 6 female rabbits after contact with the substance for 1 and 4 hours, while with occlusive application irritation was observed at both times. On guinea pig skin, undiluted diisobutyl ketone was moderately irritative both after non-occlusive and occlusive application (BIBRA 1995). A single dermal application for 8.5 hours and repeated dermal application (12 times for 5 hours within 21 days) of undiluted diisobutyl ketone to the skin of 2 rabbits led after a delay to slight erythema (no other details) (BIBRA 1995).

4.3.2 Eyes

In various studies, undiluted diisobutyl ketone caused at most mild irritation in the rabbit eye, visible as slight conjunctivitis and swelling of the cornea. The eye irritation was no longer evident after 72 hours (BIBRA 1995, Carpenter and Smyth 1946).

An atmosphere saturated with diisobutyl ketone (about 3200 ml/m^3) led in mice after 10 minutes to irritation of the eyes and nasal mucosa (no other details) (BIBRA 1995). In the experiment described in Section 5.2.1 (Dodd *et al.* 1987) lacrimation and partial eyelid closure was seen in rats exposed to 905 ml/m^3, but not in those exposed to 300 ml/m^3.

4.4 Genotoxicity

There are no data available for the genotoxic effects of the substance *in vivo*.

4.4.1 *In vitro*

In the *Salmonella* strains TA98, TA100, TA1535 and TA1537, diisobutyl ketone was not found to be mutagenic in either the standard test (max. 333 µg/plate, Mortelmans *et al.* 1986) or the preincubation test (max. 4000 µg/ml in the preincubation mixture, Brooks *et al.* 1988), either in the presence or absence of a metabolic activation system. Also in a test for differential killing with the *Escherichia coli* strains WP_2 and WP_2uvrA and in a test for mitotic gene conversion with *Saccharomyces cerevisiae*, the substance had no effect, either with or without an activation system, even at the highest concentrations of 4000 and 5000 µg/ml, respectively (Brooks *et al.* 1988). In the rat hepatocyte line RL_4, diisobutyl ketone was not found to cause chromosomal damage up to the highest concentration tested of 500 µg/ml. The experiment was carried out only without metabolic activation (Brooks *et al.* 1988).

5 Manifesto (MAK value/classification)

Diisobutyl ketone is irritative to the eyes and mucous membranes of the respiratory passages in man. Two earlier studies with volunteers yielded evidence that a concentration of 50 ml/m^3 initially causes mild irritation, but exposure for a longer period can be tolerated without complications. As a result of the small number of persons tested, this information is, however, inadequate for evaluating a threshold limit value and further studies should be carried out to determine the irritation threshold of the substance. As there is also insufficient information available for the effects of long-term exposure, a threshold limit value for the workplace cannot be derived scientifically. Therefore the previous MAK value for diisobutyl ketone has been withdrawn and the substance is listed in Section IIb of the *List of MAK and BAT Values*.

The data from experiments with animals indicate that designation with an "H" (for substances for which there is danger from cutaneous absorption) is not necessary.

The sensitizing potential cannot be evaluated because of lack of data; the substance is, therefore, not designated with an "S".

There is also no information available for the carcinogenicity, germ cell mutagenicity or toxic effects on reproduction of diisobutyl ketone.

6 References

ACGIH (American Conference of Governmental Industrial Hygienists) (1996) Diisobutyl ketone. in: *Documentation of TLVs and BEIs*, ACGIH, Cincinnati, OH, USA

BIBRA (The British Industrial Biological Research Association) (1995) *Diisobutyl ketone. Toxicity profile*, BIBRA, Surrey

Brooks TM, Meyer AL, Hutson DH (1988) The genetic toxicology of some hydrocarbon and oxygenated solvents. *Mutagenesis 3*: 227–232

Calvert GM, Fajen JM, Hills BW, Halperin WE (1990) Testicular cancer, dimethylformamide, and leather tanneries. *Lancet 336*: 1253–1254

Carpenter CP, Smyth Jr HF (1946) Chemical burns of the rabbit cornea. *Am J Ophthalmol 29*: 1363–1372

Carpenter CP, Pozzani UC, Weil CS (1953) Toxicity and hazard of diisobutyl ketone vapors. *Arch Ind Hyg Occup Med 8*: 377–381

De Ceaurriz J, Micillino JC, Marignac B, Bonnet P, Muller J, Guener JP (1984) Quantitative evaluation of sensory irritating and neurobehavioural properties of aliphatic ketones in mice. *Food Chem Toxicol 22*: 545–549

Dodd DE, Losco PE, Troup CM, Pritts IM, Tyler TR (1987) Hyalin droplet nephrosis in male Fischer-344 rats following inhalation of diisobutyl ketone. *Toxicol Ind Health 3*: 443–457

Longenecker HE, Musulin RR, Tully RH, King CG (1939) An acceleration of vitamin C excretion by feeding known organic compounds to rats. *J Biol Chem 129*: 445–453

Mortelmans K, Haworth S, Lawlor T, Speck W, Trainer B, Zeiger E (1986) *Salmonella* mutagenicity tests: II. Results from the testing of 270 chemicals. *Environ Mutagen 8, Suppl 7*: 1–119

Silverman L, Schulte HF, First MW (1946) Further studies on sensory response to certain industrial solvent vapors. *J Ind Hyg Toxicol 28*: 262–266

Smyth Jr HF, Carpenter CP, Weil CS (1949) Range-finding toxicity data, list III. *J Ind Hyg Toxicol 31*: 60–62

White RF, Feldman RG, Moss MB, Proctor SP (1993) Magnetic resonance imaging (MRI), neurobehavioral testing, and toxic encephalopathy: two cases. *Environ Res 61*: 117–123

completed 29.01.1998

Endrin

MAK value (1969)	**0.1 mg/m³ I (inhalable aerosol fraction)**
Peak limitation (1983)	**Category III**
Absorption through the skin (1969)	**H**
Sensitization	–
Carcinogenicity	–
Prenatal toxicity (1998)	**Group C**
Germ cell mutagenicity	–
BAT value	–
Synonyms	1,2,3,4,10,10-hexachloro-6,7-epoxy-1,4,4a,5,6,7,8,8a-octahydro-*endo*, *endo*-1,4:5,8-dimethanonaphthalene
Chemical name (CAS)	(1aα,2β,2aβ,3α,6α,6aβ,7β,7aα)-3,4,5,6,9,9-hexachloro-1a,2,2a,3,6,6a,7,7a-octahydro-2,7:3,6-dimethanonaphth[2,3-*b*]oxirene
CAS number	72-20-8
Structural formula	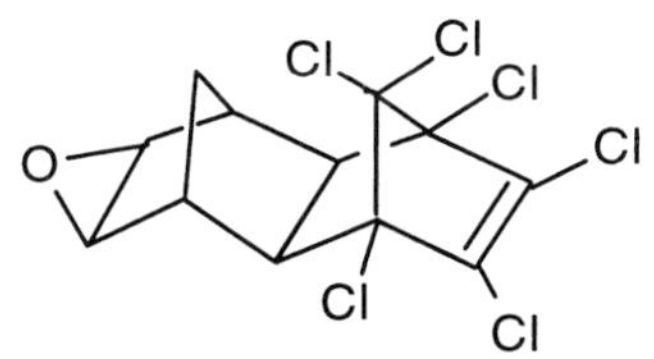
Molecular formula	$C_{12}H_8Cl_6O$
Molecular weight	380.93
Melting point	226–230°C (decomposition at 245°C)
Density at 25°C	1.64 g/cm³
$\log P_{ow}$*	5.34

* *n*-octanol/water distribution coefficient

This documentation is based mainly on a published review of the toxicological data (WHO 1992), supplemented with more recent information relevant to the evaluation.

1 Toxic Effects and Mode of Action

Endrin is a stereoisomer of dieldrin; both substances are used as pesticides. Endrin has been used since the 1950s as an insecticide and rodenticide, in particular in the growing of cotton, maize, rice and sugar cane. Its use is now banned in the EU and in numerous other countries. Technical grade endrin with a purity of 92 to 99 % can contain e.g. dieldrin (0.42 %), aldrin (0.03 %) and isodrin (0.73 %) as impurities.

Endrin is of higher acute toxicity than dieldrin, but is metabolized and eliminated more rapidly; thus endrin accumulates less in adipose tissue than do dieldrin and other substances of similar chemical structure. Endrin is absorbed well via the gastrointestinal tract and skin, and can pass the placenta. Endrin and its metabolite 12-ketoendrin can be detected in foetal tissue. The main metabolite *anti*-12-hydroxyendrin and the metabolites *syn*-12-hydroxyendrin and 12-ketoendrin are of greater acute toxicity than endrin itself. In addition to unchanged endrin, the free metabolites and their conjugates are excreted via the faeces. The glucuronic acid and sulfate conjugates of the metabolites can also be eliminated with the urine.

Endrin is highly toxic after oral and dermal uptake of single doses. In man, the severe neurotoxic effects of endrin rapidly result in symptoms such as nausea, vomiting, dizziness, stomach-ache, headache, sudden unconsciousness, convulsions and CNS depression. In animals, also paralysis, dyspnoea and cyanosis are observed. In addition to the effects on the central nervous system, the other main symptoms of acute and chronic toxicity are liver and kidney damage, probably the result of oxidative stress. After long-term exposure, a NOEL (no observed effect level) of 0.025 mg/kg body weight and day was found for the most sensitive species, the dog.

Endrin does not have irritative or sensitizing effects on the skin. Endrin is not genotoxic or carcinogenic. *In vivo*, DNA single strand breaks are induced only by exposure to high levels of endrin, which at the same time results in marked lipid peroxidation and probably the production of reactive oxygen species. Lipid peroxidation and DNA single strand breaks are also detectable in foetal tissue.

Endrin does not cause impairment of fertility. Foetotoxic effects occur only with highly maternally toxic doses or doses which are known to produce tissue damage in the dams as a result of oxidative stress after single or repeated administration.

2 Mechanism of Action

The neurotoxic effects (see Sections 4 and 5) of endrin and its metabolite 12-ketoendrin are the result of disturbances in nerve conduction following interaction with the γ-aminobutyric acid (GABA) neurotransmitter system, which has an inhibiting effect on the central nervous system. The probably non-competitive binding to an allosteric binding site of the γ-aminobutyric acid receptor of the chloride ion channels causes blocking of the chloride channels in cell membranes of the central nervous system and thus a decrease in the chloride ion flow. This facilitates synaptic transmission, which in turn leads to overstimulation of the nervous system which is associated with changes in EEG activity and the triggering of convulsions (Narahashi *et al.* 1992, WHO 1992). It is also conceivable that inhibition of the calmodulin-dependent calcium pump in the neurons is involved in the neurotoxic effects (WHO 1992).

Endrin causes the induction of cytochrome P450-dependent monooxygenases (WHO 1992). Oxidative stress probably plays a decisive role in the tissue-damaging, foetotoxic and lethal effects of endrin. Lipid peroxidation has been observed in various organs such as the liver, kidneys, brain, placenta and in foetal tissue (Bagchi *et al.* 1992a, 1992b, 1995, Hassan *et al.* 1991, Hassoun and Stohs 1996a, 1996b, Hassoun *et al.* 1996, Numan *et al.* 1990; see Section 5.1.2). Pretreatment with antioxidants led to the complete inhibition of lipid peroxidation and of the effects in the liver and kidneys (Hassan *et al.* 1991, Numan *et al.* 1990). It is still unclear whether the lipid peroxidation is initiated by a reactive metabolite or free radicals. Therefore oxidative damage to the central nervous system could also be involved in the neurotoxic effects of endrin.

3 Toxicokinetics and Metabolism

Endrin is absorbed via the gastrointestinal tract and the skin. There are no data available for absorption rates. The toxicokinetics of endrin differ in different species and sexes.

In rats endrin was stored mainly in adipose tissue and skin. In male rats which were fed endrin at a level of 1.5 mg/kg body weight and day for 8 days, the concentrations in adipose tissue reached 3 to 4 µg/g. Beagle dogs were given endrin in the diet at a level of 0.1 mg/kg body weight and day for 128 days. From day 9 the concentration in blood remained constant at 4 µg/l. 7 days after the end of exposure, concentrations less than 0.2 µg/g were found in 8 tissues and up to 0.8 µg/g in adipose tissue.

Endrin persists in the body of female rats for longer and in higher concentrations than in males. After oral doses of up to 0.064 mg/kg body weight, the elimination half-time was about 2 days. Elimination was slower with a half-time of about 6 days after a higher dose of 0.128 mg/kg body weight. After intravenous administration of 0.200 mg/kg body weight, the half-life of endrin in male rats was 2.5 to 3 days, in female rats 4 days.

While rats excreted up to about 90 % of the dose with the faeces mainly in metabolized form, rabbits excreted about half of the dose in the urine (only metabolites) over several days and in the faeces (unchanged endrin) within 24 hours. Because the molecular weight thresholds for biliary elimination are higher in rabbits than in rats, rabbits can excrete conjugates of endrin metabolites with the urine while rats cannot (WHO 1992).

Endrin can pass the placenta. After oral exposure of pregnant rats to endrin doses of 0.45 mg/kg body weight and day from days 7 to 20 of gestation, endrin concentrations of 0.327 µg/g were found in the livers of the dams and of 0.027 µg/g in the tissue of the foetuses. In hamsters exposed to higher endrin doses of 1.5 and 2.5 mg/kg body weight and day from days 5 to 14 of gestation, tissue concentrations of 1.45 and 2.55 µg/g were reached in the livers of the dams, in foetal tissue concentrations of 0.021 and 0.075 µg/g. The tissue concentrations of 12-ketoendrin were not determined quantitatively. In hamster foetuses there is much more 12-ketoendrin present than endrin (ratio of peak heights 3.2:1), while in rat foetuses more endrin is present than 12-ketoendrin (peak height ratio 1:0.4) (Chernoff *et al.* 1979, Kavlock *et al.* 1981).

Rats metabolize endrin mainly via *anti*-12-hydroxyendrin to form the sulfate and glucuronic acid conjugate. In lower concentrations *syn*-12-hydroxyendrin, 3-hydroxyendrin, 12-ketoendrin and 4,5-*trans*-dihydroisodrindiol and their conjugates are formed (metabolic pathways from WHO 1992). The transformation of endrin to *anti*-12-hydroxyendrin and of *syn*-12-hydroxyendrin to 12-ketoendrin by monooxygenases takes place rapidly. Male animals form more *anti*-12-hydroxyendrin and excrete this via the faeces, while female animals excrete this mainly as *anti*-12-hydroxyendrin sulfate in urine. In the male animals 12-ketoendrin was the main urinary metabolite. In the tissues mainly endrin and 12-ketoendrin are found; in male animals in the ratio 1:8, in female animals in the ratio 2:1. In the liver and kidneys of male animals more 12-ketoendrin is therefore found, in the kidneys of female animals, however, more endrin.

In rabbits mainly *anti*-12-hydroxyendrin is formed and excreted in the urine as *anti*-12-hydroxyendrin sulfate in addition to *syn*-12-hydroxyendrin sulfate and the glucuronic acid conjugates of 3-hydroxyendrin and 4,5-*trans*-dihydroisodrindiol. *syn*-12-Hydroxyendrin does not seem to be oxidized to 12-ketoendrin.

In the urine of workers exposed to endrin (level of exposure not specified), *anti*-12-hydroxyendrin was identified (up to 0.36 mg/g creatinine; decrease over the weekend) and in the faeces its glucuronic acid conjugate. In another study, endrin and *anti*-12-hydroxyendrin were found in the faeces of workers, and the glucuronic acid conjugate of *anti*-12-hydroxyendrin (up to 0.14 mg/l, expressed as *anti*-12-hydroxyendrin) in all urine samples.

4 Effects in Man

4.1 Single exposures

Repeatedly in the past there have been cases of intoxication as a result of the ingestion of accidentally contaminated food and also after dermal exposure. The main symptoms reported were nausea, vomiting, dizziness, stomach-ache, headache, sudden unconsciousness, seizures and CNS depression, which began 0.5 to 10 hours after uptake of endrin. Respiratory depression was occasionally fatal. Generally the condition of the patients improved within hours or days (Ferrer and Cabral 1991, 1995, WHO 1992). Changes in the EEG were still registered after several weeks and were attributed to brain stem irritation. It was not possible to deduce the exact dose-response relationship for effects in man, as the doses absorbed were not known. Estimated doses from cases of accidental intoxication are available, but the values differ greatly. Oral doses of about 0.25 and 1 mg/kg body weight are said to cause single and repeated seizures, respectively, and 10 mg/kg body weight to be lethal. In other reports, however, doses of 5 to 50 mg/kg body weight are given as toxic and a total dose of about 6 g (about 100 mg/kg body weight) as lethal (WHO 1992).

After higher-level short-term inhalation exposure (no details of exposure level), a worker from an endrin formulation plant suffered seizures, while two workers who accidentally spilt the substance on their skin had no symptoms of intoxication. For these workers a half-life of the substance in blood of about 24 hours was estimated and a LOEL (lowest observed effect level) for symptoms of intoxication of 50 to 100 µg/l blood (WHO 1992).

4.2 Repeated exposures

In several studies with workers exposed to endrin (average exposure duration 12 to 24 years, level of exposure and number of persons exposed not specified), liver function tests and determination of urine and blood parameters did not yield any unusual findings. The workers exposed to endrin were not off sick any more often than other workers. Endrin was not detectable in plasma or adipose tissue (detection limit 3 µg/l, 0.03 µg/g), presumably because little accumulation of endrin took place because of rapid metabolism and elimination (see Section 3). Significant induction of microsomal enzyme systems was observed after exposure to endrin; the exposure level was not determined. Increased excretion of the metabolite *anti*-12-hydroxyendrin was used as an indicator of enzyme induction; a urinary *anti*-12-hydroxyendrin level of 0.130 µg/g creatinine was given as the LOEL; no NOEL was given (WHO 1992). The associated air concentrations were, however, not determined.

It is not clear from the reports how inhalation and dermal absorption contribute to the body burden during exposure at the workplace. Nor are air concentrations or daily exposure durations described; thus a MAK value cannot be derived from these studies.

4.3 Genotoxicity

In peripheral lymphocytes from 8 male workers exposed to endrin (exposure level not specified) no increase in the incidence of chromosomal aberrations was observed (WHO 1992).

4.4 Carcinogenicity

Mortality and the causes of death of 232 workers exposed for 4 to 27 years (on average 11 years) to endrin were followed over a period of 4 to 29 years (on average 24 years); mortality and tumour incidence were not increased relative to the values for the general population. At the beginning of employment, the workers were exposed to high levels of endrin; there are, however, no data available for the levels of exposure (WHO 1992).

5 Animal Experiments and *in vitro* Studies

5.1 Acute toxicity

5.1.1 Inhalation

Exposure for 1 hour of groups of 10 rats to an endrin concentration of about 2000 mg/m^3 ("mist of an emulsifiable concentrate in xylene" or "25 % and 30 % endrin dust concentrates of endrin") led to the death of 3/10, and 5/10 and 3/10 animals (WHO 1992). These studies from the 1950s cannot be used in the present evaluation because the exposure atmosphere (particle size not specified) and conditions were insufficiently characterized.

5.1.2 Ingestion

Single oral doses of endrin are highly toxic. The LD_{50} values for technical grade endrin are given as 4.0 to 17.8 mg/kg body weight for rats and mice, 7 to 10 mg/kg body weight for rabbits and 16 to 36 mg/kg body weight for guinea pigs; female animals were much more sensitive. Cats and monkeys were also found to be more sensitive, with lethal doses for the cat of 3 to 6 mg/kg body weight and LD_{50} values for the monkey of 3 to 12 mg/kg body weight. With LD_{50} values of 0.8 to 2.8 mg/kg body weight, the endrin metabolites *anti*-12-hydroxyendrin, *syn*-12-hydroxyendrin and 12-ketoendrin, administered orally to rats, are more toxic than endrin itself. The symptoms of intoxication included overexcitability, tremor followed by tonic-clonic cramps, paralysis, ataxia, dyspnoea, wheezing and cyanosis. In animals that survived, these effects were completely reversible.

In addition to the neurotoxic effects of endrin, the main other effect is tissue damage, in particular of the liver and kidneys, and is probably the result of oxidative stress (see Section 2). After oral administration of endrin doses of 1.5 to 6.0 mg/kg body weight to Sprague-Dawley rats, a dose-dependent increase in lipid peroxidation (excretion of formaldehyde, malondialdehyde, acetaldehyde and acetone with the urine) was observed over a period of 12 to 72 hours after administration (maximum 24 hours after administration)

(Bagchi *et al.* 1992b). A similar course was observed after doses of 3 to 6 mg/kg body weight with evidence of oxidation products and changed membrane structures in liver mitochondria and liver microsomes (increased membrane viscosity and reduced membrane fluidity) (Bagchi *et al.* 1992a). After administration of two endrin doses of 2.25 mg/kg body weight (0.25 times the LD_{50}) 21 hours apart to rats, lipid peroxidation was detected 3 hours after the last dose, not only in liver homogenates but also in brain homogenates (Bagchi *et al.* 1995). In liver and kidney tissue, lipid peroxidation and decreased glutathione levels were found 24 hours after the administration of endrin doses of 4.0 mg/kg body weight (Numan *et al.* 1990). The extent of lipid peroxidation decreased in the following order: in the liver, rat > mouse > guinea pig (no significant increase in the hamster), in the kidneys, rat > mouse > hamster (no increase in the guinea pig) (Hassan *et al.* 1991). When endrin was administered after pre-treatment of the animals with antioxidants, the level of lipid peroxidation was found to be the same as in untreated animals or even lower; histological examination did not reveal any unusual findings in the liver or kidneys and mortality after an otherwise lethal endrin dose was greatly reduced (Hassan *et al.* 1991, Numan *et al.* 1990).

Effects on the organs were dose-dependent and time-dependent increases in relative liver and spleen weights and decreases in the relative thymus weight from doses of 3 mg/kg body weight, as for lipid peroxidation (Bagchi *et al.* 1992a, 1992b). Histopathological examination of the hepatotoxic and nephrotoxic effects in the various species 24 hours after administration of oral endrin doses of 4 mg/kg body weight was carried out with Sprague-Dawley rats, Swiss-Webster mice, hamsters and guinea pigs. Liver damage was found in all species in the form of apoptosis and focal necrosis (changes in the cell nucleus, cytoplasm and membrane fragmentation), inflammatory reactions (leukocyte infiltration and congestion), interstitial oedema, hyperplasia of Kupffer's cells and green-brown pigments. In the rats, the species which reacted most sensitively, there was also fat accumulation and bile stasis. In the kidney tubules of rats, mice and hamsters, endrin caused cell necrosis, cloudy swelling, accumulation of hyaline droplets (only rat and hamster) and narrowing of the lumen, in the interstitium oedema and infiltration of inflammatory cells (only rats and mice). The most severe effects were found in rats, the least changes (only cloudy swelling and narrowing of the tubule lumen) in guinea pigs.

5.1.3 Dermal absorption

Dermal application of technical grade endrin dissolved in a hydrophobic solvent or in oil revealed an LD_{50} of 5 to 20 mg/kg body weight for the rat; thus the acute toxicity of the substance is similar after dermal and oral application (WHO 1992). After dermal exposure to the solid, the toxicity of a minimum lethal dose of 94 mg/kg body weight was found to be much lower when rabbits were exposed for 24 hours occlusively than after oral administration (Treon *et al.* 1955).

5.2 Subacute, subchronic and chronic toxicity

5.2.1 Inhalation

In an inhalation study from the 1950s, 3 mice, 3 rats, 2 guinea pigs, 2 hamsters, 4 rabbits and one cat were exposed via inhalation to endrin concentrations of 5.44 mg/m^3, 7 hours a day, 5 days a week for up to 26 weeks. The exposure atmosphere contained endrin vapour, which, according to the authors, was produced by sublimation. In rabbits and mice, deaths occurred from the 22nd exposure (2/4 and 1/3, respectively), seizures were not observed (Treon *et al.* 1955). As a result of the unclear exposure conditions and small number of animals, this study does not meet present-day standards and therefore cannot be included in the evaluation.

5.2.2 Ingestion

The numerous studies with mice, rats and rabbits and up to life-long exposure to endrin are listed in Table 1. They were carried out between 1955 and 1980. From what is known today of the mechanism of toxicity after short-term exposure (see Sections 2 and 4) it is clear, however, that in the studies with repeated exposure the most sensitive parameters of toxic effects were not always investigated and neurotoxic symptoms not always described. Accordingly, high doses were without effect in some studies but not in others, and there are considerable differences between the NOELs (see Table 1).

For rats exposed for 16 weeks, a NOAEL (no observed adverse effect level) of 0.05 mg/kg body weight was found (reduced body weights with reduced food consumption, alkaline phosphate activity increased; WHO 1992) and in a 2-year study also a NOEL of 0.05 mg/kg body weight (Treon *et al.* 1955). The documentation of these studies from the 1950s is incomplete, but more recent studies do not yield a better-founded NOEL for the rat. A NOEL of 0.045 mg/kg body weight was determined after lifelong exposure of 2 strains of mouse to endrin. A dose of 0.45 mg/kg body weight led initially to seizures and in the long term to liver damage (Shell 1970). In dogs given endrin with the diet for 2 years, a LOEL of 0.05 mg/kg body weight was determined for seizures and liver changes, doses up to 0.025 mg/kg body weight had no effects (Shell 1969). Another study with dogs (Treon 1955) yielded similar threshold values for increased kidney and heart weights (without histopathological findings). In extensive studies, neurotoxic symptoms and histological changes in the liver were found to be the most sensitive parameters of the chronic toxicity of endrin in mice and dogs. Assuming the dog to be the most sensitive species, a NOEL of 0.025 mg/kg body weight and day can be deduced, and used to derive a MAK value.

Table 1. Toxicity after repeated oral administration of endrin

Species	Duration of administration, dose	Results	References
mouse $B6C3F_1$ 5 ♂ and 5 ♀	6 weeks treatment, 2 weeks follow-up 2.5–20 mg/kg diet (0.375–3 mg/kg body weight and day)	body weights unaffected **from 0.75 mg/kg**: over-excitability in ♂ (LOEL) **from 1.5 mg/kg**: mortality increased in both sexes **NOEL**: 0.375 mg/kg body weight and day	NTP 1979, WHO 1992
mouse $B6C3F_1$ 50 ♂ and 50 ♀; matched controls: groups of 10; pooled controls: groups of 50	80 weeks treatment, 10–11 weeks follow-up endrin (purity 97 %) ♂: average dose: 1.6, 3.2 mg/kg diet (0.24, 0.48 mg/kg body weight and day); weeks 1–25: 2.5, 5 mg/kg diet (0.375, 0.75 mg/kg body weight and day); dose reduction: weeks 26–80: 1.2, 2.5 mg/kg diet (0.19, 0.375 mg/kg body weight and day) ♀: weeks 1–80: 2.5, 5 mg/kg diet (0.375, 0.75 mg/kg body weight and day)	body weights, histology unaffected; **from 0.24 mg/kg**: clinical symptoms: e.g. over-excitability as evidence of neurotoxic effects (LOEL); **♂: 0.24 mg/kg**: accidental overdose in week 66, therefore mortality could not be evaluated; **0.48 mg/kg**: ♂: mortality increased **up to 0.75 mg/kg**: ♀: mortality not increased, therefore ♂ much more sensitive **NOEL** could not be determined	NTP 1979, WHO 1992
mouse C57B1/6J 100 ♂ and 100 ♀, controls: 200 ♂ and 200 ♀	117 weeks endrin (purity 99 %) 0.3, 3 mg/kg diet (0.045, 0.45 mg/kg body weight and day)	mortality, body weight gains, food consumption, haematology unaffected **0.45 mg/kg**: at the beginning of the study occasional seizures, effects reversible; histology: liver damage typical of chlorinated insecticides **NOEL**: 0.045 mg/kg body weight and day	Shell 1970
mouse $C32F_1/J$ 100 ♂ and 100 ♀, controls: 200 ♂ and 200 ♀	♂: 117 weeks; ♀: 78 weeks, without follow-up endrin (purity 99 %) 0.3, 3 mg/kg diet (0.045, 0.45 mg/kg body weight and day)	mortality, body weight gains, food consumption, haematology unaffected **0.45 mg/kg**: at the beginning of the study occasional seizures, effects reversible; histology: liver damage typical of chlorinated insecticides ♀: experiment terminated after 78 weeks because of high incidence of spontaneous fibroadenomas of the mamma **NOEL**: 0.045 mg/kg body weight and day	Shell 1970
rat Osborne-Mendel 5 ♂ and 5 ♀	6 weeks treatment, 2 weeks follow-up endrin (purity 97 %) 2.5–80 mg/kg diet (0.125–4.0 mg/kg body weight and day)	body weights unaffected **from 1 mg/kg**: mortality increased **NOEL**: 0.5 mg/kg body weight and day	NTP 1979, WHO 1992

Table 1. continued

Species	Duration of administration, dose	Results	References
rat Sprague-Dawley 5 ♂ and 5 ♀	16 weeks technical grade endrin 1, 5, 25, 50, 100 mg/kg diet (0.05, 0.25, 1.25, 2.5, 5 mg/kg body weight and day)	**from 0.05 mg/kg**: dose-dependent decrease in body weights with reduced food consumption; alkaline phosphatase increased **from 0.25 mg/kg**: mortality increased in ♂, ♂ much more sensitive than ♀ **from 1.25 mg/kg**: over-sensitivity to external stimuli; seizures; mortality increased in ♀ study from 1956 **NOAEL**: marginal effects at 0.05 mg/kg body weight and day	WHO 1992
rat Osborne-Mendel 50 ♂ and 50 ♀	29 months endrin (purity 98 %) 2, 6, 12 mg/kg diet (0.1, 0.3, 0.6 mg/kg body weight and day)	mortality, body weight gains, relative liver weights unaffected; **from 0.1 mg/kg**: dose-dependent symptoms of intoxication; histology: moderate, not dose-dependent, cloudy swellings in liver and kidney tubules; lungs: congestion and focal haemorrhage moderately increased **NOEL** could not be determined	Deichmann *et al.* 1970
rat Osborne-Mendel 50 ♂ and 50 ♀ matched controls: groups of 10; pooled controls: groups of 50	80 weeks treatment, 31–34 weeks follow-up endrin (purity 97 %) ♂: 2.5, 5 mg/kg diet (0.125, 0.25 mg/kg body weight and day); ♀: average dose: 3, 6 mg/kg diet (0.15, 0.30 mg/kg body weight and day); weeks 1–9: 5, 10 mg/kg diet (0.25, 0.5 mg/kg body weight and day); dose reduction: weeks 10–80: 2.5, 5 mg/kg diet) (0.125, 0.25 mg/kg body weight and day)	body weights and mortality unaffected **from 0.125 mg/kg**: clinical symptoms as evidence of neurotoxic effects **from 0.25 mg/kg**: not tolerated by ♀, dose reduced as a result **NOEL** could not be determined	NTP 1979, WHO 1992
rat Osborne-Mendel 15–23 ♂ and 15–23 ♀	104 weeks endrin (purity 99 %) 0.1, 1.0, 5, 10, 25 mg/kg diet (0.005, 0.05, 0.25, 0.5, 1.25 mg/kg body weight and day)	mortality unaffected; no other details **from 0.005 mg/kg**: liver: increase in hyperplasia and hyperplastic nodules, not dose-dependent **0.05 and 1.25 mg/kg**: nephritis increased only in ♂, not dose-dependent **NOEL** could not be determined	Reuber 1978

Table 1. continued

Species	Duration of administration, dose	Results	References
rat Carworth 20 ♂ and 20 ♀	2 years endrin (purity not specified) 1, 5, 25, 50, 100 mg/kg diet (0.05, 0.25, 1.25, 2.5, 5 mg/kg body weight and day)	**from 0.25 mg/kg**: ♂: delayed body weight gains during the first 40 weeks; relative liver weights increased **from 1.25 mg/kg**: mortality increased (♀ reacted more sensitively); in animals that died: diffuse degenerative changes in the liver, kidneys, adrenal glands and brain; in ♀: relative liver weights increased **from 2.5 mg/kg**: also in surviving animals histopathological changes; over-sensitivity to external stimuli; seizures **NOEL**: 0.05 mg/kg body weight and day	Treon *et al.* 1955
dog beagle 1–2 ♂ or 1–2 ♀	18–47 days, 6 days/week endrin (purity not specified) 5, 10, 25, 50 mg/kg diet (about 0.20–0.27, 0.49–0.81, 1.21–2.20, 2.50–4.00 mg/kg body weight and day)	**from 0.20 mg/kg**: lethal within 18–47 days, body weight gains delayed; symptoms of intoxication: vomiting of food, food refusal; lethargy, salivation; shortness of breath; tremor, seizures; histopathology: degenerative changes in the liver (fat vacuoles, cell necrosis), kidneys (necrosis of the convoluted kidney tubules), heart, brain; hyperaemia and oedema of the lungs **from 0.49 mg/kg**: weight loss **NOEL** could not be determined	Treon *et al.* 1955
dog beagle 2 ♂ and 2 ♀	80 week, 6 days/week endrin (purity not specified) 1, 3 mg/kg diet (0.045–0.12, 0.12–0.25 mg/kg body weight and day)	body weight gains, haematology unaffected; no evidence of intoxication **high dose**: relative kidney and heart weights significantly increased; histopathological examination revealed no effects **NOEL**: 0.045–0.12 mg/kg body weight and day	Treon *et al.* 1955
dog beagle 3 ♂ and 3 ♀	2 years endrin (purity 99 %) 0.1, 0.5, 1, 2, 4 mg/kg diet (about 0.0025, 0.0125, 0.025, 0.05, 0.10 mg/kg body weight and day)	body weights, food consumption, haematology, urinalysis, alkaline phosphatase, prothrombin time, clinico-chemical parameters unaffected **from 0.05 mg/kg**: seizures; in ♀: liver weights slightly increased in some animals; histopathology: liver: slight to moderate vacuolation; kidneys: no effects **NOEL**: 0.025 mg/kg body weight and day	Shell 1969

5.2.3 Dermal absorption

Dermal exposure of rabbits to endrin (solid) doses of 75 mg/animal and day for a period of up to 14 weeks, for 2 hours a day on 5 days a week, led to symptoms of intoxication such as seizures, tremor and facial twitching, and to death of 1 of 3 animals. All animals given doses of 150 mg/animal and day died within 5 weeks (Treon *et al.* 1955).

5.3 Local effects on skin and mucous membranes

Single applications to the skin of endrin doses of up to 250 mg/kg body weight in the form of a dry powder did not lead to any gross pathological or microscopic findings in rabbits. Likewise, with repeated application (75 or 150 mg/animal, 2 hours daily, 5 days weekly, up to 70 applications) of dry endrin powder to the intact or scarified skin, irritative effects on the skin were not observed (Treon *et al.* 1955).

There are no data available for the allergenic effects of endrin.

5.4 Reproductive and developmental toxicity

5.4.1 Multigeneration studies

In a 3-generation study from the year 1965, endrin doses of 0.15 mg/kg body weight and day administered to rats led in the offspring of the F_2 and F_3 generations to increased mortality; it is unclear when exactly the animals died. In the follow-up study, up to the highest dose of 0.1 mg/kg body weight and day neither mortality nor toxic effects on reproduction were observed (WHO 1992).

5.4.2 Fertility

The fertility studies available are listed in Table 2. A fertility study with the mouse and endrin doses of 0.75 mg/kg body weight and day did not yield any evidence of adverse effects on fertility or incidence of pregnancy despite high mortality of the parent animals; litter sizes were reduced. Because of the high toxicity in the parent animals, the study cannot be evaluated (Good and Ware 1969). In two 3-generation studies with rats, doses of up to 0.15 mg/kg body weight and day had no effect on litter size or number of litters (WHO 1992).

5.4.3 Developmental toxicity

An extensive review of the developmental toxicity studies can also be found in Table 2.

Table 2. Studies of toxic effects of endrin on reproduction after oral administration

Study type Species	Administration Dose	Findings, Evaluation, $NOEL_{ET}$, $NOEL_{MT}$	References
3-generation study rat Long-Evans groups of 10 ♂, 20 ♀	79 days pretreatment and during mating and lactation each generation mated once 0.1, 1, 3 mg/kg diet (0.005, 0.05, 0.15 mg/kg body weight and day)	**parent animals**: body weights, liver weights, kidney weights, brain weights, behaviour unaffected; no effects on histopathology of the weaned F_2 animals **reproductive toxicity**: litter size, number of litters unaffected; no other data 0.15 mg/kg: offspring of the F_2 and F_3 generation: mortality increased, in subsequent study with 0.1 mg/kg no evidence of comparable effects; time of mortality not specified, therefore evaluation questionable **NOEL**: could not be determined	WHO 1992
3-generation study rat Long-Evans groups of 10 ♂, 20 ♀	79 days pretreatment and during mating and lactation each generation mated twice 0.1, 1, 2 mg/kg diet (0.005, 0.05, 0.1 mg/kg body weight and day)	**parent animals**: all generations: mortality, body weights unaffected, no gross pathological findings; F_{3b}: liver weights, kidney weights, brain weights, histology, behaviour unaffected **reproductive toxicity**: litter size and number of litters unaffected; no other details; no evidence of increased mortality in the offspring of the F_2 and F_3 generations unlike with 0.15 mg/kg in the previous study **NOEL**: at least 0.1 mg/kg body weight and day	WHO 1992
fertility study mouse CFW 20 ♂, 20 ♀	120 days, beginning: 30 days before mating endrin (purity 96 %) 5 mg/kg diet (0.75 mg/kg body weight and day)	**parent animals**: mortality significantly increased (32 %) **reproductive toxicity**: fertility, number of offspring/pair, number of litters not affected; litter size significantly decreased, $p < 0.05$ study cannot be evaluated because of the high mortality of the parent animals **NOEL**: could not be determined	Good and Ware 1969
teratogenicity study mouse CD-1 10	day 9 of gestation endrin (purity 99 %) 2.5 mg/kg body weight controls: vehicle corn oil (gavage)	**maternal toxicity**: dose corresponds to half the LD_{50}; no clinical symptoms of intoxication within 24 hours **embryotoxicity**: resorptions, number of dead foetuses, foetal weights unaffected; malformations: significantly increased (5.0 %, controls: 0 %); open eyes significantly increased (2.7 %); cleft palates slightly increased (2.2 %) teratogenic effects questionable as dose high; open eyes evidence of delay in development; of only limited meaningfulness as only one dose tested **NOEL**: could not be determined	Ottolenghi *et al.* 1974

Table 2. continued

Study type Species	Administration Dose	Findings, Evaluation, $NOEL_{ET}$, $NOEL_{MT}$	References
teratogenicity study mouse CD-1 groups of 40 and 20	days 7–17 of gestation endrin (purity 99 %) 0.5, 1.0, 1.5, 2.0 mg/kg body weight controls: vehicle corn oil (gavage)	**maternal toxicity**: from 0.5 mg/kg: relative liver weights in dams significantly increased from 1.0 mg/kg: body weight gains of the dams significantly decreased from 1.5 mg/kg: mortality of the dams increased (20 % and 90 %) **embryotoxicity**: number of implantations, number of dead foetuses, incidence of malformations unaffected from 1.0 mg/kg: foetal weights significantly decreased; number of caudal vertebrae significantly decreased, delayed ossification of the supraoccipitalis significantly increased from 1.5 mg/kg: delayed development of the renal pelvis sign. increased delays in development correlate with maternally toxic effects **$NOEL_{ET}$**: 0.5 mg/kg body weight and day; **$NOEL_{MT}$**: < 0.5 mg/kg body weight and day	Kavlock *et al.* 1981
teratogenicity study mouse CD-1 21 and 40	day 8 of gestation endrin (technical grade) 7, 9 mg/kg body weight controls: vehicle corn oil (gavage)	**maternal toxicity**: mortality increased, with 14 % at 7 mg/kg, 47 % at 9 mg/kg body weight gains significantly decreased **embryotoxicity**: resorptions, number of dead and living foetuses unaffected 7 mg/kg: foetal weights significantly decreased; ossification of the sternum and caudal vertebrae significantly decreased, additional ribs significantly increased 9 mg/kg: foetal weights unaffected; additional ribs significantly increased; fused ribs in 3/184 foetuses from 1/16 litters effects with clearly maternally toxic doses, not always dose-dependent; correlation between decreased body weight gains in the dams and additional ribs **NOEL**: could not be determined	Kavlock *et al.* 1985
teratogenicity study mouse DBA/2J groups of 5–6	day 12 of gestation endrin (high purity) 4.5, 6.0 mg/kg body weight controls: vehicle corn oil (gavage)	**maternal toxicity**: body weight gains unaffected 6.0 mg/kg: mortality of the dams increased (25 %) **embryotoxicity**: implantations, number of resorptions and dead foetuses, foetal thymus weights, incidence of cleft palates and hydronephrosis unaffected from 4.5 mg/kg: foetal weights significantly decreased 6.0 mg/kg: placenta weights significantly decreased **NOEL**: could not be determined	Hassoun and Stohs 1996a

Table 2. continued

Study type Species	Administration Dose	Findings, Evaluation, $NOEL_{ET}$, $NOEL_{MT}$	References
teratogenicity study mouse C57BL/6J groups of 5–6	day 12 of gestation endrin (high purity) 4.5, 6.0 mg/kg body weight controls: vehicle corn oil (gavage)	**maternal toxicity**: body weight gains unaffected 6.0 mg/kg: mortality of the dams increased (25 %) **embryotoxicity**: implantations, incidence of cleft palates unaffected from 4.5 mg/kg: significantly decreased foetal weights, foetal thymus weights, placenta weights 6.0 mg/kg: number of resorptions and dead foetuses significantly increased; incidence of hydronephrosis significantly increased (5.6 %; controls: 0 %) **NOEL**: could not be determined	Hassoun and Stohs 1996a
teratogenicity study rat CD groups of 25	days 6–15 of gestation endrin (purity 97 %) 0.1, 0.5, 2.0 mg/kg body weight and day controls: vehicle 0.5 % methocel (gavage)	**maternal toxicity**: 2.0 mg/kg: no clinical symptoms of intoxication, salivation in 1/25; mortality 2/25, reversible decrease in body weights (day 15 9.9 %; day 20 4.9 %) **embryotoxicity**: implantations, number of resorptions, dead and living foetuses, sex ratio, incidence of malformations unaffected; variations: no significant effects: from 0.5 mg/kg delayed ossification of sternebrae and skull only slightly increased **$NOEL_{ET}$**: 2.0 mg/kg body weight and day; **$NOEL_{MT}$**: 0.5 mg/kg body weight and day	Velsicol 1978a
teratogenicity study rat CD groups of 15–32	days 7–20 of gestation endrin (purity > 99 %) 0.075, 0.15, 0.3, 0.45 mg/kg body weight and day controls: vehicle corn oil (gavage)	**maternal toxicity**: mortality, relative liver weights unaffected from 0.3 mg/kg: body weight gains significantly decreased (38 % and 87 %) **embryotoxicity**: implantations, number of resorptions and dead foetuses, foetal weights, maturation of skeleton and viscera, incidence of malformations unaffected determination of endrin after 0.45 mg/kg: liver of the dams 0.327 µg/g, foetal tissue 0.027 µg/g **$NOEL_{ET}$**: at least 0.45 mg/kg body weight and day; **$NOEL_{MT}$**: 0.15 mg/kg body weight and day	Kavlock *et al.* 1981

Table 2. continued

Study type Species	Administration Dose	Findings, Evaluation, $NOEL_{ET}$, $NOEL_{MT}$	References
teratogenicity study rat CD not specified	day 7 of gestation to day 15 of lactation endrin (purity 99 %) 0.075, 0.15, 0.3 mg/kg body weight and day controls: vehicle corn oil (gavage)	**maternal toxicity**: mortality unaffected, no other details **embryotoxicity**: litter size, birth weights, body weight gains, age at which eyes opened, vaginal opening unaffected; from 0.15 mg/kg: 16 and 20 days after birth increased motor activity ('figure 8 maze'), 90 days after birth regressed; effects interpreted as delayed adaptation to new surroundings **$NOEL_{ET}$**: 0.075 mg/kg body weight and day; **$NOEL_{MT}$**: 0.3 mg/kg body weight and day	Gray *et al.* 1981
teratogenicity study Syrian golden hamster 7 (day 7 of gestation) 24 (day 8 of gestation) 8 (day 9 of gestation)	day 7 of gestation day 8 of gestation day 9 of gestation endrin (purity 99 %) 5 mg/kg body weight controls: vehicle corn oil (gavage)	**maternal toxicity**: dose corresponds to half the LD_{50}; no clinical symptoms of intoxication within 24 hours **embryotoxicity**: day 7 of gestation / day 8 of gestation / day 9 of gestation resorptions: 23.4 %* / 30.9 %* / 15.9 % dead foetuses: 7.0 %* / 7.2 %* / 0.9 % malformations: 10.5 %* / 27.6 %* / 10.3 %* open eyes: 0.0 % / 21.6 %* / 0.0 % webbed toes: 0.0 % / 16.2 %* / 1.0 % cleft palates: 9.5 %* / 4.9 %* / 8.2 %* fused ribs: 11.6 % / 7.9 %* / 11.4 %* * parameter significantly increased, $p < 0.01$; open eyes and webbed toes interpreted as evidence of delays in development as a result of foetotoxicity; of limited meaningfulness as only one high dose tested, small number of animals in some groups **NOEL**: could not be determined	Ottolenghi *et al.* 1974
teratogenicity study Syrian golden hamster groups of 27–29	days 4–13 of gestation endrin (purity 97 %) 0.1, 0.75, 2.5 mg/kg body weight and day controls: vehicle 0.5 % methocel (gavage)	**maternal toxicity**: no abnormal findings **embryotoxicity**: implantation; number of resorptions, dead and living foetuses, post-implantation losses, foetal weights, crown-to-rump length, incidence of malformations unaffected no evidence of the effects observed by Ottolenghi *et al.* (1974) **$NOEL_{ET}$**: 2.5 mg/kg body weight and day; **$NOEL_{MT}$**: 2.5 mg/kg body weight and day	Velsicol 1978b

Table 2. continued

Study type Species	Administration Dose	Findings, Evaluation, $NOEL_{ET}$, $NOEL_{MT}$	References
teratogenicity study Syrian golden hamster (LVG) groups of 18–61, controls 87	day 8 of gestation endrin (purity 98 %) 0.5, 1.5, 5, 7.5, 10 mg/kg body weight controls: vehicle corn oil (gavage)	**maternal toxicity**: body weight gains, relative liver weights, mortality unaffected. 10 mg/kg: reversible seizures in one animal **embryotoxicity**: incidence of pregnancy, implantations, number of dead foetuses, foetal weights, skeletal maturation unaffected from 5 mg/kg: meningoencephalocoeles significantly increased, not dose-dependent from 7.5 mg/kg: significant dose-dependent increase in fused ribs endrin determination: not detected in liver of the dams and foetal tissue **$NOEL_{ET}$**: 1.5 mg/kg body weight and day; **$NOEL_{MT}$**: 7.5 mg/kg body weight and day	Chernoff *et al.* 1979
teratogenicity study Syrian golden hamster (LVG) groups of 20–38, controls 63	day 5–14 of gestation endrin (purity 98 %) 0.75, 1.5, 2.5, 3.5 mg/kg body weight and day controls: vehicle corn oil (gavage)	**maternal toxicity**: from 1.5 mg/kg: body weight gains of the dams significantly decreased, from day 8 of gestation mortality increased (37 %, 68 %, 86 %) with weight loss; hypoactivity, tremor **embryotoxicity**: number of implantation sites unaffected from 1.5 mg/kg: resorptions and dead foetuses significantly increased, foetal weights significantly decreased; delayed ossification of the sternebrae and caudal vertebrae significantly increased, ossification disturbances of the supraoccipitalis significantly increased endrin determination: at 1.5 and 2.5 mg/kg: liver of the dams 1.45 and 2.55 µg/g, foetal tissue 0.021 and 0.075 µg/g toxic effects on reproduction correlate with maternal toxicity no evidence of the effects observed by Ottolenghi *et al.* (1974) **$NOEL_{ET}$**: 0.75 mg/kg body weight and day; **$NOEL_{MT}$**: 0.75 mg/kg body weight and day	Chernoff *et al.* 1979

Table 2. continued

Study type Species	Administration Dose	Findings, Evaluation, $NOEL_{ET}$, $NOEL_{MT}$	References
teratogenicity study Syrian golden hamster (LVG) 10 and 30	days 5–14 of gestation endrin (purity 98 %) 0.75, 1.5 mg/kg body weight controls: vehicle corn oil (gavage)	**maternal toxicity**: 1.5 mg/kg: mortality of the dams increased (57 %) behavioural tests: from 0.75 mg/kg: reactive locomotor activity and rearing behaviour in an open field considerably decreased during exposure, reversible by day 27 after exposure **embryotoxicity**: litter size 15 days post partum unaffected; body weights up to weaning slightly decreased behavioural tests: rearing behaviour in an open field, behaviour in treadmill and male sexual behaviour not unusual 1.5 mg/kg: locomotive activity (reactive locomotor activity in an open field) significantly increased 15 and 20 days post partum; from day 27 post partum effects reversible, comparable with controls; 75 and 125 days post partum 'figure 8 maze' activity significantly increased effects with marked maternal toxicity, effects interpreted as delayed adaptation to new surroundings **$NOEL_{ET}$**: 0.75 mg/kg body weight and day; **$NOEL_{MT}$**: could not be determined	Gray *et al.* 1981

$NOEL_{ET}$ NOEL for embryotoxicity
$NOEL_{MT}$ NOEL for maternal toxicity

After administration of single doses of 2.5 mg/kg body weight to CD-1 mice on day 9 of gestation, increased resorptions and dead foetuses, and open eyes and cleft palates were described (Ottolenghi *et al.* 1974). Because the doses may be presumed maternally toxic (half the LD_{50}, insufficient data for maternal toxicity) and because comparable changes were not observed in this dose range in subsequent studies (Hassoun and Stohs 1996a, Kavlock *et al.* 1981, 1985), the described findings are not included in the present evaluation. Only with doses that led to mortality in the dams were effects seen in the offspring: fused ribs in CD-1 mice after 9 mg/kg body weight and hydronephrosis in C57BL mice after 6 mg/kg body weight, while no malformations were found in DBA mice (Hassoun and Stohs 1996a, Kavlock *et al.* 1985). With the last two strains of mouse, endrin was found to be foetotoxic from doses of 4.5 mg/kg body weight without it affecting the body weights or mortality of the dams; oxidative stress (evidence of increased lipid peroxidation and DNA single strand breaks in the placenta and foetal tissue) was suspected of being the cause (Hassoun and Stohs 1996a, 1996b, Hassoun *et al.* 1996). A teratogenicity study with oral administration of endrin on days 7 to 17 of gestation in doses of 0.5 to 2.0 mg/kg body weight yielded a NOEL of 0.5 mg/kg body weight for foetotoxic and developmental effects in the mouse. In the dams this dose led to increased liver weights; from doses of 1.0 mg/kg body weight reduced foetal weights and delayed development correlated with reduced weight gains in the dams (Kavlock *et al.* 1981).

In teratogenicity studies with the rat, no embryotoxic effects were observed (maternally toxic effects from 0.3 mg/kg body weight and day) up to the highest dose of 0.45 mg/kg body weight and day administered from days 7 to 15 or 20 of gestation (vehicle corn oil) (Gray *et al.* 1981, Kavlock *et al.* 1981). In another study (Velsicol 1978a), doses of 2.0 mg/kg body weight were not found to cause toxic effects on development; the aqueous methocel used, is, however, an unsuitable vehicle for the highly lipophilic endrin. For rats the NOEL for developmental toxicity is therefore given as 0.45 mg/kg body weight and day (highest dose investigated), for maternal effects (body weight changes) 0.15 mg/kg body weight and day.

In hamsters, doses of 0.1 to 2.5 mg/kg body weight and day administered in aqueous methocel on days 4 to 13 of gestation did not have embryotoxic effects, but the body weight gains of the dams were slightly reduced (Velsicol 1978b). In a teratogenicity study with hamsters given oral endrin doses of 0.75 to 3.5 mg/kg body weight and day (vehicle corn oil) on days 5 to 14 of gestation, foetotoxicity and delays in development after clearly maternally toxic doses of 1.5 mg/kg body weight and day are described, (Chernoff *et al.* 1979); the NOEL for toxic effects on development in hamsters is therefore given as 0.75 mg/kg body weight. Although no teratogenic effects were found in hamsters after repeated administration of endrin doses up to 3.5 mg/kg body weight, after single doses of 5 and 7.5 mg/kg body weight (day 8 of gestation) the same research group found significantly increased incidences of meningoencephaloceles and fused ribs, without any evidence of maternal toxicity (Chernoff *et al.* 1979). In another study, single endrin doses of 5 mg/kg body weight administered on days 7, 8 or 9 of gestation, led, as in rats, to increased incidences of malformations such as open eyes, cleft palates, fused ribs and webbed toes (Ottolenghi *et al.* 1974). As the dose was probably maternally toxic (half the LD_{50}, data for maternal toxicity inadequate), the findings cannot be evaluated. Elsewhere these results were sometimes interpreted as delayed development caused by foetotoxicity (Chernoff *et al.* 1979). Only after repeated administration could endrin and 12-ketoendrin be detected in foetal tissue at the time of autopsy (Chernoff *et al.* 1979).

After prenatal, and prenatal and perinatal exposure, increased motor activity was observed in rats and hamsters; it was interpreted as delayed adaptation to new surroundings. It was concluded from the increased motor activity of the pups of rats exposed *in utero* and via the mother's milk up to day 15 *post partum* that endrin leads to reversible behavioural changes (motor activity increased until day 20 *post partum,* then reversible) from doses of 0.15 mg/kg body weight and day with a NOEL of 0.075 mg/kg body weight and day. There was no evidence of toxic effects on reproduction or delays in development (Gray *et al.* 1981). In this study, however, maternally toxic effects are inadequately documented, and the behavioural changes cannot therefore be clearly attributed to the exposure to endrin. In the same publication comparable behavioural changes are described in hamsters exposed *in utero* (0.75 and 1.5 mg/kg body weight from days 5 to 14 of gestation) only at the high dose, which led to increased mortality and reduced motor activity of the dams. It is therefore questionable whether the increased motor activity described in the offspring, in some cases up to day 125 *post partum*, with a LOEL of 1.5 mg/kg body weight (Gray *et al.* 1981) can be regarded as evidence of toxic effects on behaviour.

Concentrations determined in foetal tissue of hamsters and rats in teratogenicity studies with repeated administration showed that despite comparable endrin concentrations of 0.021 and 0.027 μg in hamsters (after 1.5 mg/kg body weight and day) and rats (after 0.45 mg/kg body weight and day), the hamsters reacted more sensitively than the rats with marked foetotoxicity (no toxic effects on reproduction). The fact that in the hamster foetuses there was much more 12-ketoendrin present than endrin (ratio 3.2:1) while in rat foetuses more endrin than 12-ketoendrin was present (ratio 1:0.4) could play a role (see Section 3.1; Chernoff *et al.* 1979, Kavlock *et al.* 1981). In the teratogenicity studies similar NOEL values were found for toxic effects on reproduction for all species investigated: hamster 0.75 mg/kg body weight and day, rat 0.45 mg/kg body weight and day (highest dose investigated) and mouse 0.5 mg/kg body weight and day.

5.5 Genotoxicity

5.5.1 *In vitro*

Endrin was not found to be mutagenic in genotoxicity tests with numerous strains of *Salmonella typhimurium* or *Escherichia coli*, with *Bacillus subtilis*, *Saccharomyces cerevisiae* or *Serratia marescens*, neither with nor without metabolic activation. Weak mutagenic activity in a test system with 6-thioguanine-resistant mouse cells (FM3A) was reported in an abstract, which cannot be evaluated because the documentation is inadequate. In the HPRT test with cultures of rat liver epithelium cells (WHO 1992) and in the TK test with mouse lymphoma cells (McGregor *et al.* 1991), however, there was no increase in the incidence of mutations. In a modified SOS chromotest (induction of β-galactosidase in *E. coli* PQ37) with 47 test substances, there was evidence of relatively high DNA-damaging activity of endrin (Venkat *et al.* 1995). In cultures of rat or hamster liver cells, of primary rat hepatocytes or of human lung fibroblasts (with or without metabolic activation), however, endrin did not induce DNA-repair synthesis. Treatment with endrin did not increase the incidence of sister chromatid exchange in human lymphoid cells (LAZ-007), while the number of M1 metaphases increased with the dose and the number of M3 metaphases decreased. These effects occurred in parallel with reduced cell proliferation and provide evidence of cytotoxicity (WHO 1992). In cultures of PC12 cells (cell line from a phaeochromocytoma of the adrenal gland) a cytotoxic concentration of endrin (100 nM), which also induced heat shock proteins, an indicator of oxidative stress, caused a significant increase in the incidence of DNA strand breaks (Bagchi *et al.* 1995).

Overall, there is no evidence that endrin is genotoxic *in vitro* with or without metabolic activation in concentrations which are not cytotoxic.

5.5.2 *In vivo*

In the host-mediated assay with mice, no gene conversion was seen in *Saccharomyces cerevisiae*. No mutagenic activity could be detected in a dominant lethal test with mice after intraperitoneal or oral administration of endrin, or in the recessive lethal test with *Drosophila*. The treatment of mice, rats and hamsters with endrin did not cause DNA repair synthesis in liver cells.

Significant dose-dependent increases in DNA strand breaks were detected in the livers of rats 12 to 72 hours after oral administration of endrin doses of 3, 4.5 and 6 mg/kg body weight and in the liver and brain after two doses of 2.25 mg/kg body weight (0.25 times the LD_{50}) 21 hours apart (Bagchi *et al.* 1992a, 1995). Also in C57BL/6J and DBA/2J mice in foetal tissue such as foetal liver, and in the placenta, significant increases in DNA single strand breaks were observed 48 hours after oral administration of endrin doses of 4.5 mg/kg body weight on day 12 of gestation (Hassoun and Stohs 1996a, 1996b). In all these studies severe lipid peroxidation and effects on the tissues were also observed (see Section 5.1.2); the DNA strand breaks are therefore probably the result of the oxidative stress induced by the high endrin concentrations in the tissues.

5.6 Carcinogenicity

5.6.1 Short-term studies

In the liver foci test with the hepatectomized rat, endrin was administered in doses of 1.25 mg/kg body weight and day (25 mg/kg with the diet) for a period of 6 weeks after administration of *N*-nitrosodiethylamine (DEN) or *N*-2-fluorenylacetamide (2-FAA) as initiator. After the two treatments the number of foci were not increased; the area of the foci was significantly increased, but the increase was slight relative to that seen after treatment with other substances (Ito *et al.* 1980). In V79 cells, endrin in concentrations of 5 μg/ml or more caused inhibition of metabolic co-operation (Kurata *et al.* 1982). In an *in vitro* test for the stimulation of protein kinase C there was no evidence of tumour-promoting effects of endrin (Moser and Smart 1989).

5.6.2 Long-term studies

In several long-term feeding studies with mice given endrin in doses up to the maximum tolerated dose (MTD) (NTP 1979, Shell 1970; details of the experimental conditions are given in Table 1), there were no significant increases in the tumour incidences.

Administration of endrin in concentrations of 0.1 to 25 mg/kg diet for 2 years (see Table 1) led to the formation of sarcomas and carcinomas at different locations in some Osborne-Mendel rats in various dose groups, completely independent of the dose. Therefore the tumours cannot be considered as evidence of a carcinogenic effect. The highest tumour incidences were found in the male animals at the lowest dose (0.1 mg/kg diet or 0.005 mg/kg body weight and day), in the female animals at 1 mg/kg diet (0.05 mg/kg body weight and day) (Reuber 1978). In other studies with the same strain of rats and with similar doses (see Table 1; Deichmann *et al.* 1970, NTP 1979) there was also no evidence of a carcinogenic potential of endrin. In the NTP study the incidences of adrenal gland adenomas and carcinomas (male animals: 4/46, 8/44; female animals: 16/49, 7/47) were increased relative to those in the pooled controls (male animals: 2/44; female animals: 4/46) but not relative to those in the experimental controls (male animals: 2/9; female animals: 3/9). As, in addition, no dose-response relationship was observed in the female animals, these tumours were not regarded as substance-related (NTP 1979).

The data available from carcinogenicity studies therefore yield no evidence of carcinogenic effects of endrin.

6 Manifesto (MAK value/classification)

In various test systems with bacteria or mammalian cells, with and without metabolic activation, in concentrations which were not cytotoxic, endrin was not found to have genotoxic effects. *In vivo* the dominant lethal test, the recessive lethal test with *Drosophila* and the DNA repair test in liver cells yielded negative findings. DNA strand breaks are probably induced *in vivo* only as the result of oxidative stress. In several long-term feeding studies with mice and rats, endrin in doses up to the MTD was not found to have carcinogenic effects.

For observations at workplaces there are no data available for exposure concentrations or duration which could be used in the evaluation of the MAK value. In experiments with animals, short-term and long-term uptake of endrin led to neurotoxic effects and to tissue damage, e.g. in the liver and kidneys, as a result of lipid peroxidation. There are no inhalation studies available which meet present-day standards and which are suitable for deriving a MAK value. After repeated oral administration the dog was found to be the most sensitive species. Doses as low as 0.05 mg/kg body weight still led to seizures and liver damage while 0.025 mg/kg body weight and day had no effects. In the mouse the NOEL was given as 0.045 mg/kg body weight and day, in the rat 0.05 mg/kg body weight and day. The MAK value can be derived from the NOEL of 0.025 mg/kg body weight for long-term oral administration. Assuming a body weight of 70 kg and a respiratory volume of 10 m^3 per workshift, complete absorption and exposure exclusively by inhalation, this NOEL correlates with an air concentration of 0.175 mg/m^3. The previous MAK value of 0.1 mg/m^3 and Peak limitation category III can, therefore, be retained. Because of the high acute dermal toxicity of the substance, designation with an "H" has also been retained.

The teratogenicity studies with mice, rats and hamsters with oral administration are relevant for the classification of endrin in a pregnancy risk group, as there are no investigations available with inhalation exposure. These yielded a NOEL for prenatal toxicity of 0.5 mg/kg body weight for the mouse, of 0.45 mg/kg body weight for the rat and of 0.75 mg/kg body weight for the hamster. These values are higher by a factor of 20 to 30 than the NOEL of 0.025 mg/kg body weight and day from long-term studies with oral administration, which were used for evaluating the MAK value of 0.1 mg/m^3. The dose–response curves in these studies of reproductive toxicity are relatively flat and notable foetotoxic effects occurred only after much higher doses with maternally toxic effects. Therefore a factor of 20 to 30 is regarded as sufficient to exclude prenatal toxicity given observance of the MAK value. Endrin is therefore classified in Pregnancy risk group C.

7 References

Bagchi M, Hassoun DA, Bagchi D, Stohs SJ (1992a) Endrin-induced increases in hepatic lipid peroxidation, membrane microviscosity, and DNA damage in rats. *Arch Environ Contam Toxicol 23*: 1–5

Bagchi D, Bagchi M, Hassoun DA, Stohs SJ (1992b) Endrin-induced urinary excretion of formaldehyde, acetaldehyde, malondialdehyde and acetone in rats. *Toxicology 75*: 81–89

Bagchi D, Bagchi M, Hassoun DA, Stohs SJ (1995) *In vitro* and *in vivo* generation of reactive oxygen species, DNA damage and lactate dehydrogenase leakage by selected pesticides. *Toxicology 104*: 129–140

Chernoff N, Kavlock RJ, Hanisch RC, Whitehouse DA, Gray JA, Gray Jr LE, Sovocool GW (1979) Perinatal toxicity of endrin in rodents. I. Fetotoxic effects of prenatal exposure in hamsters. *Toxicology 13*: 155–165

Deichmann WB, MacDonald WE, Blum E, Bevilacqua M, Radomski J, Keplinger M, Balkus M (1970) Tumorigenicity of aldrin, dieldrin and endrin in the albino rat. *Ind Med 39*: 37–45

Ferrer A, Cabral R (1991) Toxic epidemics caused by alimentary exposure to pesticides: a review. *Food Add Contam 8*: 755–776

Ferrer A, Cabral R (1995) Recent epidemics of poisoning by pesticides. *Toxicol Letters 82/83*: 55–63

Good EE, Ware GW (1969) Effect of insecticides on reproduction in the laboratory mouse. IV. Endrin and dieldrin. *Toxicol Appl Pharmacol 14*: 201–203

Gray jr LE, Kavlock JR, Chernoff N, Gray JE, McLamb J (1981) Perinatal toxicity of endrin in rodents. III. Alterations of behavioral ontogeny. *Toxicology 21*: 187–202

Hassan MQ, Numan IT, Al-Nasiri N, Stohs SJ (1991) Endrin-induced histopathological changes and lipid peroxidation in livers and kidneys of rats, mice, guinea pigs and hamsters. *Toxicol Pathol 19*: 108–114
Hassoun EA, Stohs SJ (1996a) Comparative teratological studies on TCDD, endrin, and lindane in C57BL/6J and DBA/2J mice. *Comp Biochem Physiol 113 C*: 393–398
Hassoun EA, Stohs SJ (1996b) TCDD, endrin, and lindane induced oxidative stress in fetal and placental tissues of C57BL/6J and DBA/2J mice. *Comp Biochem Physiol 115 C*: 11–18
Hassoun EA, Bagchi D, Stohs SJ (1996) TCDD, endrin, and lindane induced increases in lipid metabolites in maternal sera and amniotic fluids of pregnant C57BL/6J and DBA/2J mice. *Res Comm Mol Pathol Pharmacol 94*: 157–169
Ito N, Tatematsu M, Nakanishi K, Hasegawa R, Takano T, Imaida K, Ogiso T (1980) The effects of various chemicals on the development of hyperplastic liver nodules in hepatectomized rats treated with *N*-nitrosodiethylamine or *N*-2-fluorenylacetamide. *Gann 71*: 832–842
Kavlock RJ, Chernoff N, Hanisch RC, Gray J, Rogers E, Gray Jr LE (1981) Perinatal toxicity of endrin in rodents. II. Fetotoxic effects of prenatal exposure in rats and mice. *Toxicology 21*: 141–150
Kavlock RJ, Chernoff N, Rogers EH (1985) The effect of acute maternal toxicity on fetal development in the mouse. *Teratogen Carcinogen Mutagen 5*: 3–13
Kurata M, Hirose K, Umeda M (1982) Inhibition of metabolic cooperation in Chinese hamster cells by organochlorine pesticides. *Gann 73*: 217–221
McGregor DB, Brown AG, Howgate S, McBride D, Riach C, Caspary WJ (1991) Responses of the L5178Y mouse lymphoma cell forward mutation assay. V: 27 coded chemicals. *Environ Mol Mutagen 17*: 196–219
Moser GJ, Smart RC (1989) Hepatic tumor-promoting chlorinated hydrocarbons stimulate protein kinase C activity. *Carcinogenesis 10*: 851–856
Narahashi T, Frey JM, Ginsburg KS, Roy ML (1992) Sodium and GABA-activated channels as the targets of pyrethroids and cyclodienes. *Toxicol Lett 64/65*: 429–436
NTP (National Toxicology Program) (1979) *Bioassay of endrin for possible carcinogenicity*. Carcinogenesis Technical report Series No. 12, NTIS PB-288461
Numan IT, Hassan MQ, Stohs SJ (1990) Protective effects of anti-oxidants against endrin-induced lipid peroxidation, glutathione depletion and lethality in rats. *Arch Environ Contam Toxicol 19*: 302–306
Ottolenghi AD, Haseman JK, Suggs F (1974) Teratogenic effects of aldrin, dieldrin, and endrin in hamsters and mice. *Teratology 9*: 11–16
Reuber MD (1978) Carcinomas, sarcomas and other lesions in Osborne-Mendel rats ingesting endrin. *Exp Cell Biol 46*: 129–145
Shell (1969) *The effects upon beagle dogs during a period of two years by the introduction of 1,2,3,4,10,10-hexachloro-6,7-epoxy-1,4,4a,5,6,7,8,8a-octahydro-1,4-endo,endo-5,8-dimethanonaphthalene into their daily diets*. Kettering Laboratory, University of Cincinnati, Cincinnati, Ohio, unpublished
Shell (1970) *The incidence of neoplasms in two strains of mice sustained on diets containing endrin*. Kettering Laboratory, University of Cincinnati, Cincinnati, Ohio, unpublished
Treon JF, Cleveland FP, Cappel J (1955) Toxicity of endrin for laboratory animals. *J Agr Food Chem 3*: 842–848
Velsicol Chemical (1978a) *Endrin. Teratology study in rats*. International Research and Development Corporation, Report 163-488, unpublished
Velsicol Chemical (1978b) *Endrin. Teratology study in hamsters*. International Research and Development Corporation, Report 163-478, unpublished
Venkat JA, Shami S, Davis K, Nayak M, Plimmer JR, Pfeil R, Nair PP (1995) Relative genotoxic activities of pesticides evaluated by a modified SOS microplate assay. *Environ Mol Mutagen 25*: 67–76
WHO (1992) *Endrin, Environmental Health Criteria 130*, IPCS (International Programme on Chemical Safety)

completed 31.03.98

Furfural

MAK value	**–**
Peak limitation	**–**
Absorption through the skin (1966)	**H**
Sensitization	**–**
Carcinogenicity (1994)	**Category 3B**
Prenatal toxicity	**–**
Germ cell mutagenicity	**–**
BAT value	**–**
Synonyms	2-furaldehyde artificial oil of ants
Chemical name (CAS)	2-furancarboxaldehyde
CAS number	98-01-1

The toxicology of furfural was reviewed in Volume 9 of the present series; this chapter is a supplement to the earlier review.

1 Allergenic Effects

1.1 Effects in man

There are no case reports of contact allergy caused by furfural. In a few reviews furfural is, however, listed as a sensitizing compound (Hamilton and Hardy 1974, Mishra 1992, Smyth 1956) or as a photosensitizer (Hamilton and Hardy 1974), although no details are given. In another review, furfural resins (furan resin) are mentioned as the possible cause of skin diseases (Schwarz *et al.* 1957). Two studies of the effects of these and similar resins on the skin do not, however, contain any information about patch tests carried out with furfural (Bonari and Martini 1971, Fregert 1981). The short summary of a Polish study does not provide any information on the type of occupational skin diseases caused by exposure to furfural vapour in a refinery (Grzegorzyk and Mazurowa 1959). Of a total of 15 patients with contact allergy to the nitrofuran derivatives, nitrofurazone

(2-[(5-nitro-2-furanyl)methylene]hydrazinecarboxamide, CAS 59-87-0) or furazolidone (3-[[(5-nitro-2-furanyl)methylene]amino]-2-oxazolidinone, CAS 67-45-8), none reacted in patch tests to 0.2 % or 1 % 2-furylmethanal in petrolatum or to 2 % 2-furylmethanal in polyethylene glycol 400 (Bleumink *et al.* 1974, de Groot and Conemans 1990). In a maximization test with 25 volunteers, no sensitizing effects on the skin could be detected in tests carried out 10 to 14 days after occlusive application of 2 % furfural in petrolatum 5 times every 48 hours over 10 days (Opdyke 1978).

There are no data available for airway sensitization caused by the substance.

1.2 Effects in animals

There are no data available from experiments with animals.

2 Manifesto (sensitization)

On the basis of the little data available, a sensitizing capacity cannot be deduced for furfural. Furfural is therefore not designated as sensitizing.

3 References

Bleumink E, te Lintum JCA, Nater JP (1974) Kontaktallergie durch Nitrofurazon (Furacin) und Nifurprazin (Carofur) (Contact allergy caused by nitrofurazone and nifurprazin) (German). *Hautarzt 25*: 403–406

Bonari R, Martini E (1971) Patologica cutanea da agente vulcanizzante. Note cliniche, sperimentali e prevenzionali (Skin pathology caused by a vulcanizing agent. Clinical, experimental and preventive notes) (Italian). *Med Lav 62*: 297–299

Fregert S (1981) Contact allergy to phenoplastics. *Contact Dermatitis 7*: 170

Grzegorczyk L, Mazurowa J (1959) Furfural poisoning (Polish). *Med Pracy 10*: 199–202, in: *Chem Abstr* (1960) 13613h

de Groot AC, Conemans JMH (1990) Contact allergy to furazolidone. *Contact Dermatitis 22*: 202–205

Hamilton A, Hardy HL (1974) *Industrial toxicology*, Publishing Sciences Group, Acton, 334

Mishra A (1992) Furfural: a toxic chemical. *Agr Biol Res 8*: 93–104

Opdyke DLJ (1978) Fragrance raw materials monographs: furfural. *Food Cosmet Toxicol 16*: 759–764

Schwartz L, Tulipan L, Birmingham DJ (1957) *Occupational diseases of the skin*, H Kimpton, London

Smyth jr HF (1956) Hygienic standards for daily inhalation. *Am Ind Hyg Assoc Q 17*: 129–185

completed 10.07.1997

Methyl acetate

MAK value (1958)	**200 ml/m³ (ppm) ≙ 610 mg/m³**
Peak limitation (2001)	**Category II, excursion factor 4**
Absorption through the skin	**–**
Sensitization	**–**
Carcinogenicity	**–**
Prenatal toxicity (1997)	**Pregnancy risk group C**
Germ cell mutagenicity	**–**
BAT value	**–**
Synonyms	–
Chemical name (CAS)	acetic acid methyl ester
CAS number	79-20-9
Structural formula	CH_3–COO–CH_3
Molecular formula	$C_3H_6O_2$
Molecular weight	74.08
Melting point	–98°C
Boiling point	56.9°C
Density at 20°C	0.93 g/cm³
Vapour pressure at 20°C	220 hPa
$\log P_{ow}$*	0.18
1 ml/m³ (ppm) ≙ 3.07 mg/m³	**1 mg/m³ ≙ 0.325 ml/m³ (ppm)**

The MAK value for methyl acetate was established at 200 ml/m³ in 1958 in analogy to the value for methanol. More recent studies relevant for evaluation of the substance are described below. A detailed review of the data can be found in the BG Chemie publication (BG Chemie 1995).

---

* *n*-octanol/water distribution coefficient

1 Toxic Effects and Mode of Action

Methyl acetate causes irritation of the mucous membranes of the respiratory tract in man after short-term exposure to concentrations of 4000 ml/m^3 and in animal experiments from about 6600 ml/m^3, and is narcotic from about 16000 ml/m^3. Liquid methyl acetate does not irritate the skin, but does irritate the eyes. During metabolism, methanol and acetic acid are formed. Methanol is further metabolized to formaldehyde, formic acid and CO_2. The toxic metabolite is formic acid. Repeated exposure of cats causes irritation of the respiratory tract and changes in haematological parameters from concentrations of 6600 ml/m^3. The substance was not found to be mutagenic in *Salmonella* mutagenicity tests. In a maximization test in man methyl acetate did not cause sensitization.

2 Mechanism of Action

The atrophy of the optic nerve described after inhalation of mixtures of methanol and methyl acetate could also be partly caused by methyl acetate, as methyl acetate is metabolized to methanol, which is known to cause such eye damage.

3 Toxicokinetics and Metabolism

Methyl acetate is degraded in the organism to form methanol and acetic acid (BG Chemie 1995). The rate of hydrolysis in the nasal homogenate of rats is 15 ± 3 mmol/mg protein and minute (Dahl *et al.* 1987). Methanol is further metabolized to formaldehyde, formic acid and CO_2 (see MAK documentation for methanol in Volume 16 of the present series). Exposure of volunteers to methyl acetate concentrations of 200 ml/m^3 for two hours twice daily on three days led to increasing urine methanol concentrations of 8 to 23 mg/l. The morning after the exposure the concentrations were once again within the normal range below 5 mg/l. Similar values were obtained after exposure for the same period to methanol concentrations of 200 ml/m^3 (Tada *et al.* 1974). The BAT value for methanol of 30 mg/l is based on the level of methanol excreted in the urine at a MAK value of 200 ml/m^3 (Henschler and Lehnert 1983).

4 Effects in Man

Inhalation abuse ("sniffing") of solvent mixtures containing methanol and methyl acetate, and inhalation of methyl acetate at the workplace (no further details), led to atrophy of the optic nerve, as is also described for methanol alone (BG Chemie 1995). As a result of metabolism, such effects can also be assumed for methyl acetate, at least with high concentrations.

In the study with volunteers exposed twice to 200 ml/m^3 for two hours (see Section 3), irritation was not reported (Tada *et al.* 1974). Exposure of volunteers for 5 minutes to methyl acetate concentrations of 325 ml/m^3 was tolerated without irritation, while from about 4050 ml/m^3 the substance caused slight irritation of the airways and throat which was reversible. Habituation was observed (Flury and Wirth 1933).

High concentrations of methyl acetate (no further details) caused irritation of the mucous membranes of the eyes and respiratory tract. Exposure of the skin for longer periods to liquid methyl acetate led to irritation (ACGIH 1992). A 10 % solution of methyl acetate did not cause skin irritation and was not found to be allergenic in a maximization test with volunteers (Opdyke 1979).

5 Animal Experiments and *in vitro* Studies

5.1 Acute toxicity

Rats (6/6) survived exposure for 4 hours to 16000 ml/m^3, while all 6 rats died after 32000 ml/m^3 (Smyth *et al.* 1962). From about 16000 ml/m^3 methyl acetate was narcotic. In cats and rabbits, inhalation of 19500 to 81250 ml/m^3 for 20 to 155 minutes led to a concentration-dependent reduction in the pH value of the blood. After 19500 ml/m^3 a reduction of 0.11 to 0.21 units was observed (Flury and Wirth 1933).

The oral LD_{50} for rats and dermal LD_{50} for rabbits were more than 5000 mg/kg body weight (Opdyke 1979). For rats the dermal LD_{50} was over 2000 mg/kg body weight. In this study, neither deaths nor clinical symptoms were recorded, and gross pathological examination did not yield any findings (BG Chemie 1995).

5.2 Subacute, subchronic and chronic toxicity

Repeated exposure of 5 cats to 6600 ml/m^3 for 6 hours on each of 8 days led to weight loss, changes in haematological parameters, salivation and eye irritation. One animal died during the exposure. Two other cats were exposed to 13000 ml/m^3, 7 hours a day for 3 days. Eye irritation and salivation were observed in all animals. Lower concentrations were not investigated (Flury and Wirth 1933).

5.3 Local effects on the skin and mucous membranes

In two valid studies with rabbits, liquid methyl acetate was not found to cause skin irritation, but did irritate the eyes (BG Chemie 1995). Irritation of the respiratory tract and of the eyes was observed after inhalation exposure to concentrations of 6600 ml/m^3 and more (Flury and Wirth 1933).

There are no data available for allergenic effects of methyl acetate.

5.4 Reproductive toxicity

There are no data available for the reproductive toxicity of methyl acetate. Methyl acetate was formerly classified in analogy to its metabolite methanol in Pregnancy risk group D. In 1995 methanol was re-classified in Pregnancy risk group C (see MAK documentation for methanol in Volume 16 of the present series). There are no valid studies available on the reproductive toxicity of the second metabolite of methyl acetate, acetic acid.

5.5 Genotoxicity

Methyl acetate was not found to be mutagenic in *Salmonella* mutagenicity tests (BG Chemie 1995). It caused mitotic aneuploidy in yeasts at a concentration of 3 % to 4 % in the medium, but not mitotic recombination (Zimmermann *et al.* 1985). The meaningfulness of these results is questionable as *in vivo* such high concentrations are hardly to be expected.

5.6 Carcinogenicity

There are no data available on the carcinogenicity of methyl acetate. The metabolites formed, methanol and acetic acid are not expected to produce carcinogenic effects.

6 Manifesto (MAK value/classification)

The critical effects are the systemic effects of the metabolites methanol and formic acid. As the amount of methanol excreted in the urine after the exposure of volunteers to methyl acetate was the same as that after exposure to equal concentrations of methanol, the MAK value for methanol of 200 ml/m^3 can also be adopted for methyl acetate. Metabolic acidosis is not to be expected at this concentration, as even after exposure to methyl acetate concentrations of 19500 ml/m^3 in animal experiments, the pH value in the blood was reduced at most by only 0.21 units (Flury and Wirth 1933).

In analogy to methanol, also methyl acetate has now been classified in Pregnancy risk group C.

As indicated above, the MAK value and the Peak limitation category for methyl acetate were established in analogy to methanol because methyl acetate is thought to be hydrolysed quantitatively to methanol and acetic acid. The Peak limitation category for methanol was changed in 1999 to II,2 (see MAK documentation for methanol in Volume 16 of the present series) with an excursion factor of five times the MAK value.

In the text which follows it is estimated whether peak exposures to methyl acetate in line with an excursion factor of 4 could result in blood methanol levels high enough to cause prenatal toxicity. Methyl acetate has a partition coefficient blood:air of 70; 53 % of the dose inhaled during a 10-minute exposure is retained by man. In comparison, methanol has a partition coefficient blood:air of 2590 and about 60 % of an inhaled dose is retained during exposure to concentrations up to 200 ml/m^3 (Kumagai *et al.* 1999). The reproductive toxicity of methanol is caused by the substance itself; the NOEL for reproductive toxicity of methanol in the mouse is 1000 ml/m^3 (see MAK documentation for methanol in Volume 16 of the present series). With a toxicokinetic model for methanol, predicted blood methanol concentrations have been determined for various exposure periods, concentrations and species (Perkins *et al.* 1995): exposure of the mouse for 8 hours to a methanol concentration of 1000 ml/m^3 is predicted to result in a blood methanol level of about 200 µg/ml at the end of exposure. The blood methanol concentration in persons exposed to methanol at the level of the MAK value for 8 hours is about 7.5 µg/ml. Exposure to 1000 ml/m^3 for 15 minutes results in a blood methanol level of about 2 to 3 µg/ml. That means that during a 15-minute exposure at 800 ml/m^3 in the worst case (in the last quarter hour of the shift) a maximum blood methanol concentration of 10 µg/ml could be attained. Because of its similar retention characteristics and rapid breakdown, after exposure to methyl acetate in a concentration of 200 ml/m^3 the maximum blood methanol level would also be 10 µg/ml. Thus, the expected maximum blood methanol level is a factor of 20 smaller than the NOEL found in the animal study. This difference is sufficient to justify classification of methyl acetate in Pregnancy risk group C, given an excursion factor of 4 for the limitation of peak exposures. Therefore for methyl acetate an excursion factor of 4 was established.

There are no suitable studies available for evaluating absorption of the substance through the skin. The dermal LD_{50} is above 5000 mg/kg body weight. Calculations by Fiserova-Bergerova *et al.* (1990) demonstrate that dermal exposure of 2 % of the body surface to methyl acetate increases internal exposure by an amount between 30 % and 200 % of that absorbed during inhalation of 200 ml/m^3. For methanol, however, which is designated with an "H" (for substances which can be absorbed in dangerous amounts through the skin), an increase in the internal exposure of at least 200 % was calculated. As a result of the high volatility of methyl acetate, the danger of persistent dermal exposure is slight. There is no reason to designate methyl acetate with an "H".

There are no data available on potential sensitization to methyl acetate.

7 References

ACGIH (American Conference of Governmental and Industrial Hygienists) (1992) Methyl acetate. in: *Documentation of TLVs and BEIs*. ACGIH, Cincinnati, OH

BG Chemie (1995) Acetic acid methyl ester. *BG Chemie Toxicological Evaluations 10, Potential health Hazards of Existing Chemicals*. Springer Verlag, pp. 31–66, c. 1966

Dahl AR, Miller SC, Petridou-Fischer J (1987) Carboxyesterases in the respiratory tract of rabbits, rats and Syrian hamsters. *Toxicol Lett 36*: 129–136

Fiserova-Bergerova V, Pierce JT, Droz PO (1990) Dermal absorption potential of industrial chemicals: criteria for skin notation. *Am J Ind Med 17*: 617–635

Flury F, Wirth W (1933) Zur Toxikologie der Lösungsmittel. *Arch Gewerbepathol Gewerbehyg 5*: 1–90

Henschler D, Lehnert G (Eds) (1983) Methanol. in: *Biologische Arbeitsstoff-Toleranz-Werte (BAT-Werte) und Expositionsäquivalente für krebserzeugende Arbeitsstoffe (EKA)*, 1st issue, VCH-Verlagsgesellschaft, Weinheim, and in English translation in *Biological Exposure Values, Volume 1*

Kumagai S, Oda H, Matsunaga I, Kosaka H, Akasaka S (1999) Uptake of 10 polar solvents during short-term respiration. *Toxicol Sci 48*: 255–263

Opdyke DLJ (1979) Monographs on fragrance raw materials: methyl acetate. *Food Cosmet Toxicol 17*: 859–861

Perkins RA, Ward KW, Pollack GM (1995) A pharmacokinetic model of inhaled methanol in humans and comparison to methanol disposition in mice and rats. *Environ Health Perspect 103*: 726–733

Smyth HF Jr, Carpenter CP, Weil CS, Pozzani UC, Striegel JA (1962) Range-finding toxicity data: list VI. *Am Ind Hyg Assoc J 23*: 95–107

Tada O, Nakasaki K, Fukabori S (1974) On the method of evaluating the exposure to methanol and methyl acetate (Japanese). *J Sci Labour 50*: 239–248

Zimmermann FK, Mayer VW, Scheel I, Resnick MA (1985) Acetone, methyl ethyl ketone, ethyl acetate, acetonitrile and other polar solvents are strong inducers of aneuploidy in *Saccharomyces cerevisiae. Mutat Res 149*: 339–351

completed 01.03.2001

4,4′-Methylene-bis(2-methylaniline)

MAK value	–
Peak limitation	–
Absorption through the skin	–
Sensitization	–
Carcinogenicity (1983)	2
Prenatal toxicity	–
Germ cell mutagenicity	–
BAT value	–
CAS number	838-88-0

1 Allergenic Effects

1.1 Effects in man

There are no reports of sensitizing effects on the skin or respiratory passages after exposure to 4,4′-methylene-bis(2-methylaniline).

1.2 Animal studies

The substance was tested with rabbits for sensitizing effects on the skin; 0.5 ml of a 15 % 4,4′-methylene-bis(2-methylaniline) solution in acetone was applied daily for 10 to 30 days to the shaved skin (5 × 5 cm) of the right flank. This treatment did not cause skin irritation. The left flank served as control. 10 days after the last application, a single dose of 0.5 ml of a 25 % solution of the test substance was applied to the left flank. This caused swelling with hyperaemia, which the authors interpreted as evidence of skin sensitizing effects (Volodchenko *et al.* 1970). As information on the duration of contact and the time of evaluation was not given, it is not possible to evaluate these results. Furthermore, it cannot be excluded that this was not an irritative effect, as the 4,4′-methylene-bis(2-methylaniline) concentration used for provocation was higher than for induction and no irritation threshold was given.

2 Manifesto (MAK value/classification)

The allergenic effects of 4,4′-methylene-bis(2-methylaniline) cannot be evaluated on the basis of the available data. There are no data available for man and it is not clear whether the skin reaction described in the only experiment with animals was of allergic genesis. 4,4′-Methylene-bis(2-methylaniline) has therefore not been designated with an “S”.

3 References

Volodchenko BA, Vasilenko NM, Sadocha ER (1970) Toxicological characteristics of new addition agents to polymere materials (Russian). *Vrach Delo 12*: 116–119

completed 05.05.1997

Molybdenum and its compounds

MAK value (2000)	**not yet established, see Section IIb of the *List of MAK and BAT Values* (except for molybdenum trioxide)**
Peak limitation	**–**
Absorption through the skin	**–**
Sensitization	**–**
Carcinogenicity (2000)	**only molybdenum trioxide: Category 3B**
Prenatal toxicity	**–**
Germ cell mutagenicity	**–**
BAT value	**–**
Chemical names	see next page

The present documentation is a review of data for metallic molybdenum and for the following molybdenum compounds: molybdenum dioxide, molybdenum trioxide, ammonium molybdate, lead molybdate, calcium molybdate, sodium molybdate, zinc molybdate, phosphomolybdic acid, molybdenum dichloride, trichloride, pentachloride and disulfide.

1 Characterization of the Substance

1.1 Occurrence

Molybdenum is one of the rarer elements. Molybdenum is the earth's 38th most abundant element, and is similar in abundance to lead. Molybdenum is a metal and occurs not only in metallic form but also as salts. There are more than 50 known inorganic molybdenum compounds and also a few organometallic compounds. In its compounds, molybdenum is found in practically all oxidation states from –4 to +6, whereby the molybdenum(+VI) compounds are the most stable. Molybdenum is soluble in nitric acid, concentrated sulfuric acid and *aqua regia* but not in hydrochloric acid, dilute sulfuric acid, hydrofluoric acid or potassium hydroxide solution. It forms alloys with aluminium, tungsten, lead, iron, nickel, manganese, chromium and many other metals.

Chemical name (CAS)	Synonyms	CAS number	Formula	Molecular weight	Melting point (°C)	Solubility
molybdenum		7439-98-7	Mo	95.95	2626	
molybdenum oxide	molybdenum(VI) oxide molybdenum trioxide molybdic anhydride	1313-27-5	MoO_3	143.93	795	0.49 g/l at 28°C
molybdenum oxide (MoO_2)	molybdenum dioxide molybdenum(IV) oxide	18868-43-4	MoO_2	127.94		
molybdenum sulfide	molybdenite molybdenum disulfide molybdenum(IV) sulfide	1317-33-5	MoS_2	160.07	1185	
molybdic acid, disodium salt	disodium molybdate molybdic(VI) acid sodium salt sodium molybdate	7631-95-0	Na_2MoO_4	205.92	687	443 g/l at 20°C
molybdic acid, calcium salt (1:1)	calcium molybdate(VI)	7789-82-4	$CaMoO_4$	200.01	965	
molybdenum zinc oxide	basic zinc molybdate molybdic acid, zinc salt, basic zinc molybdate zinc molybdenum oxide	61583-60-6	$ZnMoO_4$	225.32	>700	840 g/l at 100°C 1.57 g/l at 20°C
molybdic acid, lead(2+) salt (1:1)	lead(II) molybdate	10190-55-3	$PbMoO_4$	367.14		
molybdic acid, diammonium salt	ammonium molybdate ammonium paramolybdate diammonium molybdate	13106-76-8	$(NH_4)_2MoO_4$	196.01	decomposition	
molybdic acid, hexaammonium salt	ammonium paramolybdate hexaammonium molybdate	12027-67-7	$(NH_4)_6Mo_7O_{24}$	1163.79	decomposition	430 g/l
molybdophosphoric acid	molybdophosphoric acid hydrate phosphomolybdic acid	51429-74-4	$H_3Mo_{12}O_{40}P$	1825.25	78	
molybdenum chloride	molybdenum(II) chloride molybdenum dichloride	12184-22-4	$MoCl_2$	166.86		–
molybdenum chloride	molybdenum(III) chloride molybdenum trichloride	13478-18-7	$MoCl_3$	202.32		–
molybdenum chloride	molybdenum(V) chloride molybdenum pentachloride	10241-05-1	$MoCl_5$	273.24	194	decomposition

The most important naturally occurring molybdenum compounds are:

molybdenum disulfide	MoS_2	(molybdenite)
calcium molybdate	$CaMoO_4$	(powellite)
lead molybdate	$PbMoO_4$	(wulfenite, yellow lead ore)

Only these three molybdenum ores are mined commercially (Falbe and Regitz 1992).

Molybdenum forms several simple and many complex compounds and a large number of complex molybdate anions are known. It has not been demonstrated that molybdenum can exist on its own as a cation. Most molybdenum compounds are insoluble in water or decompose on contact with water. Readily soluble in water are the molybdates such as ammonium and sodium molybdate (443 g/l at 20°C) (Elvers *et al.* 1999).

1.2 Production and use

Pure molybdenum is extracted from ores by means of a flotation process. In this process the molybdenum is first obtained as the trioxide. Molybdenum trioxide is the main commercial product. It is the starting material for the production of molybdenum-steel and all other molybdenum compounds. Powdered molybdenum metal is produced by reduction of highly pure molybdenum compounds with hydrogen. The particle diameter of the metallic molybdenum powder produced in this way is between 2 and 10 μm. Solid molybdenum metal is produced from the powder by pressing and sintering. The final processing of the material then involves melting and recrystallization or welding (inert gas welding because molybdenum reacts rapidly with oxygen at high temperatures).

In the middle 1990s, the western world used about 90000 tonnes of molybdenum per year. The amounts of molybdenum and molybdenum compounds used in 1986 in various branches of industry are shown below in percent world molybdenum production.

mechanical engineering	35%
oil and gas industry	10%
transport	15%
chemical industry	10%
electrical industry	15%
others	15%

The substance is mainly used for the production of various kinds of steel (mild steel, stainless steel, tool steel, cast iron) and super alloys for high temperature use. Other areas in which molybdenum and its alloys are used include (HSDB 1999, Elvers *et al.* 1999):

lamps and lighting industry	resistance wire, sealing strips
electronics and semiconductor industry	semiconductor base plates, metal coolers, contacts, grids for radio valves
high temperature and vacuum ovens	heating elements, insulation
glass and ceramics industries	electrodes, stirrers and furnace parts for the glass melt, crucibles for the production of single sapphire crystals
foundry technology and metal processing	hot galvanizing, injection moulding, in isothermic forging processes

medicine	dental alloys, rotating X-ray targets, collimators
nuclear technology	muffle furnaces and reaction vessels for UO_2 sintering (production of nuclear reactor fuel cores)
chemical industry	pigments, inks, paints, rubber, in fertilizers for microbiological nitrogen fixation
others	in colloidal form in the lubricants industry, as catalyst in the oil industry, for rocket and aeroplane parts

In addition, some molybdenum compounds are important in industry. They are used in many industrial processes as catalysts, for example, cobalt molybdate in the hydrogenation of petroleum and especially in the desulfurization of petroleum, iron molybdate for the oxidation of methanol to formaldehyde. Soluble molybdates are used in corrosion control. Molybdenum oxides and molybdates have flame-retardant properties and so are added to plastics for fire prevention purposes. Molybdenum(VI) sulfide has unusual lubricant properties and is added to lubricating greases and oils. It is also used as a dry film or solid lubricant. The particle diameters in the powdered substance are around 4 µm. In addition, molybdenum compounds are used as mordants for dyeing, as reagents for identification of compounds in analytical chemistry and as fluxes or wetting agents in ceramics production.

1.3 Exposure

Workplace exposures involve mostly molybdenum fume or molybdenum dust produced in metal working (e.g. grinding, pressing, welding). Commonly used processes during which persons can be exposed to soluble molybdenum compounds include:

- working hot blocks of metal during molybdenum steel production
- welding materials containing molybdenum
- use of molybdenum electrodes
- use of molybdenum compounds in chemical laboratories (synthesis, analysis)
- use of molybdenum catalysts for desulfurization of petroleum
- use of molybdenum compounds in enamelling to improve the adhesion of glass to ceramics and metal
- use in agriculture in fertilizers
- use as intermediates in the production of corrosion inhibitors
- colouring of glass and ceramics with molybdenum pigments
- tanning of leather and furs (HSDB 1999).

1.4 Analytical procedures

Quantitative analysis of molybdenum alloys or compounds can be carried out by means of atomic absorption and emission spectroscopy, X-ray fluorescence spectroscopy, and by spectrophotometric, gravimetric and volumetric chemical procedures.

Sampling of molybdenum compounds in the workplace air is normally carried out with aerosol samplers for the inhalable dust fraction (BIA 1989). The analytical methods currently used for workplace monitoring generally determine the level of the element molybdenum in the sample. Differentiation of the kind of compound is often not practicable.

2 Toxic Effects and Mode of Action

Molybdenum is an essential trace element. It is a cofactor for the enzymes xanthine oxidoreductase, sulfite oxidase and aldehyde oxidase. High concentrations of molybdenum can have toxic effects. Antagonistic effects are known between molybdenum and copper, sulfate and tungsten.

Molybdenum and its compounds are taken up via the gastrointestinal tract and the lungs, retained transiently in the tissues, and then excreted entirely in the form of molybdates, mostly in the urine. The biological half-life in animals is of the order of magnitude of hours, in man of weeks. Molybdenum can also be eliminated in the milk.

High endogenous levels of molybdenum, which are accompanied by high uric acid levels in the serum, cause gout-like symptoms.

Molybdenum trioxide, calcium molybdate and ammonium molybdate administered to rats and guinea pigs were more toxic than molybdenum disulfide. The LD_{50} values for molybdenum trioxide were 125 mg/kg body weight after oral administration to rats and 94 mg/kg body weight after subcutaneous injection into mice. In the lungs of rats exposed for long periods by inhalation of molybdenum trioxide in concentrations of 30 mg/m^3 and more, inflammatory changes developed like those typically caused in this species by exposure to particles. In mice exposed to concentrations up to 100 mg/m^3, such effects were not seen.

There is no clear evidence that the substance causes sensitization.

In vitro test systems did not reveal genotoxic activity of molybdenum compounds. Molybdenum trioxide can induce cell transformation. In long-term inhalation studies, weak evidence of carcinogenic effects was obtained with mice but no clear evidence with rats.

3 Mechanism of Action

Although molybdenum is an essential trace element, molybdenum deficiency in man has never been reported. The daily requirements of man are estimated to be in the range of 0.1 to 0.5 mg, the amount of molybdenum in the human body is about 0.1 mg/kg body weight (NTP 1997).

The mechanism of action of molybdenum is very complex, as is that of other essential metals. In mammals, metabolic interaction takes place between molybdenum and copper, sulfate and tungsten.

Molybdenum is a cofactor for the enzymes xanthine oxidoreductase, sulfite oxidase and aldehyde oxidase. Increased molybdenum intake results in increased levels of xanthine oxidoreductase in liver, intestine and kidney tissue and consequently in increased concentrations of uric acid in the blood and urine. In mammals, a high level of molybdenum in the diet results in reduction of the level of metabolically active copper in the liver and in symptoms of copper deficiency. With increasing levels of molybdenum in blood, the binding of copper to plasma protein is increased; this leads to reduced copper uptake by liver cells and increased excretion of copper. The result is disturbance of the synthesis of copper-containing proteins such as ceruloplasmin.

The antagonism between molybdenum and copper is affected by the sulfate content of the diet. Sulfate reduces the retention of molybdenum by reducing its absorption from the gastrointestinal tract and increasing its elimination with the faeces and the urine (NTP 1997). Sulfate induces a redistribution of molybdenum in the blood, an effect characterized by reduced levels of molybdenum in the erythrocytes (Stokinger 1981). Molybdenum can inhibit the enzyme sulfite oxidase in the liver. This results in accumulation of sulfite which leads to the formation of insoluble copper sulfide and to symptoms of copper deficiency (NTP 1997).

Tungsten is a molybdenum antagonist. It inhibits the absorption of molybdenum and increases the excretion of molybdenum in the urine. It seems probable that tungsten can replace molybdenum in molybdenum-dependent oxidases and so inhibit the action of these enzymes (NTP 1997).

The mechanism by which molybdenum induces lung tumours is not yet known. Molybdenum proved not to be genotoxic *in vitro*. In the rat lung, the inflammatory changes typical of exposure to particles developed in animals of both sexes, but no tumours were found in the female animals, and in the males the tumour incidences were in the range of the historical controls. In the mouse, however, the lung tumour incidence was significantly increased but no chronic inflammatory changes were detected.

4 Toxicokinetics and Metabolism

Molybdenum and molybdenum compounds are taken up by man and animals mainly via the gastrointestinal tract and the lungs; nothing is known about percutaneous absorption of molybdenum. The amount of molybdenum taken up with the diet is very different in different geographical regions. In the USA, the average daily molybdenum intake with the diet is about 350 µg (NTP 1997), in England 128 ± 34 µg (Lener and Bibr 1984). Absorption, distribution and excretion of molybdenum are similar in man and animals. The solubility of the particular molybdenum compound determines the absorption and distribution patterns after oral, intramuscular and inhalative administration. Absorption in the lungs is also affected by the particle size.

In an early study it was demonstrated that in guinea pigs exposed to molybdenum trioxide fume (150–300 mg/m^3, 1 hour daily, 5 days weekly for 5 weeks) the concentrations of molybdenum in the lung, liver, kidney, spleen and bone tissue at the end of exposure were in the range between 2 and 27 µg/g tissue. Within 2 weeks the concentrations had decreased to about 20 % of these values. Concentrations of molybdenum in blood were determined in animals exposed for 2 years to molybdenum trioxide dust in concentrations of 10, 30 and 100 mg/m^3 in a carcinogenicity study (see Table 1). The molybdenum concentrations in the blood of the male and female mice, respectively, were 0.102, 0.208 and 0.770 µg/g and 0.066, 0.198 and 0.523 µg/g (controls 0.04 µg/g). In the rats, these levels were much higher in males (0.800, 1.774 and 6.036 µg/g) than in females (0.355, 0.655, 2.411 µg/g) (controls 0.221, 0.059 µg/g). That the systemic doses of molybdenum were much higher in the rats than in the mice was ascribed to the higher lung capacity of rats (NTP 1997).

Four hours after oral administration of 50 mg molybdenum trioxide to guinea pigs, increased levels of molybdenum were found in the kidney, spleen, blood, bile, liver and lungs. 48 hours *post applicationem* the concentrations in the blood and bile had increased further and those in the tissues had decreased (NTP 1997). Molybdenum can cross the placental barrier (Lener and Bibr 1984). After injection of radioactive molybdenum into dogs, radioactivity was detected in the liver, kidney and endocrine glands. In the brain and adipose tissue, only low levels of radioactivity were found (no other data; Stokinger 1981). Molybdenum circulating in the blood is firmly bound to erythrocytes and plasma proteins (NTP 1997).

After transient retention in the tissues, the molybdenum is excreted entirely in the form of molybdates. Accumulation of molybdenum in mammals is negligible. The biological half-life in animals is of the order of magnitude of hours, in man of weeks.

For the elimination of molybdenum, excretion via the bile and faeces is less important than that via the kidneys and urine. Twice as much molybdenum is excreted in the urine as in the faeces. Molybdenum can also be eliminated in the milk (NTP 1997).

5 Effects in Man

5.1 Repeated exposures

Of 19 persons who had been exposed to metallic molybdenum and molybdenum trioxide in concentrations of 1 to 25 mg/m^3 for 4 to 7 years, 3 suffered from symptoms such as respiratory difficulties and frequent coughing. Roentgenograms revealed early stages of pneumoconiosis (Mogilevskaya 1963).

Unspecific symptoms such as weakness, lassitude, anorexia, headaches, joint and muscle pains, and hand tremor were reported for Russian miners and metal workers exposed to molybdenum in concentrations of 60 to 600 mg/m^3 (no details of exposure situation) (Lener and Bibr 1984).

An increased incidence of gout-like disorders was found in a region of Armenia with high molybdenum levels in the soil and plants (increased 38-fold and 190-fold). The daily intake of molybdenum for a 70 kg adult was estimated to be 10 to 15 mg with a reduced copper intake of 5 to 10 mg (expected values 1–2 mg Mo, 10–15 mg Cu). Medical examination of 400 persons (100 persons who had lived in the area for a maximum of 6 years and 300 persons living in an old town) revealed gout-like symptoms such as arthralgia in the knees, hands and feet in 18 % and 32 % of individuals (incidence in an area without increased molybdenum levels 1–4 %). The increased uric acid levels in the serum correlated with increased molybdenum levels and increased xanthine oxidoreductase activities. This provides a possible explanation for the gout-like symptoms (EPA 1998).

In contrast, in a study carried out in a molybdenite roasting plant in Colorado, no gout-like symptoms were observed in the exposed workers although the molybdenum concentrations in blood and urine and the uric acid and ceruloplasmin concentrations in serum were increased. 25 persons had been exposed to an average molybdenum concentration of 9.5 mg/m^3 (soluble molybdenum compounds in total dust) for a period of 4 years. Biochemical analyses revealed only an increase in the serum ceruloplasmin concentration to 50.47 ± 1.38 mg/100 ml (control 30.50 ± 1.33 mg/100 ml) and a slight increase in serum uric acid concentration to 5.90 ± 0.24 mg/100 ml (control 5.01 ± 0.25 mg/100 ml). The molybdenum concentration in the urine was in the range between 20 and 230 µg/l in the controls and was increased to between 120 and 11000 µg/l in the exposed persons (Walravens *et al.* 1979).

5.2 Allergenic effects

In early studies carried out in Poland but available in the literature only as a summary, reactions to molybdenum sulfide (no other data) were reported in 2 of 860 persons working with metal-working fluids (Sokolowski 1976); occasional reactions to 2 % ammonium molybdate were reported in welders (Sokolowski 1972). In tests carried out with 52 patients with contact eczema thought to be caused by chromium (or chromate), 2 persons reacted to 0.5 % phosphomolybdic acid but none reacted to 2 % ammonium molybdate (Kresbach and Grabner 1969). In an abstract, weak reactions in a patch test with 20 % ammonium paramolybdate (molybdate(VI)) in water were reported for persons employed in molybdenum production, and also work-related skin disorders characterized clinically by itching and a slight tendency to eczema and secondary infections (Dueva and Stepanian 1989).

Although molybdenum levels up to 10^5 ng/g tissue were found in patients in whom metal bone implants had led to infectious or non-infectious complications, in none of the 541 patients was molybdenum allergy detected. The tests were either intracutaneous with 0.0005 % ammonium heptamolybdate in water or epicutaneous with 1 % molybdenum dichloride in petrolatum. However, the number of persons tested is not given (Hierholzer and Hierholzer 1987). Likewise in another study in which 50 patients with joint prostheses were subjected to patch tests with 1 % ammonium molybdate, no positive results were obtained (Elves *et al.* 1975). One of 128 patients who had been given a hip prosthesis produced a positive result in a test with 1 % ammonium molybdate in water (Török *et al.* 1995). A female patient who developed generalized dermatitis three months after having a broken bone fixed with an osteosynthesis screw reacted in the patch test not only to 2.5 % nickel sulfate (+++ reaction) but also to 1 % cobalt chloride, 0.5 % potassium dichromate, 0.5 % vanadium oxide and 1 % ammonium molybdate (all ++, reactions after 72 hours, no reaction after 48 hours). After the screw had been removed, the dermatitis became more severe but then regressed within 3 weeks. However, afterwards the patient produced local skin reactions after contact with products containing nickel (Oleffe and Wilmet 1980). Implantation of two metal plates into a 24-year-old woman was followed by fever and symptoms of an ANA-negative (antinuclear antibodies negative) systemic *lupus erythematodes* which became progressively more severe. The symptoms regressed after removal of the metal plates but reappeared after patch tests with the alloy and with its components (test material not specified in detail) and later also after dental treatment in which a material containing molybdenum was used. Later recurrences of the disorder also developed without provocation by contact with metals. The patient was treated with daily doses of 10 mg prednisone; in the lymphocyte transformation test with 10 mg molybdenum (no other details) a lymphocyte stimulation score of 4.1 was obtained. The scores for two healthy control persons were 1.94 and 2.1 (Federmann *et al.* 1994).

In a study of the effects of a low-metal diet and of the removal of dental metal on the course of moderate to severe atopic dermatitis, positive results in tests with 1 % molybdenum pentachloride in water were obtained for 6/27 patients (Adachi *et al.* 1997). None of 63 patients produced a positive result with 1% ammonium molybdate in water in patch tests with salts of many of the metals used in dental prostheses (Van Loon *et al.* 1986).

In patch tests with 18 metal salts carried out on 80 volunteers with healthy skin, three erythematous and 2 simple reactions to 1 % molybdenum pentachloride in water were obtained (application for 48 hours, reading after 72 hours; reading after 7 days still revealed a reaction) (Hattori *et al.* 1996).

Between 1981 and 1984 a total of 1200 patients were subjected to patch tests with numerous metal salts in various test blocks. Positive results with 2 % ammonium molybdate and 2 % molybdenum pentachloride were obtained for 3/208 (+ reactions) and 6/211 (including 3 ++/+++ reactions), respectively. In only one case were positive results obtained with both test preparations. Three of the eight patients who reacted worked with metals, and all also produced reactions to nickel or cobalt; seven persons reacted to at least 3 metals. Data for the clinical relevance of the reactions are, however, lacking (Rammelsberg 1985).

In the clinics of the IVDK (Informationsverbund dermatologischer Kliniken: Information Network of German Departments of Dermatology) in the years between 1992 and 1999, simple reactions to 1 % ammonium heptamolybdate in water were recorded for 3/787 patients. In addition, 7 questionable or irritant reactions were observed. In tests with 1 % of the substance in petrolatum, reactions were not observed in any of the 275 patients (IVDK 2000).

5.3 Genotoxicity

In peripheral blood lymphocytes from persons occupationally exposed to molybdenum, molybdenite (MoS_2), ammonium paramolybdate and molybdenum trioxide, an increased incidence of chromosomal aberrations was detected (no other data; Babaian *et al.* 1980).

5.4 Carcinogenicity

For a case-control study, 478 cases of confirmed lung carcinoma and 536 hospital patients (mostly with cardiovascular disorders) were recruited in Antwerp and the surrounding area (Droste *et al.* 1999). Information as to occupation, potential exposures and smoking habits were obtained by means of questionnaires. There are no data available for levels of exposure to molybdenum. Workplaces with exposure to molybdenum included processing of mineral oils, the chemical industry, production of non-metallic mineral substances, production and processing of metals, the mechanical and electrical industries, and the production of leather goods. Persons at these workplaces were generally exposed to mixtures of substances including chromium and mineral oils. 21 cases and 13 controls reported exposure to molybdenum themselves. For these, the lung cancer risk was not significantly increased (odds ratio (OR) adjusted for smoking habits, age and some demographic data: 2.1; 95 % confidence interval (CI): 0.9–5.1). When the persons were classified as exposed to molybdenum on the basis of their occupation (52 cases, 34 controls), the risk of developing a lung carcinoma was found to be significantly increased (2.1-fold; 95 % CI: 1.2–3.7). Analysis of exposure duration in this collective revealed that the risk increased with duration (1–4 years: OR = 1.6, 4–21 years: OR = 1.8, > 21 years: OR = 3.3) but was significantly increased only after 21 years of exposure or more. This is the first epidemiological study in which a significant association between exposure to molybdenum and the development of lung cancer has been described. Significant associations were also found for chromium and mineral oils. Therefore the increased lung cancer risk cannot be ascribed entirely to the molybdenum exposure and cannot be used to assess the carcinogenic effects of molybdenum.

6 Animal Experiments and *in vitro* Studies

6.1 Acute toxicity

During the four weeks after exposure, no effects were seen in rats which had inhaled for one hour metallic molybdenum in concentrations of 25000 to 30000 mg/m^3, molybdenum dioxide at 10000 to 12000 mg/m^3 or molybdenum trioxide at 12000 to 15000 mg/m^3 (estimated doses per animal: 50 mg Mo, 21 mg MoO_2, 26 mg MoO_3). Exposure of the animals to ammonium paramolybdate in concentrations of 3000 to 5000 mg/m^3 (5 to

6 mg/animal) caused slight signs of irritation of the upper airways and conjunctiva. In rats exposed for two hours to a condensation aerosol of molybdenum trioxide (64 mg/m^3) and observed for two weeks, dystrophic changes in the heart, kidneys and liver were observed; body weights were unaffected (no other data; Mogislevskaya 1963).

Molybdenum is more toxic for ruminants than for monogastric animals, presumably because of the formation of thiomolybdate in the sulfide-rich milieu in the rumen (NTP 1997).

Molybdenum doses of 1200 to 6000 mg/kg body weight in the form of molybdenum trioxide, calcium molybdate and ammonium molybdate were lethal after oral administration to rats and guinea pigs; in the groups given doses between 120 and 600 mg/kg body weight, mortality was lower. Oral molybdenum doses of 6000 mg/kg in the form of molybdenum disulfide were not lethal for any of the animals (ACGIH 2000). The LD_{50} values for molybdenum trioxide were 125 mg/kg body weight after oral administration to rats and 94 mg/kg body weight after subcutaneous injection into mice. The symptoms included diarrhoea, coma and death from heart failure (NTP 1997).

Molybdenum trioxide doses of 400 mg/kg body weight administered to guinea pigs by intraperitoneal injection were lethal within 4 days for 6 of 8 animals. All (12/12) guinea pigs given parenteral ammonium molybdate doses of 800 mg/kg body weight died within a few hours. An ammonium molybdate dose of 80 mg/kg body weight was, however, not lethal. Lower mortality was observed in guinea pigs given molybdenum sulfide in doses of 800 mg/kg body weight by injection (2/12 animals within 4 days, 3/12 within 4 months; no other details) (ACGIH 2000).

6.2 Subacute, subchronic and chronic toxicity

6.2.1 Inhalation

As part of an NTP carcinogenicity study (NTP 1997), groups of 5 female and 5 male F344 rats were exposed to molybdenum trioxide dust in concentrations of 0, 3, 10, 30, 100 and 300 mg/m^3 for 6 hours daily, 5 days weekly, 10 times during 14 days. All animals survived the treatment. The body weights of the male animals in the 100 mg/m^3 group and those of the male and female animals in the 300 mg/m^3 group were significantly lower (31 % and 13 % reduction, respectively) than those in the control group. Likewise, in groups of 5 female and 5 male B6C3F$_1$ mice exposed under the same conditions, all the animals survived. Body weights of animals of both sexes were significantly lower (about 10 %) in the group exposed to 300 mg/m^3 than in the control group. Histological and biochemical parameters were unaffected in both species.

Groups of 10 female and 10 male B6C3F$_1$ mice and F344 rats were exposed by inhalation of molybdenum trioxide in concentrations of 0, 1, 3, 10, 30 or 100 mg/m^3, 6.5 hours daily, 5 days weekly for 13 weeks; in the exposed mice the copper levels in the liver were increased (in females from 30 mg/m^3, in males from 100 mg/m^3). Body and organ weights, clinicochemical, haematological and sperm parameters, and the results of the histopathological examination were not different from the control values.

For the long-term study, groups of 50 male and 50 female B6C3F_1 mice and F344 rats were exposed to molybdenum trioxide in concentrations of 0, 10, 30, or 100 mg/m^3, 6.5 hours/day, 5 days/week for 105 and 106 weeks, respectively. The results are shown in Table 1; the tumour findings are presented and discussed in Section 6.7.

Survival, body weight development and clinical parameters for the exposed rats were similar to the control values. There were no changes in bone density or form and no effects on liver or kidneys. In the male rats, the molybdenum concentrations in the blood were very much higher than in the female rats. The incidence of chronic pulmonary inflammation in the alveolar region was significantly increased in both sexes after exposure to 30 mg/m^3 or more; both incidence and severity increased in a dose-dependent manner. The changes were qualitatively like those seen after long-term inhalation of other particulate substances. However, the effects were much less severe. The inflammatory changes were accompanied by hyperplasia and metaplasia. Exposure to molybdenum trioxide caused hyaline degradation of the nasal olfactory epithelium of practically all females and all males exposed to 30 mg/m^3 or more. The incidence of squamous cell metaplasia in the larynx was significantly increased in all groups in a dose-dependent manner. The changes observed in the nose and larynx were considered to be an unspecific irritative reaction.

In the mice, survival was slightly reduced only in the male animals of the 30 mg/m^3 group. Body weights were unaffected in the males but in the exposed females they were higher (at most 15 %) than in the control animals. In the lungs of the male animals the incidence of histiocyte infiltration was significantly increased but not in a dose-dependent manner. The incidence of mild metaplasia of the alveolar epithelium was significantly increased in all groups; the increase was dose-dependent. Unlike in the rats, in mice there were no chronic inflammatory changes in the lungs. In the noses of animals exposed to 100 mg/m^3, the incidence of slight hyaline degeneration was significantly increased in the respiratory epithelium of male and female animals and in the olfactory epithelium only in the females. The incidence of squamous cell metaplasia in the larynx was significantly increased in all groups in a dose-dependent manner. These effects were considered to be an unspecific irritative reaction. In the long-term study, effects were observed in both species at the lowest molybdenum trioxide concentration of 10 mg/m^3, and so no NOEL can be derived from the results of this study.

In the NTP publication several inhalation studies were reviewed (NTP 1997).

In one study from the year 1945, 24 guinea pigs were exposed to molybdenum trioxide dust in a concentration of 205 mg/m^3, 1 hour daily, 5 days weekly for 5 weeks. The mortality was 51 %. Irritation of the eyes and nose, anorexia, weight loss, diarrhoea, impairment of muscular coordination, and hair loss were recorded. In addition, vacuolation and necrosis in the liver, cell changes in the spleen, and alveolar and bronchiolar exudate in the lungs were observed. Molybdenum trioxide vapour in a concentration of 191 mg/m^3 was less toxic under the same exposure conditions. Mortality was 8.3 %.

Inadequately documented studies in which rabbits and rats were exposed to molybdenum trioxide dust in a concentration of 30 mg/m^3, 4 hours/day for 5.5 months, revealed reduced activities of alkaline and acid phosphatases in the serum and a reduced level of inorganic phosphorus in the tibia. No changes were found in the liver or spleen. The ascorbic acid levels in serum and urine were increased only in the rabbits.

Exposure of rats to condensation aerosols of molybdenum trioxide in concentrations of 3 to 10 mg/m^3 for 2 hours every second day during two months led to changes in the liver (dystrophy, focal fatty degeneration, binucleate cells with hyperchromic nuclei), lungs (thickening of the alveolar walls, interstitial pneumonia, emphysema), kidneys and heart (dystrophy) (Mogilevskaya 1963).

6.2.2 Intratracheal administration

Four to eight months after intratracheal instillation of 50 mg molybdenum trioxide or 50 mg molybdenum into 8 and 17 rats, respectively, fibrotic changes were detected in the lungs and the neighbouring lymph nodes (Mogilevskaya 1963).

6.2.3 Ingestion

Cattle and sheep given feed containing increased molybdenum levels (20–100 ppm) developed within a few days a disease called molybdenosis which can be lethal. The high molybdenum uptake causes copper deficiency in the organism (see Section 3). The symptoms of this disease are weight loss, diarrhoea, discoloration of the fur, anaemia, reduced conception rate, changes in the testis, the bones and the joints (Davis 1991, NTP 1997).

Molybdenum disulfide, molybdenum trioxide, calcium molybdate and ammonium molybdate were fed for up to 232 days to albino rats in concentrations in the diet corresponding to molybdenum doses of 10 to 500 mg/animal and day. In animals given molybdenum disulfide, no toxic effects were seen. All the other compounds caused anorexia, weight loss, lethargy and ruffled fur in the exposed animals. LD_{50} values (expressed as mg Mo per kg body weight and day) for repeated daily doses were found to be 125 mg/kg for molybdenum trioxide, 100 mg/kg for calcium molybdate and 333 mg/kg for ammonium molybdate (ACGIH 2000).

Oral administration of molybdenum trioxide doses of 25 to 200 mg per animal and day for one month by gavage to guinea pigs resulted in dose-dependent changes in the liver (fatty degeneration) and kidneys (focal necrosis and granulomatosis) (NTP 1997).

Ammonium molybdate was administered orally (gavage) for 8 weeks to Sprague-Dawley rats in amounts corresponding to molybdenum doses of 40 or 80 mg/kg body weight and day; in the high dose group body weight gains were delayed (by 35 %), absolute kidney weights were reduced, but because of the effects on body weight, relative kidney weights were increased. In addition, diuresis and the excretion of creatinine and kallikrein were increased significantly. These findings were interpreted as mild chronic renal failure caused by a glomerular dysfunction (Bompart *et al.* 1990).

In rabbits given oral ammonium molybdate doses of 5 mg/kg body weight and day for 4 to 6 months, the spleen weights were increased and liver weights decreased (no other data; NTP 1997).

6.3 Local effects on skin and mucous membranes

Calcium and zinc molybdate were not irritating for the skin or mucous membranes of the eye. Sodium molybdate caused slight irritation of skin and eyes which regressed within 72 hours (no other details; Stokinger 1981).

6.4 Allergenic effects

A maximization test with molybdenum pentachloride yielded positive results. After intradermal induction with 0.1 % molybdenum pentachloride in petrolatum and epicutaneous induction with 15 %, provocation 2 weeks after the epicutaneous induction with 5 %, 2.5 %, 1 % and 0.5 % molybdenum pentachloride in petrolatum produced after 24 hours reactions in 17, 13, 4 and 1 of the 20 animals and after 48 hours in 16, 10, 2 and 2/20. Of the 20 control animals pretreated with Freund's complete adjuvant (FCA), reactions to 5 % molybdenum pentachloride were seen in only 3 animals at the 24 and 48 hour readings (Boman *et al.* 1979). A second maximization test carried out by the same authors with sodium molybdate (induction intradermal: 10 %, epicutaneous: 25 %) yielded negative results. However, this test cannot be evaluated because the number of animals for which the results were positive was not given exactly and the data for the provocation concentrations are inconsistent.

In an auricular lymph node assay, C58Bl/6 mice were treated with 25 µl of a solution of 0.1 %, 1 %, 2.5 % or 5 % molybdenum pentachloride in acetone/olive oil (4:1) on the backs of both ears daily for three consecutive days. One day after the last treatment, dose-dependent increases in lymph node weights and in total cell count per lymph node were observed. However, after analysis of the cell populations affected, the authors concluded that molybdenum pentachloride is rather an unspecific contact irritant (Abdouh *et al.* 1995).

6.5 Reproductive and developmental toxicity

In an NTP study, groups of 10 female and 10 male $B6C3F_1$ mice and F344 rats were exposed, 6.5 hours daily, 5 days weekly for 13 weeks, by inhalation of molybdenum trioxide in concentrations of 0, 10, 30 or 100 mg/m^3; no evidence of toxic effects was found (see Section 6.2.1) and male fertility was unaffected. The weights of testis, cauda and epididymis, the sperm count and the motility and concentration of spermatozoa in the epididymis were in the control range (NTP 1997).

Male and female rats (strain not specified) were given sodium molybdate in the diet in a concentration of 80 or 140 mg/kg feed in the period from week 3 to week 11 of life. The animals were then mated and the following results obtained: reduction in the number of litters, degeneration of the seminiferous tubules, reduced growth of the progeny (no other details; NTP 1997).

In a 3-generation study, CD mice were exposed to 10 ppm molybdenum in the drinking water in the form of a soluble molybdenum salt (no other details). The exposure led to an increase in early deaths among the progeny (in the F_1 generation 15/238, control 0/209; F_2 generation 7/242, control 6/248; F_3 generation 34/123, control 1/230). In the F_1 and F_2 generations the litter parameters were unchanged in the absence of maternal toxicity. The results obtained for the F_3 generation were increased mortality of the dams, increased number of pairs without offspring, reduced number of litters (14, control 22), increased number of litters with only stillbirths (4/14, control 0/22) and underdeveloped pups (11/123, control 0/230) (Schroeder and Mitchener 1971).

6.6 Genotoxicity

6.6.1 *In vitro*

In the *Salmonella* mutagenicity test with the *S. typhimurium* strains TA97, TA98, TA100, TA102, TA104, TA1535, TA1537 and TA1538, molybdenum trioxide in concentrations between 10 and 10000 µg/plate was not found to be mutagenic either in the presence or absence of a metabolic activation system (NTP 1997, Zeiger *et al.* 1992). Molybdenum trioxide (0.5–15 µg/ml) did not increase the incidence of sister chromatid exchange or of chromosomal aberrations in CHO cells either in the presence or absence of a metabolic activation system (NTP 1997).

A co-mutagenic effect of sodium molybdate (100–30000 µM) was seen as an increase in the number of mutants induced by UV light in the presence of sodium molybdate in *Escherichia coli* WP2 and, to a lesser extent, in the repair-deficient strain *E. coli* $WP2_S$ (uvrA) (Rossman and Molina 1986).

In vitro genotoxicity tests with molybdenum trioxide in the rec-assay with *Bacillus subtilis* yielded negative results (NTP 1997).

6.6.2 *In vivo*

In the *Drosophila* wing spot test, molybdenum trichloride (10–50 mM) was shown to be genotoxic (Ogawa *et al.* 1994).

6.7 Carcinogenicity

6.7.1 Short-term studies

Molybdenum trioxide in concentrations of 75 to 200 µg/ml induced concentration dependent cell transformation in Syrian hamster embryo cells (Kerckaert *et al.* 1996).

In a short-term carcinogenicity study with Strain A mice (Strong Research Foundation), molybdenum trioxide administered by intraperitoneal injection (3× weekly, a total of 19 injections, end of study after 30 weeks) in total doses of 950, 2735 and 4750 mg/kg body weight induced an increase in the incidence of lung tumours only at the highest dose (1.13± 0.20, control 0.42 ± 0.1 tumours/animal). The carcinogenic activity in this test system was considered to be weak (Stoner *et al.* 1976).

6.7.1 Long-term studies

In a long-term carcinogenicity study in which mice and rats inhaled molybdenum trioxide (details in Table 1, non-neoplastic findings in Section 6.2.1), evidence of carcinogenic effects in the lung was found (NTP 1997).

Groups of 50 male and 50 female B6C3F$_1$ mice and 50 male and 50 female F344/N rats were exposed to molybdenum trioxide in concentrations of 0, 10, 30, or 100 mg/m^3, 6 hours/day, 5 days/week for 105 weeks.

In the lungs of the male mice, the incidence of alveolar/bronchiolar carcinomas was increased significantly in all exposure groups (2/50, 16/50, 14/49, 10/50) but a dose-response relationship was not apparent. Likewise when the alveolar/bronchiolar adenomas and carcinomas were taken together, the incidence was increased (11/50, 27/50, 21/49, 18/50) but the increase was significant only in the 10 and 30 mg/m^3 groups. In the female mice, the incidence of alveolar/bronchiolar adenomas (1/50, 4/50, 8/49, 9/49) was increased significantly in the 30 and 100 mg/m^3 groups. When the alveolar/bronchiolar adenomas and carcinomas were taken together (3/50, 6/50, 8/49, 15/49) the incidence was increased significantly only in the female animals of the highest dose group. All these significantly increased incidences of lung tumours were above the range of the historical control values (see Table 1). The results were considered by the NTP to be some evidence of a carcinogenic effect of molybdenum trioxide in male and female mice. No tumours were detected apart from lung tumours. In all exposure groups in the lungs of both male and female mice, metaplasia of the alveolar epithelium was observed and in the male animals histiocyte infiltration (see Table 1 and Section 6.2.1). Unlike in the rats, in mice there were no chronic inflammatory changes in the lungs. The changes observed in the nose and larynx (see Table 1 and Section 6.2.1) were considered to be an unspecific irritative reaction.

When the alveolar/bronchiolar adenomas and carcinomas in the male rats were taken together, the incidence was increased with a marginally significant positive trend (0/50, 1/49, 1/49, 4/50) but was within the range of the historical controls (see Table 1). The results were considered by the NTP to be equivocal evidence of a carcinogenic effect of molybdenum trioxide in male rats. In tissues apart from the lung no tumours were found in either sex, in female rats not in the lungs either. The incidence of chronic alveolar inflammation was significantly increased in both male and female animals exposed to 30 mg/m^3 or more. This kind of effect is a typical reaction of the rat to exposure to particles (see Section 6.2.1). The changes observed in the nose and larynx (see Table 1 and Section 6.2.1) were considered to be an unspecific irritative reaction.

Table 1. Long-term inhalation studies with molybdenum trioxide (NTP 1997)

Species:	rat, F344, 50 ♂, 50 ♀
Concentration:	MoO_3: 0, 10, 30, 100 mg/m^3; average MMAD between 1.5 and 1.8 μm
Duration:	106 weeks, 5 days/week, 6.5 h/day
Toxicity:	no effects on mortality or body weight development, no abnormal clinical findings; no effects on liver, kidneys, bones (density, form); maximum tolerated dose (MTD) not attained

		MoO_3 concentration (mg /m^3)			
		0	10	30	100
Non-neoplastic changes:					
lung:					
chronic alveolar inflammation	♂	2/50	3/50	25/50*	47/50*
	♀	14/50	13/50	43/50*	49/50*
nose:					
hyaline degeneration of the respiratory epithelium	♂	2/50	7/49	48/49*	49/50*
	♀	1/48	13/49*	50/50*	50/50*
degeneration of the olfactory epithelium	♀	39/48	47/49	50/50*	50/50*
larynx:					
metaplasia in the squamous epithelium of the epiglottis	♂	0/49	11/48*	16/49*	39/49*
	♀	0/49	18/49*	29/49*	49/50*
Lung tumours:					
carcinomas:	♂	0/50	1/49	1/49	1/50
	♀	0/48	1/49	0/50	0/50
alveolar/bronchiolar squamous cell carcinoma	♀	1/48	0/49	0/50	0/50
adenomas	♂	0/50	0/49	0/49	3/50
	♀	0/48	1/49	0/50	2/50
alveolar/bronchiolar adenomas, carcinomas or squamous cell carcinomas	♂ [a]	0/50	1/49	1/49	4/50
	♀	1/48	2/49	0/50	2/50

* p = 0.01
historical control incidences: [a] carcinoma or adenoma ♂: 23/654 (3.5% ± 3.7%); range: 0–10%

Species:	mouse, B6C3F$_1$, 50 ♂, 50 ♀
Concentration:	MoO_3: 0, 10, 30, 100 mg/m^3; average MMAD between 1.3 and 1.8 μm
Duration:	105 weeks, 5 days/week, 6.5 h/day
Toxicity:	♂: only at 30 mg/m^3 from week 75 mortality slightly reduced; body weight development unaffected ♀: mortality unaffected; from week 11 to week 105 body weights of exposed animals greater than control values all groups: no effects on clinical findings, liver, kidneys, bones (density, form)

Table 1. continued

		MoO₃ concentration (mg /m³)			
		0	10	30	100
Non-neoplastic changes:					
lung:					
histiocyte infiltration	♂	2/50	16/50*	9/49*	9/50*
metaplasia of the alveolar epithelium	♂	0/50	32/50*	36/49*	49/50*
	♀	2/50	26/50*	39/49*	46/49*
nose:					
hyaline degeneration of the respiratory epithelium	♂	11/50	13/50	11/49	41/50*
	♀	22/49	14/50	14/49	36/49*
atrophy of the olfactory epithelium	♂	3/50	5/50	3/49	10/50*
hyaline degeneration of the olfactory epithelium	♀	26/49	23/50	28/49	48/49*
larynx:					
hyperplasia	♂	1/50	3/49	6/48	41/50*
	♀	1/49	1/50	7/49	35/50*
metaplasia in the squamous epithelium of the epiglottis	♂	0/50	26/49*	37/48*	49/50*
	♀	1/49	36/50*	43/49*	49/50*
Lung tumours:					
carcinomas:	♂ [a]	2/50 (4%)	16/50* (32%)	14/49* (29%)	10/50* (20%)
	♀	2/50 (4%)	2/50 (4%)	0/49 (0%)	6/49 (12%)
alveolar/bronchiolar adenomas	♂	9/50 (18%)	14/50 (28%)	10/49 (20%)	9/50 (18%)
	♀ [b]	1/50 (2%)	4/50 (8%)	8/49* (16%)	9/49* (18%)
alveolar/bronchiolar adenomas : or carcinomas	♂ [c]	11/50 (22%)	27/50* (54%)	21/49* (43%)	18/50 (36%)
	♀ [d]	3/50 (6%)	6/50 (12%)	8/49 (16%)	15/49* (31%)

*p = 0.05
historical control incidences:
[a] carcinomas ♂: 75/947 (7.9% ± 5.7%); range: 0–16%
[b] adenomas ♀: 61/939 (6.5% ± 3.2%); range: 0–14%
[b] adenomas or carcinomas: ♂: 205/947 (21.7% ± 8.0%); range: 10–42%
[d] adenomas or carcinomas: ♀: 97/939 (10.3% ± 3.7%); range: 0–16%

7 Manifesto (MAK value/classification)

Results of studies of exposed persons which would be suitable for the derivation of a MAK value are not available.

In a long-term carcinogenicity study in which experimental animals inhaled molybdenum trioxide for a period of 104 weeks, the number of male mice with lung tumours (mice with carcinomas or sum of mice with adenomas or carcinomas) was increased significantly from the lowest exposure concentration of 10 mg/m^3 but the increase was not dose-dependent. The incidence of lung adenomas was significantly increased in female mice exposed to 30 and 100 mg/m^3, that of adenomas plus carcinomas only in the 100 mg/m^3 group. In male rats there was only a marginally significant positive trend towards an increase in the incidence of adenomas plus carcinomas and the incidences were within the range of the historical control values. In female rats, no increases in lung tumour incidence were observed.

In the lungs of rats exposed to 30 mg/m^3 and more, not only tumours but also chronic inflammatory changes were observed, like those typically seen in this species after exposure to particles. In the mice from 10 mg/m^3, metaplasia was seen in the alveolar epithelium but no chronic inflammatory changes in the lungs. In both species the incidence of degenerative and proliferative changes in the nose and larynx was increased, but these effects were considered to be unspecific reactions to the chronic irritation. A NOAEL for the effects in the lung, nose and larynx was not established in these studies. The *in vitro* genotoxicity tests yielded negative results. The mechanism which results in an increased lung tumour incidence in animals exposed to molybdenum trioxide is not understood. Therefore molybdenum trioxide has been classified in Carcinogen category 3B in Section III of the *List of MAK and BAT Values*. The available data indicate, however, that the MAK value of 4 mg/m^3 I (inhalable fraction) which applied until 2000 is too high. A new MAK value cannot be established on the basis of the available data; the MAK value has therefore been withdrawn.

Because the data are inadequate, a MAK value cannot be established for the other molybdenum compounds. Molybdenum and its compounds, with the exception of molybdenum trioxide, are therefore listed in Section IIb of the *List of MAK and BAT Values*.

The potential of molybdenum and its compounds to cause germ cell mutations cannot be assessed because of lack of data. Therefore the substances are not classified as germ cell mutagens.

To date no information is available as to the ability of molybdenum or its compounds to enter the human organism through the intact skin. Therefore these substances are not designated with an “H”.

A significant sensitizing effect of molybdenum compounds in man has not been proved in the available reports which are mostly inadequately documented. Patch tests carried out in large collectives yielded only very few positive results whose allergic genesis is often questionable and relevance unclear. The results of experimental animal studies are contradictory and in some cases do not make sense. There are no data available for sensitizing effects on the respiratory passages. Therefore for the present molybdenum and its compounds have not been designated with an “S”.

8 References

Abdouh M, Krzystyniak K, Flipo D, Therien HM, Fournier M (1995) Cytometric profile of molybdenum-induced contact sensitization versus a strong allergen reaction to oxazolone in murine auricular lymph node (ALN) test. *Int J Immunopharmacol 17*: 545–554

ACGIH (American Conference of Governmental Indutrial Hygienists) (2000) *Molybdenum and compounds*. ACGIH, Cincinnati, OH, USA; draft

Adachi A, Horikawa T, Takashima T, Komura T, Komura A, Tani M, Ichihashi M (1997) Potential efficacy of low metal diets and dental metal elimination in the management of atopic dermatitis: an open clinical study. *J Dermatol 24*: 12–19

Babaian EA, Bagramian SB, Pogosian AS (1980) Effect of some chemical hazards involved in molybdenum production on the chromosome apparatus of experimental animals and of man (Russian). *Gig Tr Prof Zabol 9*: 33–36

BIA (Berufsgenossenschaftliches Institut für Arbeitssicherheit des Hauptverbandes der gewerblichen Berufsgenossenschaften e.V.) (Ed.) (1989) *Messung von Gefahrstoffen—BIA-Arbeitsmappe—Expositionsermittlung bei chemischen und biologischen Einwirkungen* (Analysis of hazardous substances) (German). Erich Schmidt Verlag, Bielefeld

Boman A, Wahlberg JE, Hagelthorn G (1979) Sensitizing potential of beryllium, copper and molybdenum compounds studied by the guinea pig maximization method. *Contact Dermatitis 5*: 332–333

Bompart G, Pécher C, Prévot D, Girolami JP (1990) Mild renal failure induced by subchronic exposure to molybdenum: urinary kallikrein excretion as a marker of distal tubular effect. *Toxicol Lett 52*: 293–300

Davis KG (1991) Molybdenum. in: Merian E (Ed.) *Metals and their compounds in the environment. Occurrence, analysis, and biological relevance.* VCH Verlagsgesellschaft, Weinheim, 1089–1100

Droste JHJ, Weyler JJ, Van Meerbeeck JP, Vermeire PA, van Sprundel MP (1999) Occupational risk factors of lung cancer: a hospital based case-control study. *Occup Environ Med 56:* 322–327

Dueva LA, Stepanian SS (1989) Clinico-immunologic characteristics and prevention of occupational allergic dermatoses due to molybdenum exposure (Russian). *Vestn Dermatol Venerol (10)*: 47–50

Elvers B, Hawkins S, Schulz G (Eds) (1999) Molybdenum and molybdenum compounds. in: *Ullmann's Encyclopedia of Industrial Chemistry, Vol 16, fifth, completely revised edition,* 664–669

Elves MW, Wilson JN, Scales JT, Kemp HBS (1975) Incidence of metal sensitivity in patients with total joint replacements. *Brit Med J 4*: 376–378

EPA (Environmental Protection Agency) (1998) IRIS (Integrated Risk Information System) *Molybdenum; CAS No. 7439-98-7*

Falbe J, Regitz M (Eds) (1992) *Römpp Chemie Lexikon, 10. Auflage*, Georg Thieme, Stuttgart

Federmann M, Morell B, Graetz G, Wyss M, Elsner P, von Thiessen R, Wüthrich B, Grob D (1994) Hypersensitivity to molybdenum as a possible trigger of ANA-negative systemic *lupus erythematosus. Ann Rheumatic Dis 53*: 403–405

Hattori M, Yokota A, Kako M, Katoh K, Abe Y, Kawaguchi T, Gotoh H, Wakita K, Suzuki T (1996) Epidemiological survey of metal allergy. *Aichi-Gakuin Dent Sci 9*: 15–21

Hierholzer S, Hierholzer G (1987) Klinische Relevanz von Metallallergien in der Unfallheilkunde (Clinical relevance of metal allergies in traumatology) (German). *Hefte Unfallheilkunde 189*: 871–879

HSDB (Hazardous Substances Data Bank) (1999) *Molybdenum*. Databank HSDB, National Library of Medicine, Bethesda, USA

IVDK (Informationsverbund Dermatologischer Kliniken) (2000) *Auswertung der zwischen 1992 und 1999 vom IVDK erfassten Daten* (Evaluation of data recorded by the IVDK between 1992 and 1999) (German). 20.03.2000, IVDK, Göttingen

Kerckaert GA, LeBoeuf RA, Isfort RJ (1996) Use of the Syrian hamster embryo cell transformation assay for determining the carcinogenic potential of heavy metal compounds. *Fundam Appl Toxicol 34*: 67–72

Kresbach H, Grabner K (1969) Untersuchungen zur Ätiopathogenese der Kontaktekzeme. 3. Mitteilung: Zur Technik und zur Brauchbarkeit der Intrakutantestung bei Fällen mit Chromat-, Nickel- und/oder Kobaltallergien (The aetiopathogenesis of contact dermatitis. III. Procedures for and usefulness of intracutaneous tests in cases of chromate, nickel and/or cobalt allergies) (German). *Berufs-Dermatosen 17*: 66–82

Lener L, Bibr B (1984) Effects of molybdenum on the organism (a review). *J Hyg Epidemiol Microbiol Immunol 28*: 405–419

Maltoni C (1976) Predictive value of carcinogenesis bioassays. *Ann NY Acad Sci 271*: 431–443

Mogilevskaya OY (1963) Experimental studies on the effect on the organism of rare, dispersed and other metals and their compounds used in industry. in: Izrael'son ZI (Ed.) *Toxicology of the rare metals.* U.S. Atomic Energy Commission and National Science Foundation, Washington DC, 12–28

NTP (National Toxicology Program) (1997) *Toxicology and carcinogenesis studies of molybdenum trioxide in F344/N rats and B6C3F$_1$ mice (inhalation studies). Technical Report Series No. 462*, NIH Publication No. 97-3378

Ogawa HI, Shibahara T, Iwata H, Okada T, Tsuruta S, Kakimoto K, Sakata K, Kato Y, Ryo H, Itoh T, Fujikawa K (1994) Genotoxic activities *in vivo* of cobaltous chloride and other metal chlorides as assayed in the *Drosophila* wing spot test. *Mut Res 320*: 133–140

Oleffe J, Wilmet J (1980) Generalized dermatitis from an osteosynthesis screw. *Contact Dermatitis 6*: 365

Rammelsberg P (1985) *Metallallergien. Epikutantestergebnisse von 1981–1984* (Metal allergies. Patch test results from 1981–1984) (German). Doctoral thesis, Medizinische Fakultät der Julius-Maximilians-Universität, Würzburg

Rossmann TG, Molina M (1986) The genetic toxicology of metal compounds: II. Enhancement of ultraviolet light-induced mutagenesis in *Escherichia coli* WP2. *Environ Mutagen 8*: 263–271

Schroeder HA, Mitchener M (1971) Toxic effects of trace elements on the reproduction of mice and rats. *Arch Environ Health 23*: 102–106

Sokolowski F (1972) Beurteilung der Schweißarbeiten auf die Entstehung von Dermatosen in der Schiffbauindustrie (lecture) (Assessment of the effects of welding on the development of dermatosis in ship building workers) (German). *Berufs-Dermatosen 20*: 44–45

Sokolowski F (1976) Wirkung mineralischer Kühllösungen auf die Haut (lecture) (Effects on the skin of salt solutions used for cooling) (German). *Berufs-Dermatosen 24*: 141

Stoner GD, Shimkin MB, Troxell MC, Thompson TL, Terry LS (1976) Test for carcinogenicity of metallic compounds by the pulmonary tumor response in strain A mice. *Cancer Res 36*: 1744–1747

Stokinger HE (1981) Molybdenum. in: Clayton GD, Clayton FE (Eds) *Patty's Industrial Hygiene and Toxicology, Vol 2A,* John Wiley & Sons, New York, 1807–1819

Török L, Greczy I, Ocsai H, Czako J (1995) Investigation into the development of allergy to metal in recipients of implanted hip protheses: a prospective study. *Eur J Dermatol 5*: 294–295

Van Loon LAJ, van Elsas PW, van Joost T, Davidson CL (1986) Test battery for metal allergy in dentistry. *Contact Dermatitis 14*: 158–161

Walravens PA, Moure-Eraso R, Solomons CC, Chappell WR, Bentley G (1979) Biochemical abnormalities in workers exposed to molybdenum dust. *Arch Environ Health 34*: 302–307

Zeiger E, Anderson B, Haworth S, Lawlor T, Mortelmans K (1992) *Salmonella* mutagenicity tests: V. Results from the testing of 311 chemicals. *Environ Mol Mutagen 19, Suppl 21*: 2–141

completed 14.01.2000

Wood dust
(except beech and oak wood dust)

MAK value	**–**
Peak limitation	**–**
Absorption through the skin	**–**
Sensitization	**–**
Carcinogenicity (1983)	**Category 3**
Prenatal toxicity	**–**
Germ cell mutagenicity	**–**

Beech wood dust
Oak wood dust

MAK value	**–**
Peak limitation	**–**
Absorption through the skin	**–**
Sensitization	**–**
Carcinogenicity (1985)	**Category 1**
Prenatal toxicity	**–**
Germ cell mutagenicity	**–**

The toxicology of wood dust was reviewed by the German Commission for the Investigation of Health Hazards of Chemical Compounds in the Work Area in 1982 and 1985 (see *Occupational Toxicants* Volume 4). The present document is a supplement to the earlier documentation.

New data made it necessary to re-examine the 1985 classification of oak and beech wood dust in Section IIIA1 of the *List of MAK and BAT Values* (corresponding to Carcinogen category 1). The present documentation is based mainly on a published review of the toxicological data (IARC 1995), supplemented with the results of more recent studies.

1 Mechanism of Action

The mechanism of carcinogenesis of the tumours induced by wood dust in the nose and nasal sinuses has still not been clarified satisfactorily in spite of intensive research. The hypothesis that the cancer is induced by physical-mechanical processes has not been confirmed by the results of animal studies (Pott *et al.* 1989) or dust analysis (Schmitt *et al.* 1996). There is no doubt that extracts prepared from hardwoods contain substances which have genotoxic effects in various target cells *in vitro* and *in vivo* (IARC 1995). These results have been supplemented by the demonstration of chromosome and chromatid breaks in cultured human lung cells treated with hardwood extracts (Zhou *et al.* 1995). To date it is not known whether the test systems used, including the only animal study which has yielded positive results (application of extracts of beech wood to mouse skin induced mainly skin tumours; Mohtashamipur *et al.* 1989), are representative for the principle that is active in wood dust carcinogenesis in man. It is relevant in this context that genotoxic substances have been isolated to date only from so-called hardwoods and not from softwoods (IARC 1995). The structures of the compounds involved have not yet been elucidated.

In persons exposed to hardwood dust, allergic and non-allergic rhinitis developed (Ahman *et al.* 1995, De Zotti and Gubian 1996) and, in addition to other adverse effects on the olfactory organ, mucociliary clearance was inhibited (Ahman *et al.* 1996, IARC 1995). Both effects could be involved in the pathogenesis of the tumours. Inflammatory stimuli cause increased cell proliferation, and impaired mucociliary clearance prolongs the retention of incorporated dust particles in the tissue. Mucosal alterations during chronic exposure are reported consistently. They include dysplasia and metaplasia of the columnar epithelium and less consistently also of the squamous epithelium (IARC 1995, Pisaniello *et al.* 1995).

2 Effects in Man

2.1 Allergenic effects

Occupational exposure to wood has been shown not only to cause direct acute and chronic damage in the respiratory tract (ACGIH 1992, Ahmann *et al.* 1995, De Zotti and Gubian 1996) but also to have sensitizing effects on the skin and airways (1983 MAK documentation for Wood dust in *Occupational Toxicants* Volume 4, and also ACGIH 1992, Ahmann *et al.* 1995, De Zotti and Gubian 1996) and this makes it necessary to assess woods individually. The reader is referred to the separate documentation on the sensitizing effects of individual woods (in this volume and in Volumes 13 and 14 of the present series).

2.2 Genotoxicity

Analysis of the genetic changes in adenocarcinomas of intestinal type in the nose and nasal sinuses revealed no mutations in the K-ras gene and mutations in the p53 gene in only 18 % of tumours (Wu *et al.* 1996). In contrast, in histologically poorly demarcated colorectal adenocarcinomas, mutations in the p53 suppressor gene are detected in about 75 % of cases and mutations in the K-ras gene in 50 % of cases. Apparently the histologically poorly demarcated colorectal adenocarcinomas differ from the tumours in the inner nose and nasal sinuses in their molecular biology although they are histologically similar.

The incidence of micronuclei in the peripheral lymphocytes of 298 persons employed in a match factory was increased after exposure to poplar and linden wood. However, the validity of this study is limited because the micronucleus incidence was not correlated with the exposure level, the age of the persons was not taken into account as a confounder, and the data were not given as numbers of micronuclei but only as the percentage of workers for whom the results were positive and negative (Jiang *et al.* 1994).

2.3 Carcinogenicity

2.3.1 Tumours of the inner nose and the nasal sinuses

The results of the numerous case-control studies and cohort studies which were published before the end of 1994 confirmed a causal relationship between occupational exposure to so-called hardwood dusts (beech, oak) and the development of tumours in the inner nose and the nasal sinuses (IARC 1995). Evaluation of the published results demonstrated that the cancer risk is correlated with the exposure level. The conclusion that the above-mentioned cancers were induced by wood dust itself and not by concomitant exposures is based on the considerations discussed below. Studies carried out in various countries during various periods of time and involving various occupations have not revealed any other substances specifically associated with a cancer risk which is even approximately as great as that associated with wood dust (IARC 1995).

More recently, the association has been confirmed again in an analysis of data from the cancer registry in Belgium (Van den Oever 1996).

In 1994 it was not possible to decide whether also squamous cell carcinomas of the inner nose and the nasal sinuses can be caused by exposure to wood dust (IARC 1995). However, in the meta-analysis carried out in this context by Demers *et al.* (1995a), only some of the relevant studies were included in the calculations. Almost all the studies which could not be included in the evaluation of the pooled data, because the data were not available, were studies demonstrating an increased risk of developing squamous cell carcinoma. The first demonstration of an increased risk associated with very early first exposure (1945 or earlier) was published later in a case-control study (Leclerc *et al.* 1994). The increase in risk was statistically significant (odds ratio (OR) 2.5 for softwood and 2.2 for hardwood). In another evaluation of published data, Demers *et al.* (1995b) did not distinguish between tumours with differing histology. Supported by the tendency to increased risk (but not a statistically significant increase) revealed by the calculations of Demers *et al.* (1995a), the findings of Leclerc *et al.* (1994) provide good evidence that softwood dust has carcinogenic effects and that squamous cell carcinomas may be induced. In a reanalysis of the pooled data from 5 cohort studies, Demers *et al.* (1995a) found a significant excess of malignant tumours of the nasopharynx, and involvement of the squamous epithelium. The hypothesis that exposure to wood dust can induce squamous cell carcinomas is supported by these data. The hypothesis is also supported by the fact that other organic dusts are known to be causally involved in the development of nasal adenocarcinomas and of nasal squamous cell carcinomas. In particular, leather dust (IARC 1987) and textile dusts (IARC 1990) are specified here. But publications with such data for cereal dusts and peat dust are also available (Maier 1993, Van den Oever 1996). The proposal that only tumours of a single histological type can be caused by exposure to wood dust is the same as the suggestion that the carcinogenic effect has a particularly high target cell specificity. Such specificity is very rare in cancer epidemiology. The observed dysplasia and metaplasia of the squamous epithelium of the nasal mucosa of exposed persons is evidence that wood dust does not have such a high target cell specificity.

2.3.2 Tumours in other organs

Especially because of contradictory and inconsistent findings, it has been concluded up till now that there is no association between exposure to wood dust and the development of cancer at sites apart from the inner nose and the nasal sinuses (IARC 1995). This conclusion refers to publications describing cancer of the oropharynx, the hypopharynx, the lungs, the lymphatic and haematopoietic systems, the stomach, the colon and the rectum. However, evidence for an increased risk of developing cancer of the nasopharynx and multiple myeloma was obtained in the reanalysis of the data from 5 cohort studies by Demers *et al.* (1995a). Evidence for an increased lung cancer risk was found by both Spitz *et al.* (1995) and Wu *et al.* (1995), but only for Americans of African origins. In a case-control study, Pollan and Lopez-Abente (1995) found a high degree of association between larynx cancer and occupation in the Spanish wood industry. The cancer risk increased with the duration of exposure and decreased with the number of years since leaving the job. For persons employed longer than 20 years in the wood industry, the risk was significantly increased (OR = 5.63). This value was estimated after adjustment for consumption of tobacco and alcohol. In an analysis of 15 published case-control studies and 5 additional cohort studies, Deitmer (1995) considered that the relationship had not been shown to be causal.

2.3.3 The effects of substances used for wood treatment and co-exposures

According to the results of Wolf *et al.* (1988, 1994), the cancer risk is increased when certain substances used for wood treatment are also present during the work process. The authors point out that the average latency periods are shorter for persons also exposed to paints, paint solvents, dusts from chipboard and plywood, wood preservatives and leather dust. Woods treated with preservatives containing chromium compounds contain not only chromium(III) but also chromium(VI) compounds. The latter are not completely reduced to chromium(III) compounds until several weeks after application. The use of these substances was common especially between 1910 and 1970. In artificial nasal secretion, chromium(VI) is stable for several weeks. The natural level of chromium(VI) in oak and beech woods was shown by Bachg (1990) to be about 0.02 µg/g and higher than the level of chromium(III). Softwoods contain up to 5 times higher levels of chromium(VI). However, the hypothesis that substances used in the treatment of woods are the real cause of the carcinogenesis induced by wood dust must be dismissed (see 1983 MAK documentation and IARC 1995). In addition to the counter-arguments already mentioned, in April 1985 Acheson sent the Commission a communication which states that neither wood preservatives containing arsenic nor formaldehyde have been used in the English furniture industry in Buckinghamshire.

2.3.4 A potential threshold for the carcinogenic effects of wood dust

A study by Blot *et al.* (1997) investigated why adenocarcinomas of the inner nose are seen less often in persons exposed to wood dust in the United States of America than in other continents. The authors suggest that differences in the levels of exposure to wood dust could be the explanation and propose that the cancer risk associated with wood dust could be reduced if a time-weighted 8-hour average concentration of 5 mg/m^3 were to be observed. Although this is speculation, it must be emphasized that little is known about the dose-dependence of the tumour incidence in persons exposed to wood dust. Because it is currently not possible to generate such dose-response curves, and because the real mechanism of action is not known, the possibility that there is a threshold for the tumour induction cannot be excluded.

3 Manifesto (MAK value/classification)

The new data reviewed above confirm that beech and oak wood dusts are properly classified in Carcinogen category 1 (Section III of the *List of MAK and BAT Values*, formerly IIIA1). The analyses and results of recent years suggest that exposure to wood dust can induce not only adenocarcinomas of intestinal type but probably also sinonasal squamous cell carcinomas. The available data also suggest that the dusts of other hardwoods could also be potential human carcinogens, but this has not yet been demonstrated definitively in epidemiological studies. It is also probable, but slightly less probable, that dusts of so-called softwoods are carcinogenic. Therefore the classification of these softwood dusts and also of the dusts of hardwoods apart from beech and oak in Carcinogen category 3 (Section III of the *List of MAK and BAT Values*, formerly IIIB) has been retained. The data reviewed above also suggest that exposure to wood dusts can probably cause other cancers apart from adenocarcinomas of the inner nose and the nasal sinuses; however, for the various tumours which have been observed in epidemiological studies, the quality of the evidence varies. The best evidence to date is available for tumours of the nasopharynx and larynx and for multiple myeloma.

4 References

ACGIH (American Conference of Governmental Industrial Hygienists) (1992) Wood Dust. in: *Documentation of the Threshold Limit Values and Biological Exposure Indices*, 6th issue, Cincinnati, USA

Ahman M, Holmstroem M, Cynkier I, Soederman E (1996) Work related impairment of nasal function in Swedish woodwork teachers. *Occup Environ Med 53*: 112–117

Ahman M, Holmstroem M, Ingelman-Sundberg H (1995) Inflammatory markers in nasal lavage fluid from industrial arts teachers. *Am J Ind Med 28*: 541–550

Bachg D (1990) *Quantitative Differentialanalyse von Cr(III) und Cr(VI) in Holzstäuben als potentielle Cancerogene in der Arbeitswelt.* Doctoral thesis, Essen

Blot WJ, Chow W-H, McLaughlin J K (1997) Wood dust and nasal cancer risk. *J Occup Environ Med 39*: 148–156

Deitmer T (1995) Kehlkopfkarzinom durch Holzstaubexposition? (Laryngeal carcinoma caused by exposure to wood dust?) (German). *Laryngol- Rhinol- Otol 74*: 365–370

Demers PA, Kogevinas M, Boffetta P, Leclerc A, Luce D, Gerin M, Battista G, Belli S, Bolm-Audorff U, Brinton LA (1995a) Wood dust and sino-nasal cancer: pooled reanalysis of twelve case-control studies. *Am J Ind Med 28*: 151–66

Demers PA, Boffetta P, Kogevinas M, Blair A, Miller BA, Robinson CF, Roscoe RJ, Winter PD, Colin D, Matos E (1995b) Pooled reanalysis of cancer mortality among five cohorts of workers in wood-related industries. *Scand J Work Environ Health 21*: 179–190

De Zotti R, Gubian F (1996) Asthma and rhinitis in wooding workers. *Allergy Asthma Proceedings 17*: 199–203

IARC (International Agency for Research on Cancer) (1987) *IARC Monographs on the evaluation of carcinogenic risks to humans, Volumes 1 to 42, Overall evaluations of carcinogenicity; an updating of IARC Monographs*, 236–237

IARC (International Agency for Research on Cancer) (1990) *IARC Monographs on the evaluation of carcinogenic risks to humans, Volume 48, Some flame retardants and textile chemicals, and exposures in the textile manufacturing industry*, Lyon, 215–280

IARC (International Agency for Research on Cancer) (1995) IARC (International Agency for Research on Cancer) (1995) Wood dust. *IARC Monographs on the evaluation of the carcinogenic risk of chemicals to humans 62*: 32–215

Jiang ZC, Su YL, Zhang J, Deng YF, Ma ZH, Dong QL (1994) Study on micronucleus frequency in peripheral lymphocytes in workers of match factories. *Biomed Environ Sci 7*: 150–153

Leclerc A, Cortes MM, Gerin M, Luce D, Brugers J (1994) Sinonasal cancer and wood dust exposure: Results from a case-control study. *Am J Epidem 140*: 340–349

Maier A (1993) *Mitteilungen über Nasentumore, verursacht durch organische Stäube bei Berufen ohne Holzstaubexposition* (Nasal tumours caused by organic dusts in occupations not involving exposure to wood dust) (German). Doctoral thesis, Essen

Mohtashamipur E, Norpoth K, Ernst H, Mohr U (1989) The mouse-skin carcinogenicity of a mutagenic fraction from beech wood dusts. *Carcinogenesis 10*: 483–487

Pisaniello DL, Gun RT, Tkaczuk MN, Schultz MR, Stevens MW (1995) Nasal cytology in Australian furniture woodworkers. *Aust J Otolaryngology 2*: 137–141

Pollan M, Lopez-Abente G (1995) Wood-related occupations and laryngeal cancer. *Cancer Detect Prev 19*: 250–257

Pott F, Roller M, Ziem U, Reiffer F-J, Bellmann B, Rosenbruch M, Huth F (1989) Carcinogenicity studies on natural and man-made fibres with the intraperitoneal test in rats. in: Bignon J, Peto J, Saracci R (Eds) *Non-occupational exposure to mineral fibres, IARC Scientific Publications No. 90*, IARC, Lyon 173–179

Schmitt U, Peek RD, Rapp AO (1996) Malignant nasal tumors in employees of the wood industry: SEM of wood dust particles. *Microsc Anal (September)*: 15–17

Spitz MR, Wu X, Hsu TC (1995) Mutagen sensitivity and wood dust exposure in lung cancer: a case control analysis (Meeting abstract). *Proc Ann Meet Am Assoc Cancer Res*: 36

Van den Oever R (1996) Occupational exposure to dust and sinonasal cancer. An analysis of 386 cases reported to the N.C.C.S.F. Cancer Registry. *Acta Otol Rhinol Laryngol Belg 50*: 19–24

Wolf J, Hartung M, Schroeder H-G, Kleinsasser O, Compes PC, Valentin H (1988) Bösartige Tumoren der Nase im Bereich der Holzwirtschaft. Empirisch-kasuistische Studie zur Belastungssituation (Malignant tumours of the nose in the wood industry. Empirical casuistic study of the exposure situation) (German). *Arbeitsmed Sozialmed Praeventivmed, Sonderheft 10*, Gentner, Stuttgart

Wolf J, Schmezer F, Kuchenmeister F, Klein RG, Schröder HG, Pool-Zobel BL, Ziegler H, Detering B, Fengel G, Stehlin J, Kleinsasser O (1994) Zur Ätiologie von malignen Nasentumoren bei Beschäftigten aus der Holzwirtschaft (On the etiology of malignant nasal tumours in persons employed in the wood industry) (German). *Arbeitsmed Sozialmed Umweltmed, Sonderheft 21*: 3–7

Wu X, Delclos GL, Annegers JF, Bondy ML, Honn SE, Henry B, Hsu TC, Spitz MR (1995) A case-control study of wood dust exposure, mutagen sensitivity and lung cancer risk. *Cancer Epidem Biomarkers Prevention 4*: 583–588

Wu TT, Barnes L, Bakker A, Swalsky PA, Finkelstein SD (1996) K-Ras-2 and P53 genotyping of intestinal-type adenocarcinoma of the nasal cavity and paranasal sinuses. *Modern Pathology 9*: 199–204

Zhou ZC, Norpoth KH, Nelson E (1995) Genotoxicity of wood dust in a human embryonic lung cell line. *Arch Toxicol 70*: 57–60

completed 21.10.1997

Woods: *Entandrophragma angolense* C.DC., *Entandrophragma candollei* Harms, *Entandrophragma cylindricum* Sprague, *Entandrophragma utile* Sprague and other *Entandrophragma* spp.

Sensitization		–
Family		Meliaceae
Synonyms	*E. angolense*	edinam gedu nohor tiama
	E. candollei	kosipo heavy mahogany heavy sapele omu
	E. cylindricum	sapele sapeli
	E. utile	utile sipo
Chemical name (CAS) and trivial name of the allergenic component		2,6-dimethoxy-2,5-cyclohexadiene-1,4-dione [1] 2,6-dimethoxy-1,4-benzoquinone
CAS number		530-55-2
Structural formula		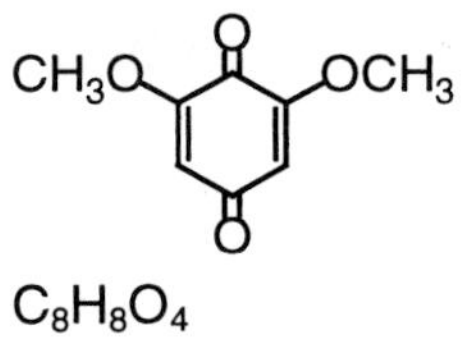
Molecular formula		$C_8H_8O_4$

[1] uncertain whether the substance is allergenic in the wood; to date only detected by thin layer chromatography in extracts of *E. utile*

The *Entandrophragma* species listed above come from central and central west Africa. Their woods are like the mahogany woods from *Swietenia* and *Khaya* species. Tiama is used as solid wood or veneer, for example, in marine plywood for boat building. Kosipo is mostly used for veneers. Sapele is used mainly for surface veneers, for example, for furniture, doors and panelling, and as solid wood, for example, for stairs and in boat building; utile is used as solid wood, for example, for windows, doors and parquet floors, and for many purposes in plywood or as veneer.

1 Allergenic Effects

The botanically variable species *E. angolense* has been split into a number of subspecies; their constituents include gedunin, an epoxidized triterpene lactone like the allergenic anthothecol from *Khaya anthotheca*, and methyl angolensate (Akisanya *et al.* 1960).

Entandrophragmin has been isolated from both *E. cylindricum* and *E. utile*, and from the latter also utilin, another epoxidized terpene lactone (Akisanya *et al.* 1960). *E. cylindricum* also contains several sapelins (e.g. sapelin C) (Chan *et al.* 1971). There are no data available for potential allergenic effects of these substances. In extracts of *E. utile* wood, 2,6-dimethoxy-1,4-benzoquinone has been detected by thin layer chromatography (Hausen 1978).

1.1 Effects in man

One of 12 woods including utile was shown to be responsible for the observed skin and mucosal disorders in 15 of 50 employees (no other details) (Oleffe *et al.* 1973). In spite of the very extensive use of *Entandrophragma* woods, there are only few reports of irritative or allergenic effects of the individual species (Table 1). None of the available study reports can be considered to demonstrate definite allergenic effects of any of the *Entandrophragma* species listed above. There is only one study of allergenic effects of the woods on the airways. It describes positive results in 4 of 157 persons tested by intracutaneous injection of an aqueous extract of sapele. A provocation test with the extract (no other details) yielded positive results for 2 persons (Kersten and von Wahl 1994).

1.2 Results of animal studies

After subcutaneous induction treatment with 0.01 M 2,6-dimethoxy-1,4-benzoquinone in Freund's complete adjuvant (FCA), epicutaneous provocation with a 0.02 M solution did not produce any reactions in 10 Pirbright White guinea pigs (Hausen 1970, Schulz *et al.* 1979). It is claimed in inadequately documented studies that epicutaneous sensitization of guinea pigs was achieved with an extract of *E. utile* (Hausen 1978). In a modified maximization test (intracutaneous induction with 0.005 M 2,6-dimethoxy-1,4-benzoquinone in FCA, epicutaneous induction with 0.2 M), non-occlusive epicutaneous provocation with 0.1 M and 0.01 M 2,6-dimethoxy-1,4-benzoquinone produced reactions in all animals after 48 hours (6× strong positive, 3× positive, 1× weak positive) and in 5/10 animals (weak to clearly positive), respectively (Buske 1974). As positive results were obtained in this maximization test also with substances which have been shown in other studies not to be allergenic, the results described above are not sufficient to document a sensitizing effect of the wood dust of *E. utile* or of the wood constituent 2,6-dimethoxy-1,4-benzoquinone in animals.

Table 1. Case reports of dermatitis caused by contact with *Entandrophragma* species

Patients with reactions	Material applied in the patch test	Observations	References
2	not specified	2 patients with dermatitis caused by utile tested (no other details)	Anonymous 1975
–		1 case report of dermatitis caused by utile listed (no other details)	
3	ethanolic extract (10 %)	positive results with *E. utile* in 3/60 patients with suspected wood allergy	Bleumink and Nater 1974
1	wood dust (5% in petrolatum)	clearly positive results with *E. cylindricum* after 48 h and 96 h	Correia 1992
4	damp wood dust from steamed veneers	1× weak, 3× clearly positive result with reaction to tiama (*E. angolense*) persistent for 72 h ;	Krille 1980
3		3× weak positive result after 72 h with kosipo (*E. candollei*) and	
3		3× weak positive result with sapele (*E. cylindricum*); no reactions with any of the woods in 50 control persons	
–	–	in the anamnesis 1× itching and 1× dermatitis caused by *E. angolense*, and mucosal symptoms caused 2× by *E. utile* and 3× by *E. angolense*	Ngangu and Foussereau 1982
1	damp wood dust	contact dermatitis caused by *E. utile*	Schleicher 1974
2	wood dust	positive results with *E. candollei* after 48 h in 2/137 employees of a veneer factory; anamnesis not suggestive	Schürkämper 1972
2	wood dust	positive results with *E. angolense* in 2/424 patients with a variety of jobs with and without wood contact, 111 of which were in furniture production	Wilkinson *et al.* 1980
1	wood dust	reaction to *E. utile*	Woods and Calnan 1976
2	wood dust	1× weak reaction and 1× reaction to *E. cylindricum*	Woods and Calnan 1976

2 Manifesto (sensitization)

The wood of the *Entandrophragma* species listed above appears to have only very weak sensitization potential. Allergenic constituents of the wood have not yet been demonstrated convincingly. Because an immunological genesis of the skin reactions which have been reported in man has not been supported by the results of sensitization tests with animals, woods of the *Entandrophragma* species are not designated as sensitizing.

3 References

Akisanya A, Bevan CWL, Hirst J, Halsall TG, Taylor DAH (1960) West African timbers. Part III. Petroleum extracts from the genus *Entandrophragma. J Chem Soc*: 3827–3829

Anonymous (1975) Fiches d'allergologie et dermatologie professionnelles: bois exotiques (Occupational allergology and dermatology datasheets: exotic woods) (French). *Dec. 1975, 2. Doc Med Trav*

Bleumink E, Nater JP (1974) Allergic reactions to (tropical) wood. *Contact Dermatitis Newslett 16*: 436–437

Buske H-H (1974) *Untersuchungen über die Sensibilisierungsfähigkeit natürlich vorkommender Chinone* (Sensitization potential of natural quinones) (German). Doctoral thesis, Hautklinik der Universität Hambug

Chan WR, Taylor DR, Yee TH (1971) Triterpenoids from *Entandrophragma cylindricum* Sprague. Part II. The structures of sapelins C, D, E, and F. *J Chem Soc (C)*: 2662–2667

Correia O, Barros MA, Mesquita-Guimaraes J (1992) Airborne contact dermatitis from the woods *Acacia melanoxylon* and *Entandrophragma cylindricum. Contact Dermatitis 27*: 343–344

Hausen BM (1970) *Untersuchungen über gesundheitsschädigende Hölzer* (Woods injurious to health) (German). Doctoral thesis, Fachbereich Biologie der Universität Hamburg

Hausen BM (1978) Sensitizing capacity of naturally occurring quinones. V. 2,6-Dimethoxy-*p*-benzoquinone: Occurrence and significance as a contact allergen. *Contact Dermatitis 4*: 204–213

Kersten W, von Wahl P-G (1994) Allergische Atemwegserkrankungen in der holzverarbeitenden Industrie (Occupational disease in the wood-processing industry) (German). *Allergologie 17*: 55–60

Krille M (1980) *Zur arbeitsdermatologischen Bedeutung exotischer Hölzer unter besonderer Berücksichtigung der Holzindustrie der DDR* (The dermatological significance of exotic woods at the workplace, in particular in the wood industry of the GDR) (German). Doctoral thesis, Zentralinstitut für Arbeitsmedizin der DDR

Mitchell J, Rook A (1979) *Botanical dermatology – Plants and plant products injurious to the skin*, Greengrass, Vancouver, 465

Ngangu Z, Foussereau J (1982) Tropische Hölzer und Kontaktekzem (Tropical woods and contact dermatitis) (German). *Dermatosen Beruf Umwelt 30*: 193–195

Oleffe J, Sporq J, Hublet P, Heylen CH (1973) Troubles cutanés et muqueux provoqués par le travail des bois tropicaux en Belgique (Disorders of skin and mucosa provoked by working with tropical wood in Belgium) (French). *Rev Fr Allergol 13*: 65–68

Schleicher H (1974) Über phytogene allergische Kontaktekzeme (Phytogenic allergic contact dermatitis) (German). *Dermatol Monatsschr 160*: 433–444

Schürkämper H (1972) *Allergien durch tropische Hölzer – Eine Betriebsuntersuchung in einem Werk der Furnierindustrie* (Allergies to tropical woods – a study in a veneer production factory) (German). Doctoral thesis, Fakultät für Medizin der Technischen Universität München

Schulz KH, Garbe I, Hausen BM, Simatupang MH (1979) The sensitizing capacity of naturally occuring quinones. Experimental studies in guinea pigs. II. Benzoquinones. *Arch Dermatol Res 264*: 275–286
Wilkinson DS, Budden MG, Hambly EM (1980) A 10-year review of an industrial dermatitis clinic. *Contact Dermatitis 6*: 11–17
Woods B, Calnan CD (1976) Toxic woods. *Br J Dermatol 95, Suppl 13*: 1–97

completed 05.05.1997

Woods: *Gonystylus bancanus* (Miq.) Baill.

Sensitization	–
Family	Gonystylaceae (formerly Thymelaeaceae)
Synonyms	melawis ramin telur ramin

The almost colourless wood of *Gonystylus bancanus* from south east Asia, especially that from northern Borneo, Malaysia and the Philippines, is used as a wood of moderate strength for indoor construction work, as a veneer, in furniture, and for wooden rods, mouldings, toys and many other purposes, in particular because of its regular structure and smooth surface. Some other *Gonystylus* species yield similar woods which are also sold as ramin.

1 Allergenic Effects

Constituents of the wood with demonstrated allergenic activity have not yet been isolated or characterized.

1.1 Effects in man

The potential allergenic effects of the wood on the skin have been described in only few and not always adequately documented reports (Hausen 1970, Woods and Calnan 1976). A positive result after 48 and 96 hours in a patch test with damp and dry wood dust was obtained for one patient (Beck and Roberts 1981). In another case the positive result obtained after 48 hours in the patch test with damp ramin dust was accompanied by a resurgence of the patient's facial dermatitis. A scratch test with wood dust yielded negative results in this case. Further patch tests carried out on this patient 5 months later with a 10 % ethanolic extract of ramin and with 2,6-dimethoxybenzoquinone (5 % and 10 % in petrolatum) yielded negative results as did the tests with wood dust on 20 control persons (Bruynzeel and de Haan 1987). In a third case it was not possible to exclude the possibility that the wood was contaminated with polymerized epoxy resin (Woods and Calnan 1976). Two other cases of occupational skin disorders caused by ramin have been reported to the Swiss Accident Insurance Association (SUVA) (Hartmann and Schlegel 1980).

Likewise, the effects of the wood dust on the respiratory passages have been described in only few inadequately documented reports. When 220 employees with respiratory symptoms were tested with a wood extract, positive results were obtained for 14 persons, but the reaction was considered to be unspecific in 13 of the cases (Oehling 1963). In a patient with bronchial hyperreactivity (dyspnoea, coughing and sweating after exposure to high levels of ramin dust), but not in 8 asymptomatic employees, precipitating antibodies against components of an aqueous extract of ramin were detected. A prick test with the extract yielded negative results for this patient. In a bronchial provocation test with wood dust, which was repeated several times, the FEV_1 was reduced by up to about 17 % within 4 to 6 hours (Howie *et al.* 1976). Precipitating antibodies were found in 3 of 96 tested persons. However, details of clinical symptoms are not given (Girard *et al.* 1980). Of 4 patients who were allergic to abachi wood (*Triplochiton scleroxylon*) 2 were also exposed to ramin dust. The reactions of the persons exposed to ramin in prick tests with a 1 % and 10 % ramin extract were much less marked than their reactions to 1 % and 0.1 % abachi extract; similar reactions were seen in the persons who were not exposed to ramin. In the bronchial provocation test, the same ramin extracts produced less marked and less prolonged reactions in both groups of patients than did the abachi extracts. Inhalation of 40 mg sodium cromoglycate before the test prevented the pulmonary reaction. Provocation tests with extracts (up to 10 %) yielded negative results for 2 non-exposed asthma patients and 4 exposed symptom-free control persons; prick tests with 10 % extracts yielded negative results for 30 control persons. Specific IgE against ramin was not determined; precipitating antibodies against components of the extracts could not be detected. In a reverse enzyme immunoassay (REIA), a ten times higher concentration of ramin extract inhibited the abachi REIA to about the same extent as did an abachi extract (Hinojosa *et al.* 1986). The authors concluded from this finding and from the results of the provocation and prick tests that cross-reactions between abachi and ramin are conceivable, but further studies are necessary to support or reject this hypothesis.

Another report of a positive result in a bronchial provocation test has been published without further details (Fasani *et al.* 1982). In addition, there is just one report of 4 cases of asthma for which positive results in intracutaneous tests with ramin extracts are listed without further details (Nava 1974).

2 Manifesto (sensitization)

It is not possible to conclude from the inadequate data available that ramin has allergenic effects on the skin. Likewise, the data are not sufficient to classify the pulmonary reactions to the dust from this wood as allergic reactions.

3 References

Beck MH, Roberts MM (1981) A case of ramin wood sensitivity. *Contact Dermatitis 8*: 74–75

Bruynzeel DP, de Haan P (1987) Sensitivity to ramin wood. *Contact Dermatitis 17*: 318–319

Fasani F, Pisati A, La-Manna A (1982) Fattori di rischio allergico nella lavorazione del legno (Factors of allergic risk in wood working) (Italian). *G Ital Med Lav 4*: 211–214

Girard JP, Surber R, Guberan E (1980) Allergic manifestations due to wood dusts. in: CA Frazier (Ed.) *Occupational asthma*, Van Nostrand Reinhold, New York, 91–101

Hartmann A, Schlegel H (1980) Durch Holz verursachte Gesundheitsschäden in der Schweiz (Health hazards caused by wood in Switzerland) (German). *Schweiz Med Wochenschr 110*: 278–281

Hausen BM (1970) *Untersuchungen über gesundheitsschädigende Hölzer* (Woods injurious to health) (German). Doctoral thesis, Fachbereich Biologie der Universität Hamburg

Hinojosa M, Losada E, Moneo I, Dominguez J, Carrillo T, Sanchez-Cano M (1986) Occupational asthma caused by African maple (obeche) and ramin: evidence of cross reactivity between these two woods. *Clin Allergy 16*: 145–153

Howie AD, Boyd G, Moran F (1976) Pulmonary hypersensitivity to ramin (*Gonystylus bancanus*). *Thorax 31*: 585–587

Nava C (1974) Patologia da polveri di legno (Pathology caused by wood dust) (Italian). *Med Lav 65*: 1–7

Oehling A (1963) Berufsallergie im Holzgewerbe (Occupational allergy in the wood industry). (German) *Allerg Asthma 9*: 312–322

Woods B, Calnan CD (1976) Toxic woods. *Br J Dermatol 95, Suppl 13*: 1–97

completed 15.11.1996

Woods: *Khaya* spp.

Family	Meliaceae
Synonyms	African mahogany African acajou *Khaya* mahagony

Khaya anthotheca C. DC.

Sensitization (1997)	**Sh**
Synonyms	white acajou

Khaya grandifoliola C. DC., *Khaya ivorensis* A. Chev., *Khaya senegalensis* A. Juss. and other *Khaya* spp.

Sensitization		–
Synonyms	*Khaya grandifoliola*[1]	big leaf mahogany Zaire mahagony
	Khaya ivorensis	Grand Bassam mahagony Lagos mahogany
	Khaya senegalensis	dry zone mahogany Gambia mahogany Senegal mahogany

[1] In the literature this species is also called *Khaya grandifolia.*

Chemical names (CAS) and trivial names of the allergenic components	I:	(11α,13α,14β,15β,17α)-11-(acetyloxy)-14,15:21,23-diepoxy-6-hydroxy-4,4,8-trimethyl-24-norchola-1,5,20,22-tetraene-3,7-dione [2] anthothecol
	II:	2,6-dimethoxy-2,5-cyclohexadiene-1,4-dione [3] 2,6-dimethoxy-1,4-benzoquinone
CAS number	I:	10410-83-0
	II:	530-55-2

Structural formulae

I:

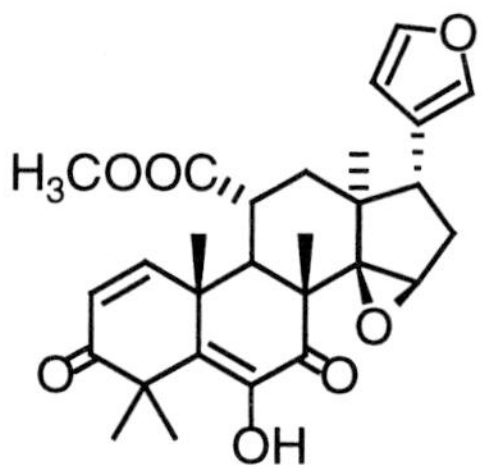

II:

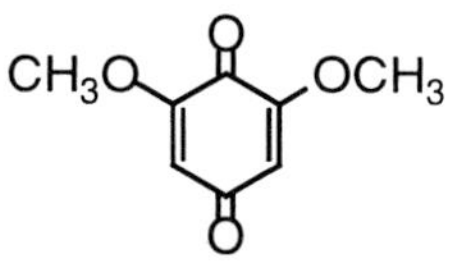

Molecular formulae	I:	$C_{28}H_{32}O_7$
	II:	$C_8H_8O_4$

Khaya species from western or central Africa yield a reddish brown wood with little grain which is weather-resistant and is used solid or as veneer, for example, for musical instruments, furniture, (window-)frames, mouldings and in boat-construction. The individual woods commercially available under the name *Khaya* (*Khaya* mahogany) are difficult to distinguish macroscopically. The particularly valuable wood of *Khaya grandifoliola* is, however, heavier and darker in colour than the wood of *Khaya anthotheca* or *Khaya ivorensis*. The wood of *Khaya* species is similar to the genuine American mahogany (*Swietenia mahagoni*, *Swietenia macrophylla*) and to sapele (*Entandrophragma cylindricum*) and utile (*Entandrophragma utile*).

---

[2] detected to date only in the wood of *Khaya anthotheca*

[3] allergenicity of the substance in wood uncertain; detected to date only by thin layer chromatography in extracts of *Khaya anthotheca* and *Khaya ivorensis*

1 Allergenic Effects

Khaya anthotheca may contain the allergenic substance anthothecol (Bevan *et al.* 1963a, 1963b), which has not yet been detected in other *Khaya* species. It is conceivable that the levels of anthothecol vary in various batches of wood, as do those of desoxylapchol in *Tectona grandis*. 2,6-Dimethoxy-1,4-benzoquinone has been detected by thin layer chromatography in extracts of *Khaya anthotheca* and *Khaya ivorensis* wood (Hausen 1978). The wood is thought to contain other allergenic components as well (Morgan and Wilkinson 1965). These could perhaps include the epoxide 3-deacetylkhivorin and khayanthone. From *Khaya ivorensis* up to 0.1 % khivorin has also been isolated (Bevan *et al.* 1962, Taylor 1977). The substances isolated from or detected by thin-layer chromatography in extracts of *Khaya grandifoliola* include khivorin, grandifolione, grandifoliolenone and 6-hydroxyangolensic acid methyl ester and in *Khaya senegalensis* 7-deacetoxy-7-oxokhivorin (Bevan *et al.* 1963a, 1965, Connolly and McCrindle 1971). The allergenicity of these substances has not been studied.

1.1 Effects in man

The first reports of skin disorders caused by *Khaya* wood appeared in the year 1907 (Hausen 1970) and African mahogany was listed in review articles at a very early date as a wood which could have adverse effects on health (there it was called "*Swietenia senegalensis*") (Czimatis and Hagemann 1910, Dantin-Gallego 1952, Herxheimer 1912). It was not until 1965 that the first results of patch tests with wood dust from *Khaya anthotheca* and its constituent anthothecol were published (Table 1) (Morgan and Wilkinson 1965, Morgan *et al.* 1968). By means of comparative tests with several control persons, the allergenic effects of anthothecol were demonstrated in patients (see Table 1). However, it appears that the only *Khaya* wood with marked sensitization potential is that of *Khaya anthotheca*; only few reactions to the woods of other *Khaya* species have been described and in these woods anthothecol has not been detected (Table 2) (Bleumink and Nater 1974, Wilkinson *et al.* 1980).

Occasional airway reactions have been thought to be caused by the wood of *Khaya anthotheca* but it has not been demonstrated that these are allergic reactions (Table 3). In other reports in which several cases of airway reactions to 'mahogany wood' were described (Table 3), the woods were not classified botanically.

1.2 Results of animal studies

Animal studies have not been carried out with anthothecol or extracts of *Khaya anthotheca* wood. In inadequately documented studies it is claimed that epicutaneous sensitization of guinea pigs was achieved with an extract of *Khaya ivorensis* (Hausen 1978). With another test system, however, the results were not positive: after subcutaneous induction treatment with 0.01 M 2,6-dimethoxy-1,4-benzoquinone in Freund's complete adjuvant (FCA), epicutanous provocation with a 0.02 M solution did not produce any reactions in 10 Pirbright White guinea pigs (Hausen 1970, Schulz *et al.* 1979). In a modified maximization test (intracutaneous induction 0.005 M with FCA, epicutaneous induction 0.2 M), non-occlusive epicutaneous provocation with 0.1 M and 0.01 M solutions of 2,6-dimethoxy-1,4-benzoquinone produced reactions in all animals after 48 hours (6× strong positive, 3× positive, 1× weak positive) and in 5/10 animals (weak to clearly positive), respectively (Buske 1974). As in this maximization test positive results were also obtained with substances which have been shown in other studies not to be allergenic, the results described above are not sufficient to document a sensitizing effect of the wood dust of *Khaya ivorensis* or of the wood constituent 2,6-dimethoxy-1,4-benzoquinone in animals.

Table 1. Case reports of dermatitis caused by contact with *Khaya anthotheca*

Patients with reactions	Material applied in the patch test	Observations	References
2	not specified	2 patients with dermatitis caused by "acajou" (*Khaya anthotheca* or *Khaya ivorensis*)	Anonymous 1975
2	not specified	2 cases with skin disorders caused by *Khaya anthotheca* (no other details)	Hartmannn and Schlegel 1980
5	wood dust	weak positive results for 5/38 patients from the wood-processing industry after 48 h and 72 h in the test with damp wood dust from steamed *Khaya anthotheca* veneers; clinical relevance unclear	Krille 1980
8	wood dust, extract, anthothecol (1 % and 0.1 % in acetone)	marked reactions to *Khaya anthotheca* wood dust in 8/9 patients with skin disorders caused by *Khaya* wood, in 3 cases reaction still marked after 94 h; in 5/6 patients also reaction to wood extract (petroleum ether, ether, acetone), no reaction to extracted wood dust; in 6/6 patients also reaction to anthothecol; in 10 and 4 control persons no reactions to wood dust or to 0.1 % anthothecol	Morgan and Wilkinson 1965
7	wood dust, anthothecol (0.1 % in methyl ethyl ketone)	in 5/55 exposed employees 4× positive, 1× weak positive result with dust from *Khaya anthotheca*, of these persons 1× marked reaction and 1× reaction to 0.1 % anthothecol; clinical symptoms mild, rapid recovery after the end of exposure; in 2/20 other persons positive results with *Khaya anthotheca* and anthothecol with unclear clinical relevance	Morgan *et al.* 1968
20	wood dust	positive results with *Khaya anthotheca* in 20/424 patients with a variety of jobs with and without wood contact, of which 111 were in furniture production	Wilkinson *et al.* 1980

Table 2. Case reports of dermatitis caused by contact with other *Khaya* species

Patients with reactions	Material applied in the patch test	Observations	References
4	ethanolic wood extract	in 4/60 patients positive results with *Khaya ivorensis*; clinical relevance unclear	Bleumink and Nater 1974
5	wood dust	weak positive results in 5/38 patients from the wood-processing industry after 48 h and 72 h in the patch test with damp wood dust from steamed *Khaya ivorensis* veneers; clinical relevance unclear	Krille 1980
1	wood dust	in 1/55 exposed employees positive results with *Khaya ivorensis*; clinical relevance unclear	Morgan *et al.* 1968
5	wood dust	dermatitis in 15/68 employees who had worked with 'mahogany' * for 3–18 months; symptoms much more severe in summer than in winter; in 5/21 tested persons positive results with wood dust	Shevlyakov 1969
1	wood dust	positive results with *Khaya ivorensis* in 1/424 patients with a variety of jobs with and without wood contact, 111 of which were in model-making;	Wilkinson *et al.* 1980
4		4× positive results in tests with other *Khaya* species	

* The author calls the wood "Suretenia mahogany". However, as the wood came from the Congo and Ghana, it is likely to have been a species of *Khaya.*

Table 3. Case reports of the effects of dust from "*Khaya*" or "mahogany" on the airways

Patients with reactions	Test, test material	Observations	References
1	intracutaneous, extract (no other details)	marked reaction to "mogano"	Fasani 1982
1	intracutaneous, provocation, extract (no other details)	patient with increased total IgE; immediate and delayed reactions in the intracutaneous test; provocation with "mahogany" or "acajou" extract resulted in 14–19 % decrease in FEV_1;	Girard *et al.* 1980, Surber *et al.* 1977
11	intracutaneous, extract (no other details)	in 11/96 patients positive results in the intracutaneous test with "mahogany" or "acajou"; clinical relevance unclear	
15	intracutaneous, wood dust	11× weak positive, 4× positive results with "mogano" among the 60 exposed employees who were tested (no other details)	Granati and Nucci 1959
1	not specified	1 case of a registered respiratory disease caused by *Khaya anthotheca*	Hartmann and Schlegel 1980

Table 3. continued

Patients with reactions	Test, test material	Observations	References
1	intracutaneous, provocation, extract (no other details)	positive results in 1/101 tested persons, no reaction in provocation test with "*Khaya*"	Kirsten 1985
3	intracutaneous, aqueous extract	2× asthma, 1× rhinitis caused by "mogano"	Nava 1974
4	intracutaneous, extract (no other details)	4× positive results in intracutaneous tests in persons with suggestive anamnesis; in addition in 11/220 persons positive results in tests with "mahogany" without clinical relevance	Oehling 1963
2	intracutaneous, provocation, extract (no other details)	positive results in 2 patients with current allergy and in 7 without current clinical relevance; 2× positive and 2× negative results of provocation test with *Khaya anthotheca*	Oertmann and Bergmann 1993
1	intracutaneous, wood dust and aqueous extract, provocation, aqueous extract	in the intracutaneous test and after provocation immediate and delayed reactions to "mahogany", after about 1 h and 7 h about 30 % decrease in FEV_1	Sosman 1969
10	not specified	10 cases of a registered respiratory disease caused by "mahogany"	Wolf 1991

2 Manifesto (sensitization)

The wood of *Khaya anthotheca* has weak to moderate sensitizing potential. Although the immunological genesis of the skin reactions seen in man has not yet been supported by the results of sensitization studies with animals, because of the relatively large number of documented cases and the detection of an allergenic constituent in the wood, *Khaya anthotheca* is designated with "Sh".

Adequate data for sensitizing effects of other *Khaya* species are not available. Because the immunological genesis of the reported skin reactions has not been supported by the results of sensitization tests with animals, the other *Khaya* species are not designated as sensitizing.

The studies of the effects of "mahogany" or "*Khaya*" on the airways do indicate allergenic potential, but in most studies the woods tested were not classified botanically and therefore it is not possible to classify *Khaya anthotheca* or any other *Khaya* species as respiratory allergens.

3 References

Anonymous (1975) Fiches d'allergologie et dermatologie professionnelles: bois exotiques (Occupational allergology and dermatology datasheets: exotic woods) (French) *Dec. 1975, 2. Doc Med Trav*

Bevan CWL, Halsall TG, Nwaji MN, Taylor DAH (1962) West African timbers. Part V. The structure of khivorin, a constituent of *Khaya ivorensis. J Chem Soc:* 768–771

Bevan CWL, Powell JW, Taylor DAH (1963a) West African timbers. Part VI. Petroleum extracts from species of the genera *Khaya, Guarea, Carapa,* and *Cedrela. J Chem Soc*: 980–982

Bevan CWL, Rees AH, Taylor DAH (1963b) West African timbers. Part VII. Anthothecol, an extractive from *Khaya anthotheca. J Chem Soc*: 983–989

Bevan CWL, Ekong DEU, Taylor DAH (1965) Extractives from West African members of the family Meliaceae. *Nature 206*: 1323–1325

Bleumink E, Nater JP (1974) Allergic reactions to (tropical) wood. *Contact Dermatitis Newslett 16*: 436–437

Buske H-H (1974) *Untersuchungen über die Sensibilisierungsfähigkeit natürlich vorkommender Chinone* (Studies of the sensitization potential of natural quinones) (German). Doctoral thesis, Hautklinik der Universität Hamburg

Connolly JD, McCrindle R (1971) Tetranortriterpenoids and related substances. Part XIII. The constitution of grandifoliolenone, an apo-tirucallol derivative from *Khaya grandifoliola* (Meliaceae). *J Org Chem (C)*: 1715–1718

Czimatis, Hagemann (1910) Ueber Erkrankungen durch Holzstaub (Diseases caused by wood dust) (German). *Hyg Med 20*: 761–771

Dantin-Gallego J, Armayor AF, Riesco J (1952) Some new toxic woods. Some new manifestations of toxicity. *Ind Med Surg 21*: 41–46

Fasani F, Pisati A, La-Manna A (1982) Fattori di rischio allergico nella lavorazione del legno (Factors of allergic risk in wood working) (Italian). *G Ital Med Lav 4*: 211–214

Girard JP, Surber R, Guberan E (1980) Allergic manifestations due to wood dusts. in: CA Frazier (Ed.) *Occupational asthma*, Van Nostrand Reinhold Company, New York, 91–101

Granati A, Nucci F (1959) Sul meccanismo d'azione lesiva delle polveri di legno (Mechanism of the harmful action of wood dusts) (Italian). *Folia Med 42*: 132–137

Hartmann A, Schlegel H (1980) Durch Holz verursachte Gesundheitsschäden in der Schweiz (Adverse effects on health caused by wood in Switzerland) (German). *Schweiz Med Wochenschr 110*: 278–281

Hausen BM (1970) *Untersuchungen über gesundheitsschädigende Hölzer* (Studies of woods injurious to health) (German). Doctoral thesis, Fachbereich Biologie der Universität Hamburg

Hausen BM (1978) Sensitizing capacity of naturally occurring quinones V. 2.6-Dimethoxy-*p*-benzoquinone: Occurrence and significance as a contact allergen. *Contact Dermatitis 4*: 204–213

Hausen BM (1981) *Woods injurious to human health – A manual*, Walter de Gruyter & Co, Berlin

Herxheimer K (1912) Über die gewerblichen Erkrankungen der Haut (Occupational skin diseases) (German). *Dtsch Med Wochenschr 1*: 18–22

Kirsten D, Liebetrau G, Meister W (1985) Holzstaub als inhalative Noxe (Wood dust as inhalative noxious agent) (German). *Z Erkr Atmungsorgane 165*: 235–241

Krille M (1980) *Zur arbeitsdermatologischen Bedeutung exotischer Hölzer unter besonderer Berücksichtigung der Holzindustrie der DDR* (The dermatological significance of exotic woods at the workplace, in particular in the wood industry of the GDR) (German). Doctoral thesis, Zentralinstitut für Arbeitsmedizin der DDR

Morgan JWW, Orsler RJ, Wilkinson DS (1968) Dermatitis due to wood dusts of *Khaya anthotheca* and *Machaerium scleroxylon. Br J Ind Med 25*: 119–125

Morgan JWW, Wilkinson DS (1965) Sensitization to *Khaya anthotheca. Nature 207*: 1101

Nava C (1974) Patologia da polveri di legno (Pathology caused by wood dust) (Italian). *Med Lav 65*: 1–7

Oehling A (1963) Berufsallergie im Holzgewerbe (Occupational allergy in the wood industry) (German). *Allerg Asthma 9*: 312–322

Oertmann CH, Bergmann KC (1993) Atemwegserkrankungen bei Arbeitern im Holzgewerbe (Airway diseases in wood workers) (German). *Allergologie 16*: 334–340

Schulz KH, Garbe I, Hausen BM, Simatupang MH (1979) The sensitizing capacity of naturally occuring quinones. Experimental studies in guinea pigs. II. Benzoquinones. *Arch Dermatol Res 264*: 275–286

Shevlyakov LV (1969) On dermatitis caused by tropic species of wood (Russian). *Vestn Dermatol Venerol 43*: 73–75

Sosman AJ, Schlueter DP, Fink JN, Barboriak JJ (1969) Hypersensitivity to wood dust. *N Engl J Med 281*: 977–980

Surber R, Guberan M, Girard JP (1977) Allergies respiratoires aux poussières de bois – Cas cliniques et études épidémiologiques (Respiratory allergies due to wood dust – Clinical cases and epidemiologic study) (French). *Rev Fr Allergol 17*: 193–198

Taylor DAH (1977) The structure of an extractive from *Khaya ivorensis*. *Phytochemistry 16*: 1847–1849

Wilkinson DS, Budden MG, Hambly EM (1980) A 10-year review of an industrial dermatitis clinic. *Contact Dermatitis 6*: 11–17

Wolf J (1991) Gesundheitsgefahren vermeiden (Avoiding health hazards) (German). *HGB-Mitteilungen 65*: 3–22

Woods B, Calnan CD (1976) Toxic woods. *Br J Dermatol 95, Suppl 13*: 1–97

completed 05.05.1997

Woods: *Quercus petraea* (Mattuschka) Liebl., *Quercus robur* L., *Quercus rubra* L. and other *Quercus* spp.

MAK value (oak wood dust)		–
Sensitization		–
Carcinogenicity (1985)		**Category 1**
Family		Fagaceae
Synonyms		
white oaks	*Quercus petraea*	Durmast oak *Quercus sessiliflora* Salisb. *Quercus sessilis* Ehr. sessile oak
	Quercus robur	common oak pedunculate oak *Quercus pedunculata* Ehrh.
red oaks	*Quercus rubra*	American red oak northern red oak *Quercus borealis* Michx. red oak

There are more than 250 species of oak [1], of which about 20 are found in Europe, and for which numerous hybrids and varieties are known. The species which are used commercially in Europe[2] are subdivided into two groups, white oaks and red oaks. The white oaks include the European species *Q. petraea* and *Q. robur*, the (north) American species *Q. alba* L. (American white oak), *Q. bicolor* Willd. (swamp white oak), *Q. garryana* Dougl. (Oregon white oak), *Q. macrocarpa* Michx. (bur oak), *Q. muehlenbergii* Engelm. (Chinquapin oak) and the east Asian species *Q. mongolica* Fisch. ex Turcz. (var. *grosseserrata*) (Mongolian oak, Japanese oak). The woods of the American species are

[1] Bog oak is the designation for partially fossilized ancient oak wood which has been preserved underground.

[2] Some woods used as substitutes for oak are also commonly designated as "oak" but are not related botanically to the *Quercus* species, for example, *Eucalyptus delegatensis* and other *Eucalyptus* species (Tasmanian oak) and *Grevillea robusta* (silver or silky oak).

often mixed in commercial wood lots and are difficult to distinguish (Wagenführ 1996). Likewise, the woods of *Q. petraea* and *Q. robur* cannot be distinguished with any certainty from each other nor from the woods of the American and Japanese white oaks (Gottwald 1958). Red oaks which are endemic in Europe and the Near East include *Q. cerris* L. (Turkey oak) and *Q. castaneifolia* C.A. Mey. (chestnut-leaved oak). Red oaks which are endemic to America but cultivated in Europe include *Q. rubra* L., *Q. falcata* Michx. (American red oak, southern red oak) and *Q. palustris* Münchh. (swamp oak, pin oak).

The wood of the various oak species is mainly used for quality veneers and, especially the more weather-resistant white oak, for furniture, panelling and parquet flooring and often as load-bearing construction wood. The holm oak (*Q. ilex* L.), which is endemic to the Mediterranean coastal regions, has only regional significance and the other Mediterranean species, the cork oak (*Q. suber*), is used practically only for the production of cork.

1 Allergenic Effects

1.1 Effects in man

In reviews of allergenic effects of woods on the skin, oaks are hardly ever mentioned, and in the more recent literature there are also only occasional reports of contact dermatitis often not caused by the wood, however, but by the bark and the mosses or lichen growing on it (Campolmi *et al.* 1986). Such cases appeared relatively early in the literature and were described as "woodcutter's eczema" or "woodcutter's disease" (Mitchell and Rook 1979). Two cases of skin disorders after contact with oak wood dust (no other details) are included in an analysis of 165 cases reported to the Swiss Accident Insurance Association (SUVA) (Hartmann and Schlegel 1980). In 2 other publications, a total of 4 cases with positive results in the patch test are listed without details of the relevance of the test results (Bleumink and Nater 1974, Woods and Calnan 1976). The data in these publications are, however, not sufficient to demonstrate an allergenic effect of oak wood.

On the other hand, in several reviews of the effects of wood dust on the airways, oak species are listed as allergenic woods without any other details (Baur 1990, Enarson and Chan-Yeung 1990, Grammer 1993). In addition, there are a number of reports of positive results in prick tests and intracutaneous tests or provocation tests with wood dust or wood extracts. In these publications, however, an exact or reliable botanical classification of the woods used in the tests is practically never given, nor details of the symptoms (Table 1).

More details of the effects of the dust are given in a report in which the symptoms in three Canadian patients were said to have been caused by dust from *Q. robur*. The anamneses of the three patients did not reveal an atopic genesis of the symptoms. Prick tests with an alcoholic extract (no other details) yielded negative results in all cases. In one patient, the asthmatic symptoms returned after 5 days at work following a 10-day break. Measurements carried out at the workplace showed that the peak expiratory flow rate was 350 l/min, markedly reduced below the control value of 640 l/min, and the FEV_1 was reduced by 25 %. After a 2-hour provocation exposure to a dust-aerosol and subsequent exposure for 20 minutes to wood dust, a dual reaction was observed (FEV_1 reduced by 20 % after 20 minutes and by 32 % after 7 hours). Similar provocation of an asthma patient as control resulted in a decrease in FEV_1 of less than 10 %. Control tests with wood dust from 3 coniferous woods and formaldehyde also yielded negative results for the patient. In a second case, two hours exposure to dust also provoked a dual reaction (33 % decrease in FEV_1, after 6 hours 19 %) and on the next day unspecific bronchial hyperreactivity was detected with the histamine test (0.04 mg/ml compared with 0.44 mg/ml). Control provocations with wood dust from fir and abachi did not cause significant decreases in FEV_1. In the third patient, an immediate 30 % decrease in FEV_1 was seen after exposure to wood dust for 3 hours but not for 2 hours; the bronchial hyperreactivity determined after 6 hours was increased (0.31 mg/ml compared with 4.57 mg/ml). The patient did not react significantly to the control provocation with wood dust from fir and meranti (Table 1) (Malo *et al.* 1995).

There are also publications which list individual cases of respiratory disease without more details or with only anamnestic findings (Girard *et al.* 1980, Hartmann and Schlegel 1980, de Zotti *et al.* 1995). In one list 10 cases of a recognized occupational disease are mentioned (Wolf 1991). However, because documentation is inadequate, these data cannot be used in the present assessment.

There are no studies available of the effects of *Quercus* wood on animals.

2 Manifesto (sensitization)

The available data do not indicate that oak wood dust has marked skin sensitization potential. Allergenic constituents of the wood have not yet been demonstrated. Therefore these woods are not designated with “Sh”. There are only a few well-documented reports of cases where oak wood dust has been shown to cause reactions of the airways which are most probably allergic reactions. In the other reports of airway reactions to oak wood dust, an allergic genesis of the reactions cannot be concluded with any certainty from the described clinical findings or the sometimes contradictory results of the allergological tests with preparations of oak wood from inexactly determined oak species. Therefore, and because the number of observed cases is very small when the extensive use of these woods is considered, oak wood is not designated with “Sa”.

Table 1. Case reports of the effects of oak wood dust on the airways

Patients with reactions	Test, test material	Observations	References
4	RAST, extract of the wood dust from the factory (no other details)	RAST yielded positive results for 4/33 persons (1× class 4, 3× class 2) and 4× questionably positive (class 1). In the follow-up examination 3 years later, the RAST result was no longer pathological in 2 of the patients whose values had been increased initially (1× class 4, also class 3 against fungal spores; 1× class 2). No correlation of results and symptoms.	Abendroth 1990, Abendroth *et al.* 1992
1	prick test, extract of oak wood dust (no other details)	positive result in the prick test in 1/90 exposed persons (35 of whom had respiratory symptoms)	Carosso *et al.* 1987
24 10	intracutaneous test; provocation, aqueous extract of oak wood (no other details)	in all 157 persons examined (medical reports 1991 to 1993); controls not mentioned	Kersten and von Wahl 1994
3	intracutaneous test;	115 patients with suspected respiratory disease examined (1979 to 1983)	Kirsten *et al.* 1985
–	provocation, extract of oak wood (no other details)	provocation tests yielded negative results	
–	prick test, alcoholic extract;		Malo *et al.* 1995
3	provocation, wood dust from *Q. robur*	after 2–3 h exposure 2× dual reaction (maximum decrease in FEV_1 20 %/32 % and 33 %/19 %) and 1× immediate reaction persisting for 4–6 hours (maximum decrease in FEV_1 30 %) after 3-hour exposure	
3	intracutaneous test, aqueous extract (no other details)	positive results for 2 patients with asthma and 1 patient with rhinitis among 13 tested exposed persons	Nava 1974
2	intracutaneous test, extract of oak wood (no other details)	220 asthmatics from woodworking factories tested; in 15 persons weak positive test result without suggestive anamnesis	Oehling 1963
1	intracutaneous test	positive results in the intracutaneous test for 1 patient with current symptoms and 8 patients without symptoms	Oertmann and Bergmann 1993
1	provocation, extract of oak wood (no other details)	acute airway obstruction in 1/3 patients	
1	intracutaneous test, provocation, 1 ml extract of oak wood (1:10)	in the acetylcholine test bronchial hyperreactivity; after provocation the pneumometer value was reduced from 4.5 to 2.1 l/sec; controls not mentioned	Scheven *et al.* 1963

Table 1. continued

Patients with reactions	Test, test material	Observations	References
1	provocation, wood dust and 10 % alcoholic extract of oak wood	1 patient with rhinorrhoea, coughing and dyspnoea; symptoms regressed about 1 h after the end of exposure; immediate reaction to provocation with extract (about 50 % decrease in FEV_1)	Sosman *et al.* 1969
1	prick test, provocation, extract of oak wood (no other details)	23/298 questioned workers described nasal symptoms; test with dialysed aqueous extract (exclusion point 10000 Dalton) with 0.5 mg protein/ml yielded positive results for 1/23 persons	Wilhelmsson *et al.* 1984

3 References

Abendroth RR (1990) Neue Wege im Arbeitsschutz (New trends in protection at work) (German). *Staub – Reinhalt Luft 50*: 19–21

Abendroth RR, Kalveram CM, Kalveram KJ (1992) Holzstauballergie: Klinik – Diagnostik – Verlauf. Einsatz eines mobilen Untersuchungssystems (Wood dust allergy: Clinical findings, diagnosis, prognosis, new trends in protection at work) (German). *Allergologie 15*: 300–303

Baur X (1990) Inhalative Allergene und Irritantien am Arbeitsplatz (Inhalative allergens and irritants at the workplace) (German). *Allergologie 13*: 134–139

Bleumink E, Nater JP (1974) Allergic reactions to (tropical) wood. *Contact Dermatitis Newslett 16*: 436–437

Campolmi P, Lotti T, Francalanci S, Sertoli A (1986) Contact sensitivity to oak. *Contact Dermatitis 14*: 314

Carosso A, Ruffino C, Bugiani M (1987) Respiratory diseases in wood workers. *Br J Ind Med 44*: 53–56

Enarson DA, Chan-Yeung M (1990) Characterization of health effects of wood dust exposures. *Am J Ind Med 17*: 33–38

Grammer LC (1993) Occupational asthma. *Immunol Allergy Clinics North Am 13*: 769–783

Girard JP, Surber R, Guberan E (1980) Allergic manifestations due to wood dusts. in: Frazier CA (Ed.) *Occupational asthma*, Van Nostrand Reinhold, New York, 91–101

Gottwald H (Ed.) (1958) *Handelshölzer – Ihre Benennung, Bestimmung und Beschreibung* (Commercial timber – naming, classification and description) (German). Ferdinand Holzmann Verlag, Hamburg

Hartmann A, Schlegel H (1980) Durch Holz verursachte Gesundheitsschäden in der Schweiz (Health hazards caused by wood in Switzerland) (German). *Schweiz Med Wochenschr 110*: 278–281

Kersten W, von Wahl P-G (1994) Allergische Atemwegserkrankungen in der holzverarbeitenden Industrie (Occupational disease in the wood-processing industry) (German). *Allergologie 17*: 55–60

Kirsten D, Liebetrau G, Meister W (1985) Holzstaub als inhalative Noxe (Wood dust as inhalative noxious agent) (German). *Z Erkr Atmungsorgane 165*: 235–241

Malo JL, Cartier A, Desjardins A, Vande Weyer R, Vandenplas O (1995) Occupational asthma caused by oak wood dust. *Chest 108*: 856–858

Mitchell J, Rook A (Eds) (1979) *Botanical dermatology – plants and plant products injurious to the skin*, Greengrass, Vancouver
Nava C (1974) Patologia da polveri di legno (Pathology caused by wood dust) (Italian). *Med Lav 65*: 1–7
Oehling A (1963) Berufsallergie im Holzgewerbe (Occupational asthma in the wood industry) (German). *Allerg Asthma 9*: 312–322
Oertmann C, Bergmann K-C (1993) Atemwegserkrankungen bei Arbeitern im Holzgewerbe (Sensitization of the respiratory tract due to wood dust) (German). *Allergologie 16*: 334–340
Scheven H, Gronemeyer W, Fuchs E (1963) Zum Antigenkatalog gewerblicher Inhalationsallergien (Catalogue of antigens causing occupational inhalative allergies) (German). *Allerg Asthma 9*: 286–290
Sosman AJ, Schlueter DP, Fink JN, Barboriak JJ (1969) Hypersensitivity to wood dust. *N Engl J Med 281*: 977–980
Wagenführ R (Ed.) (1996) *Holzatlas*. (Wood Atlas) (German) 4. Auflage, Carl Hanser Verlag, München
Wilhelmsson B, Jernudd Y, Ripe E, Holmberg K (1984) Nasal hypersensitivity in wood furniture workers. *Allergy 39*: 586–595
Wolf J (1991) Gesundheitsgefahren vermeiden (Avoiding health hazards) (German). *HGB-Mitteilungen 65*: 3–22
Woods B, Calnan CD (1976) Toxic woods. *Br J Dermatol 94, Suppl. 13*: 68
de Zotti R, Gubian F, Peresson M, Molinari S (1995) Asthma and rhinitis in woodworkers. *Allergy 50*: 38

completed 10.07.1997

Woods: *Swietenia macrophylla* King, *Swietenia mahagoni* (L.) Jacq. and other *Swietenia* spp.

Sensitization	–
Family	Meliaceae
Synonyms	American mahogany
Swietenia macrophylla	Amazon mahogany baywood Belize mahogany Brazilian mahogany broadleaf mahogany Honduras mahogany Nicaragua mahogany *Swietenia candollei* Pittier
Swietenia mahagoni	Cuban mahogany Dominican mahogany Jamaica mahogany
Chemical name (CAS) and trivial name of the allergenic component	2,6-dimethoxy-2,5-cyclohexadiene-1,4-dione [1] 2,6-dimethoxy-1,4-benzoquinone
CAS number	530-55-2
Structural formula	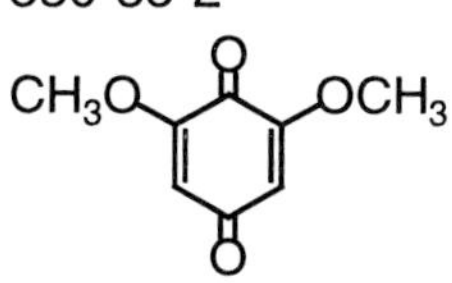
Molecular formula	$C_8H_8O_4$

The very valuable American mahogany of *Swietenia macrophylla* from Central America and central and northern South America is used today much less, just as the wood from *Swietenia mahagoni* has been for many years, because the few remaining natural

[1] uncertain whether the substance is allergenic in the wood; to date only detected by thin layer chromatography in extracts of *S. macrophylla*

resources are increasingly being protected. In practice, the wood is still used only in cabinet-making and furniture production, and as either solid wood or veneer for applications such as boat furnishings, as well as for musical instruments and scientific apparatus. The two *Swietenia* species can form hybrids with each other and with *S. humilis*. As a substitute for *Swietenia* mahogany, woods from *Khaya* species, and sapele and utile (*Entandrophragma cylindricum* and *E. utile*) are used.

1 Allergenic Effects

From the wood of *S. mahagoni* a number of steroid derivatives have been isolated; these are not epoxidized like the allergenic anthothecol from *Khaya anthotheca* but are condensed with cyclopropyl residues: cycloswietenol (Anjaneyulu *et al.* 1978), norcycloswietenol (Anjaneyulu *et al.* 1979) and cycloeucalenol (Amorós-Marín *et al.* 1959). Data for potential allergenic or irritative effects of these substances are not available. 2,6-Dimethoxy-1,4-benzoquinone has been detected in extracts of the wood by thin layer chromatography (Hausen 1978).

1.1 Effects in man

Mahogany is listed in several early reviews as a wood which causes skin damage, but often without differentiation of American and African (*Khaya*) mahogany (Bonnevie 1939, Czimatis and Hagemann 1910, Herxheimer 1912, Schwartz and Russell 1941, Touraine and Gole 1943, Touton 1929, 1932). Probably because use of these woods is becoming less common, in the more recent literature there are relatively few reports of contact dermatitis caused by *Swietenia* species (Table 1) and comparative tests with control persons to establish whether the reactions are of allergic genesis have not been carried out. In most reports in which 'mahogany wood' was claimed to have been responsible for airway reactions in several persons (Tables 2, 3), the woods involved were not classified botanically.

1.2 Results of animal studies

After induction treatment with 0.01 M 2,6-dimethoxy-1,4-benzoquinone in Freund's complete adjuvant (FCA), provocation with a 0.02 M solution did not produce any reactions in 10 Pirbright White guinea pigs (Hausen 1970, Schulz *et al.* 1979). In inadequately documented studies it is claimed that epicutaneous sensitization of guinea pigs was achieved with an extract of *S. macrophylla* (Hausen 1978). In a modified maximization test (intracutaneous induction 0.005 M 2,6-dimethoxy-1,4-benzoquinone with FCA, epicutaneous induction 0.2 M), non-occlusive epicutaneous provocation with 0.1 M and

0.01 M solutions of 2,6-dimethoxy-1,4-benzoquinone produced reactions after 48 hours in all animals (6× strong positive, 3× positive, 1× weak positive) and in 5/10 animals (weak to clearly positive), respectively (Buske 1974). As in this maximization test positive results were also obtained with substances which have been shown in other studies not to be allergenic, the results described above are not sufficient to document a sensitizing effect of the wood dust of *S. macrophylla* or of the wood constituent 2,6-dimethoxy-1,4-benzoquinone in animals.

Table 1. Case reports of dermatitis caused by contact with *Swietenia* species

Patients with reactions	Material applied in the patch test	Observations	References
1	wood dust, damp	after 48 h weak positive result, after 96 h weak to clear positive result with wood dust from Honduras mahogany (probably *S. macrophylla*)	Fowler 1985
–	not specified	5 registered cases of skin disease caused by *S. macrophylla*	Hartmann 1978, Hartmann and Schlegel 1980
1	wood dust	positive result with wood dust from Honduras mahogany (*S. macrophylla*)	Hjorth 1979
1	wood dust, damp; alcoholic extract	positive result in the patch test with wood dust and extract of *S. mahagoni*	Schleicher 1974
3	wood dust	dermatitis, in some cases vesicular, in 8 of about 40 employees of a woodworking factory, 1 positive result in the patch test with "mahogany", positive result with Honduras mahogany and with one other "mahogany" in 1/3 tested persons and 1 positive result with an extract of "mahogany wood"	Steiner and Schwartz 1944
1	wood dust	veneer maker, positive result with wood dust from Honduras mahogany (*S. macrophylla*), negative result with anthothecol (no other details)	Woods and Calnan 1976

Table 2. Case reports of the effects of *Swietenia* wood dust on the airways

Patients with reactions	Test test material	Observations	References
21	intracutaneous extract (no other details)	21 positive results in tests on 157 persons with airway disorders caused by American mahogany; provocation tests seem to have yielded negative results (no other details)	Kersten and von Wahl 1994
1	intracutaneous not specified	3 registered cases of respiratory disease, of these 1 person with positive results in a patch test with *S. macrophylla*	Hartmann 1978, Hartmann and Schlegel 1980

Table 3. Case reports of the effects of "mahogany" wood dust on the airways

Patients with reactions	Test, test material	Observations	References
1	intracutaneous, extract (no other details)	marked reaction to "mogano" (no other details)	Fasani 1982
1	intracutaneous, provocation, extract (no other details);	immediate and delayed reactions in the intracutaneous test; increased total IgE; provocation with "mahogany/acajou extract" resulted in 14–19 % decrease in FEV_1	Girard *et al.* 1980, Surber *et al.* 1978
11	intracutaneous, extract (no other details)	for 11/96 patients positive results in the intracutaneous test with "mahogany/acajou extract"; no details of exposure	
15	intracutaneous, wood dust	11× weak positive result, 4× positive result with "mogano" among the 60 exposed employees who were tested (no other details)	Granati and Nucci 1959
3	intracutaneous, aqueous extract	2× asthma, 1× rhinitis caused by "mogano" (no other details)	Nava 1974
4	intracutaneous, extract (no other details)	4× positive results for persons with suggestive anamnesis; in addition for 11/220 persons positive results in tests with "mahogany" without clinical relevance	Oehling 1963
1	intracutaneous, wood dust and aqueous extract, provocation, aqueous extract	in the intracutaneous test and after provocation immediate and delayed reactions to "mahogany": after about 1 h and 7 h about 30 % decrease in FEV_1	Sosman 1969
–	not specified	10 registered cases of respiratory disease caused by "mahogany"	Wolf 1991

2 Manifesto (sensitization)

The wood of the *Swietenia* species listed above appears to have only very weak sensitization potential. Because of the very small number of cases described in man and because the immunological genesis of the skin reactions reported in man has also not been supported by the results of sensitization tests with animals, the woods of the *Swietenia* species are not designated as sensitizing. Although the studies of the effects of "mahogany" on the airways do indicate allergenic potential, it is not possible to classify the woods of *Swietenia* species as respiratory allergens because in most studies the woods tested were not classified botanically.

3 References

Amorós-Marín L, Torres WI, Asenjo CF (1959) Isolation of cycloeucalenol from West Indian mahogany wood. *J Org Chem 24*: 411–413

Anjaneyulu ASR, Murthy YLN, Row LR (1978) Novel triterpenes from *Swietenia mahagoni* Linn.: Part I – Isolation and structure determination of 9-19 cycloswietenol. *Indian J Chem 16 B*: 650–654

Anjaneyulu ASR, Lakshminarayana V, Murthy YLN, Row LR (1979) Novel triterpenes from *Swietenia mahagoni* Linn.: Part II – Structure of 31-norcycloswietenol. *Indian J Chem 17 B*: 423–426

Bonnevie P (1939) *Aetiologie und Pathogenese der Ekzemkrankheiten* (Aetiology and pathogenesis of eczema) (German). Nordisk Forlag, Kopenhagen/JA Barth, Leipzig

Buske H-H (1974) *Untersuchungen über die Sensibilisierungsfähigkeit natürlich vorkommender Chinone* (Sensitization potential of natural quinones) German. Doctoral thesis, Hautklinik der Universität Hamburg

Czimatis, Hagemann (1910) Ueber Erkrankungen durch Holzstaub (Diseases caused by wood dust) (German). *Hyg Med 20*: 761–771

Fasani F, Pisati A, La-Manna A (1982) Fattori di rischio allergico nella lavorazione del legno (Factors of allergic risk in wood working) (Italian). *G Ital Med Lav 4*: 211–214

Fowler JF (1985) Occupational dermatitis to Honduran Mahogany. *Contact Dermatitis 13*: 336–337

Girard JP, Surber R, Guberan E (1980) Allergic manifestations due to wood dusts. in: CA Frazier (Ed.) *Occupational asthma*, Van Nostrand Reinhold, New York, 91–101

Granati A, Nucci F (1959) Sul meccanismo d'azione lesiva delle polveri di legno (Mechanism of the harmful action of wood dusts) (Italian). *Folia Med. 42*: 132–137

Hartmann AL (1978) *Das Holz als gesundheitliche Noxe – Eine Untersuchung anhand des Krankengutes der SUVA* (Wood as a health hazard – a study of SUVA cases) (German). Doctoral thesis, Medizinische Fakultät der Universität Zürich

Hartmann A, Schlegel H (1980) Durch Holz verursachte Gesundheitsschäden in der Schweiz (Adverse effects on health caused by wood in Switzerland) (German). *Schweiz Med Wochenschr 110*: 278–281

Hausen BM (1970) *Untersuchungen über gesundheitsschädigende Hölzer* (Studies of woods injurious to health) (German). Doctoral thesis, Fachbereich Biologie der Universität Hamburg

Hausen BM (1978) Sensitizing capacity of naturally occurring quinones. V. 2.6-Dimethoxy-*p*-benzoquinone: Occurrence and significance as a contact allergen. *Contact Dermatitis 4*: 204–213

Herxheimer K (1912) Über die gewerblichen Erkrankungen der Haut (On occupational skin disease) (German). *Dtsch Med Wochenschr 1*: 18–22

Hjorth N (1979) Contact dermatitis from sawdust. *Contact Dermatits 5*: 339–340

Kersten W, von Wahl P-G (1994) Allergische Atemwegserkrankungen in der holzverarbeitenden Industrie (Occupational disease in the wood-processing industry) (German). *Allergologie 17*: 55–60

Nava C (1974) Patologia da polveri di legno (Pathology caused by wood dust) (Italian). *Med Lav 65*: 1–7

Oehling A (1963) Berufsallergie im Holzgewerbe (Occupational allergy in the wood industry) (German). *Allerg Asthma 9*: 312–322

Schleicher H (1974) Über phytogene allergische Kontaktekzeme (Phytogenic allergic contact dermatitis) (German). *Dermatol Monatsschr 160*: 433–444

Schulz KH, Garbe I, Hausen BM, Simatupang MH (1979) The sensitizing capacity of naturally occuring quinones. Experimental studies in guinea pigs. II. Benzoquinones. *Arch Dermatol Res 264*: 275–286

Schwartz L, Russell JP (1941) Skin hazards in airplane manufacture. *Public Health Rep 56*: 1581–1593

Sosman AJ, Schlueter DP, Fink JN, Barboriak JJ (1969) Hypersensitivity to wood dust. *N Engl J Med 281*: 977–980

Steiner SD, Schwartz L (1944) Dermatitis from mahogany wood. *Ind Med 13*: 234–235

Surber R, Guberan M, Girard JP (1977) Allergies respiratoires aux poussières de bois – Cas cliniques et études épidémiologiques (Respiratory allergies due to wood dust – Clinical cases and epidemiologic study) (French). *Rev Fr Allergol 17*: 193–198

Touraine, Golé (1934) Les dermatites par bois toxiques (Dermatitis from toxic wood) (French). *Paris Med 91*: 62–68

Touton K (1929) Nutzhölzer und Hautreizung (Commercial timbers and irritation of the skin) (German). *Naturwissenschaften 17*: 371–375

Touton K (1932) Hauterkrankungen durch phanerogamische Pflanzen und ihre Produkte (Toxicodermia et Allergodermia phytogenes) (Skin disorders caused by phanerogamic plants and their products). in: Jadassohn (Ed.) *Handbuch der Haut- und Geschlechtskrankheiten, Band IV/1* (Handbook of skin and venereal diseases) (German), Springer, Berlin, 487–697

Wolf J (1991) Gesundheitsgefahren vermeiden (Avoiding health hazards) (German). *HGB-Mitteilungen 65*: 3–22

Woods B, Calnan CD (1976) Toxic woods. *Br J Dermatol 95, Suppl 13*: 1–97

completed 05.05.1997

Woods: *Terminalia ivorensis* A. Chev.

Sensitization	–
Family	Combretaceae
Synonyms	black afara framiré idigbo

Terminalia superba Engl. & Diels

Sensitization (1999)	**Sa**
Family	Combretaceae
Synonyms	afara fraké limba ofram white afara

Terminalia superba (limba) comes from the tropical rainforests of western and central Africa and has been used in large quantities in Europe since about 1925. The wood is not very durable; it is used for veneers, for panelling and furniture, as construction wood where only moderate strength is necessary, and it is processed to plywood and sandwiched panelling. Wood from *Terminalia ivorensis* is more durable and weather-resistant. It comes from the same geographical area as *Terminalia superba* where it is also produced in plantations.

Wounds containing splinters of limba wood can become severely inflamed and heal very slowly (Sandermann and Barghoorn 1956).

1 Allergenic Effects

The effects of limba dust on the skin have been described in only few, mostly poorly documented case reports (Bleumink and Nater 1974, Ngangu and Foussereau 1982, Preißler 1959) which do not yield clear evidence for a sensitization potential of the wood. Ten cases of skin disorders (no other details) have been reported to the Swiss Accident Insurance Association (SUVA) (Hartmann and Schlegel 1980). For *Terminalia ivorensis*, which is listed as a toxic wood in some reviews (e.g. Zafiropoulo *et al.* 1968), neither case reports nor positive results from patch tests are available.

Limba wood dust is described in older literature as a cause of the asthma which is common among carpenters (Scheven *et al.* 1963). Statistics include 5 cases of respiratory disease registered by the SUVA (no other details) (Hartmann and Schlegel 1980). The positive results of intracutaneous tests at least summarized in other publications and another case in which a Prausnitz-Küstner test was carried out suggest that limba wood could have sensitizing effects on the airways (Oehling 1963). In 3 cases specific IgE was demonstrated with the enzyme allergosorbent test (EAST; RAST class 2 and 3); this result also suggests that an immunological mechanism was the cause of the respiratory symptoms (Wirtz *et al.* 1997a, 1997b). However, intracutaneous tests in these three patients yielded negative results. The bronchial provocation test carried out in one case cannot be evaluated because limba and abachi wood dusts (*Triplochiton scleroxylon*, see 1997 MAK documentation, this volume) were used simultaneously.

In addition, positive results of 4 bronchial provocation tests with extracts of limba wood have been reported in the literature without further details (Table 1).

There are no studies available of the effects of *Terminalia* wood in animals.

2 Manifesto (sensitization)

Allergenic effects of the wood of *Terminalia ivorensis* or *Terminalia superba* (limba) on the skin have not been documented convincingly. There are also no data available for allergenic effects of the wood of *Terminalia ivorensis* on the respiratory passages. The wood of *Terminalia ivorensis* is therefore not designated as sensitizing. The wood of *Terminalia superba*, on the other hand, seems likely to have allergenic effects on the airways as indicated not only by early reports but also by the results of more recent studies in which specific IgE against the wood of *Terminalia superba* was detected. The wood of *Terminalia superba* is therefore designated with "Sa".

Table 1. Case reports of effects of *Terminalia superba* (limba wood) on the airways

Patients with reactions [1]	Test, test material (concentration)	Observations and comments	References
7 / 2	intracutaneous and provocation, wood extract (no other details)	in 157 cases intracutaneous test: 7× positive, provocation test: 2× positive	Kersten and von Wahl 1994
3 / –	intracutaneous and provocation, wood extract (no other details)	positive results in intracutaneous test in 3/101 patients with suspected respiratory disease; provocation tests with wood extracts in 26/29 patients with reactions to various wood extracts in the intracutaneous test but no positive results with limba wood	Kirsten *et al.* 1985
15 / 1	intracutaneous and provocation, wood extract (no other details)	positive results in intracutaneous tests in 10 cases with suggestive anamnesis; in addition in 5/220 persons reaction to limba wood without current clinical relevance; provocation test yielded positive results in one patient and with the serum from this patient; several positive results were obtained in Prausnitz-Küstner tests	Oehling 1963
9 / 1	intracutaneous and provocation, wood extract (no other details)	9/55 patients produced a reaction in the intracutaneous test without current clinical relevance; 1/3 provocation tests yielded positive results	Oertmann and Bergmann 1993
1	EAST (no other details) and provocation by simultaneous sanding of the wood of abachi (*Triplochiton scleroxylon*) and limba (total dust determined after personal sampling: 315 mg/m^3)	specific IgE: RAST class 2 [1]; a few minutes after provocation runny nose, shortness of breath and nausea, increase in specific airway resistance from 0.85 to 2.15 kPa × s, reduction in peak expiratory flow (PEF), forced expiratory volume in the first second (FEV_1) and in forced vital capacity (FVC) by at least 30 %; in the methacholine test no unspecific hyperreactivity detectable [1]; patient atopic; cross-inhibition experiments with abachi yielded positive results	Wirtz *et al.* 1997a
3 (including case described above)	intracutaneous test and EAST, extract of 2.5 g wood dust with 35 ml sodium carbonate buffer (no other details)	IgE: RAST class 2 × 2, 1 × 3 [1]; all patients atopic; all patients with unspecific bronchial hyperreactivity [2]	Wirtz *et al.* 1997b

[1] RAST values and unspecific bronchial hyperreactivity of a patient inconsistently documented

References

Bleumink E, Nater JP (1974) Allergic reactions to (tropical) wood. *Contact Dermatitis Newslett 16*: 436

Hartmann A, Schlegel H (1980) Durch Holz verursachte Gesundheitsschäden in der Schweiz (Health hazards caused by wood in Switzerland) (German). *Schweiz Med Wochenschr 110*: 278–281

Kersten W, von Wahl PG (1994) Allergische Atemwegserkrankungen in der holzverarbeitenden Industrie (Occupational disease in the wood-processing industry) (German). *Allergologie 17*: 55–60

Kirsten D, Liebetrau G, Meister W (1985) Holzstaub als inhalative Noxe (Wood dust as inhalative noxious agent) (German). *Z Erkr Atmungsorgane 165*: 235–241

Ngangu Z, Foussereau J (1982) Tropische Hölzer und Kontaktekzem (Tropical woods and contact dermatitis) (German). *Dermatosen Beruf Umwelt 30*: 193–195

Oehling A (1963) Berufsallergie im Holzgewerbe (Occupational allergies in the wood industry) (German). *Allerg Asthma 9*: 312–322

Oertmann C, Bergmann KC (1993) Atemwegserkrankungen bei Arbeitern im Holzgewerbe (Airway diseases in woodworkers) (German). *Allergologie 16*: 334–340

Preißler R (1959) *Reihenuntersuchungen in Exotenhölzer verarbeitenden Betrieben infolge Verdacht auf eine schädigende Einwirkung von Gurjun- und Makoré-Staub* (Serial tests in factories working with exotic woods following suspected adverse effects of gurjun and makoré dust) (German). Doctoral thesis, Medizinische Akademie Carl Gustav Carus, Dresden

Sandermann W, Barghoorn AW (1956) Gesundheitsschädliche Hölzer (Woods hazardous for health) (German). *Holz Roh- Werkstoff 14*: 87–94

Scheven H, Gronemeyer W, Fuchs E (1963) Zum Antigenkatalog gewerblicher Inhalationsallergien (Catalogue of antigens causing occupational inhalative allergies) (German). *Allerg Asthma 9*: 286–290

Wirtz C, Korn M, Chen Z, Raulf-Heimsoeth M, von Kampen V, Baur X (1997a) Asthmaanfälle durch Abachi- und Limbaholz (Fallbericht) (Asthma caused by abachi and limba wood (case report)) (German). *Arbeitsmed Sozialmed Umweltmed 32*: 257–260

Wirtz C, Chen Z, Raulf-Heimsoeth M, von Kampen V, Papenfuß F, Baur X (1997b) Atemwegssensibilisierung durch Holzstaub (Sensitization of the respiratory tract due to wood dust) (German). *Zentralbl Arbeitsmed 47*: 336–342

Zafiropoulo A, Audibert A, Charpin J (1968) À propos des accidents dus a la manipulation des bois exotiques (Accidents due to the handling of exotic woods) (French). *Rev Fr Allergol 8*: 155–170

completed 02.03.1999

Woods: *Thuja* species

Family	Cupressaceae

Thuja plicata (D. Don) Donn[1]

MAK value	–
Sensitization (1997)	**Sah**
Synonyms	(giant) arborvitae Beisugi giant cedar Pacific red cedar red cedar shinglewood *Thuja gigantea* Nutt. western red cedar

Thuja occidentalis L.

MAK value	–
Sensitization	–
Synonyms	eastern white cedar northern white cedar white cedar

---

[1] sometimes also called *Thuja plicata* D. Don

Chemical names and common names of the allergenic components

I: [1*S*-(1*α*,2*α*,3*β*)]-1-(3,4-dihydroxy-5-methoxyphenyl)-1,2,3,4-tetrahydro-2,3,7-trihydroxy-3-(hydroxymethyl)-6-methoxy-2-naphthalene carboxylic acid
plicatic acid

II: 2-hydroxy-4-(1-methylethyl)-2,4,6-cycloheptatriene-1-one [2]
4-isopropyltropolone
hinokitiol
β-thujaplicin

III: 2-hydroxy-5-(1-methylethyl)-2,4,6-cycloheptatriene-1-one [2]
5-isopropyltropolone
γ-thujaplicin

CAS number

I: 16462-65-0
II: 499-44-5
III: 672-76-4

Molecular formulae

I: $C_{20}H_{22}O_{10}$
II: $C_{10}H_{12}O_2$
III: $C_{10}H_{12}O_2$

Structural formulae

I:

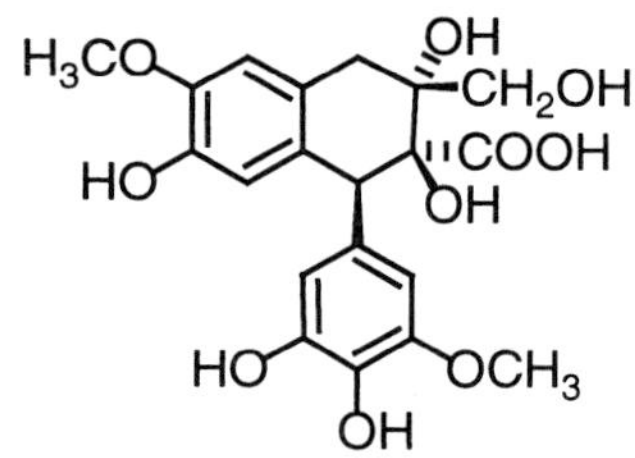

II, III:

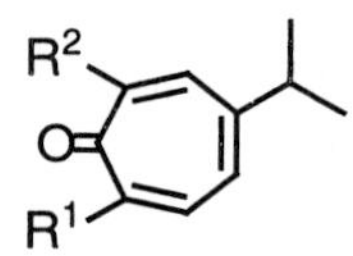

II: R^1 = H, R^2 = OH
III: R^1 = OH, R^2 = H

---

[2] allergenic effects questionable

Thuja plicata, one of the (northwest) American species of cedar which is also cultivated in Great Britain, is not related to the "true cedars" [3]. The wood is resistant to attack by fungi, insects and weather and is used in house building as a non-loadbearing construction material for windows, doors and panelling of walls and ceilings, and as veneer. To date, only insignificant amounts have been imported into Europe and into Germany.

The wood of *Thuja plicata* contains not only thujic acid and methyl thujate but also highly variable amounts of several tropolones (the main terpenes being β-thujaplicin and γ-thujaplicin with a little α-thujaplicin and 7-hydroxy-4-isopropyltropolone) (Gardner *et al.* 1957, Gardner and Barton 1958). Old wood can contain up to 1.8 % tropolone (Nault 1987, 1988). The main allergen in the wood (Gardner *et al.* 1959) is plicatic acid (about 3–4 %). Lower levels of its lactone plicatin and other polyphenols may be found in the wood (MacLean and MacDonald 1967, 1969a, 1969b, MacLean and Murakami 1966).

Thuja occidentalis, also a highly resistant wood, originates mainly from south eastern Canada or north eastern USA; in central Europe its many cultivars are used as ornamental trees. Extracts or distillates of fresh twigs and leaves of *Thuja occidentalis* which contain thujone and thujaplicin are used in homeopathic medicine as immune system stimulants and for treatment of warts. In *Thuja occidentalis* wood, plicatic acid has been identified so far only by thin layer chromatography; immunological methods indicate that the levels are much lower than in wood from *Thuja plicata* (Cartier *et al.* 1986).

1 Allergenic Effects

1.1 Effects in man

Thuja plicata is listed in several reviews as a wood causing contact allergy or skin irritation (Lovell 1993, Orsler 1969). However, the literature contains only few adequately documented reports of allergic contact dermatitis caused by *Thuja plicata* wood (Table 1). Often this disorder, which is also called "cedar poisoning", is apparently caused by allergenic components (usnic acid derivatives, frullanolide) of lichens and liverworts (*Frullania* sp.) which can grow on the bark of the trees (Mitchell and Chan-Yeung 1974). Recent studies of contact allergy caused by the wood of *Thuja occidentalis* are not available. Occasionally, however, allergic reactions to extracts or distillates of fresh twigs and leaves of *Thuja occidentalis* which contain thujone and thujaplicin have been reported (Grimm 1991, Puig *et al.* 1994).

[3] The true cedars (cedar of Lebanon, *Cedrus libani*) and deodar (*Cedrus deodara*) are members of the family Pinaceae, and the Cedrela species (*Cedrela odora*, *C. angustifolia*, *C. fissilis*: the Central and South American cedars, sometimes also called "red cedar", "cedar" or cigar box cedar) are members of the Meliaceae. Other species which are bought and sold as "cedar" such as African cedar (*Juniperus procera*), incense cedar (*Calocedrus decurrens*, Californian cedar, incense cedar, pencil cedar) eastern red cedar (*Juniperus virginiana*) and yellow cedar (*Chamaecyparis nootkatensis*, white cedar), on the other hand, are classified like *Thuja* in the family Cupressaceae.

Table 1. Case reports of the effects of *Thuja plicata* and thujaplicins on the skin

Patients with reactions	Material applied in the patch test	Observations	References
1	wood dust, aqueous and ethanolic extracts of the wood (10 %)	marked reaction to wood dust and extracts after 48 h, 72 h and 96 h; positive results also with 1 % γ-thujaplicin, no reaction with 1 % β-thujaplicin	Bleumink *et al.* 1973
4	wood dust, ethanolic extract of the wood (10 %)	positive results listed for 4/20 patients with suspected wood allergy, also for 3/21 patients with reaction to wood tar	Bleumink and Nater 1974
2	not specified	2 cases with positive results listed (no other details)	Burry *et al.* 1973
1	petroleum ether extract of the wood (1 % in yellow petrolatum)	marked reaction after 48 h, no reactions to 1 % β-thujaplicin or γ-thujaplicin	Calnan 1972
1	β-thujaplicin, (0.1 % in ethanol)	after 48 h marked reaction of one patient who had probably become sensitized after using a hair lotion containing 0.01 % β-thujaplicin; questionable or weak reaction to 5 % and 10 % β-thujaplicin in petrolatum; 10 % β-thujaplicin in petrolatum caused slight erythema in 5 control persons, 0.1 % β-thujaplicin in ethanol produced no reaction in 3 control persons	Fujita and Aoki 1983
–	–	59 cases of dermatitis among 1320 employees handling *Thuja plicata* listed without details of causes (no other data)	Ishizaki *et al.* 1973

The fine dust produced during sawing and working the wood can cause especially asthmatic complaints; forestry workers seem to be affected very rarely because they are exposed to less wood dust (Mitchell and Chan-Yeung 1974). Prevalences between 1.2 % (Ashley *et al.* 1978, Chan-Yeung *et al.* 1978) and 13.5 % (Brooks *et al.* 1981) have been reported in epidemiological studies of persons working with *Thuja plicata* (see Table 3).

The first reports of asthma caused by *Thuja plicata* ("Beisugi asthma") came from Japan where large amounts of red cedar were imported after the earthquake in 1923. In Japan, unlike in cases described later, the symptoms induced by the wood dust mostly developed after the persons had been exposed for many years, sometimes more than 10 years. Further reports from Japan and other regions, for example, Australia (Milne and Gandevia 1969) did not appear until the 1960s (Hausen 1981, Mue *et al.* 1975). The initial symptoms such as irritation of the eyes and nose, rhinitis or nasal obstruction are followed often weeks later by coughing which is most severe after the end of the working day or at night and regresses at weekends and during work-free periods. The diagnosis was often supported by the observation that the symptoms became worse on re-exposure but could not be confirmed with skin tests.

In later studies too, skin tests did not prove useful for diagnostic purposes. Provocation tests with aqueous extracts or plicatic acid, on the other hand, mostly yielded positive results. Immediate reactions in skin tests or after provocation with extracts have been reported only relatively rarely (see Table 2).

An atopic diathesis has been suggested to predispose persons to *Thuja plicata* asthma (Ishizaki *et al.* 1973) but most studies have not yielded evidence to support this hypothesis. There is evidence for a synergistic effect in smokers (Chan-Yeung *et al.* 1978), although in the collectives of patients with *Thuja plicata* asthma there tend to be less (ex)-smokers than in the control collectives. In several follow-up studies it has been shown that in up to one half of cases, variously severe respiratory symptoms occurred even after years without further exposure. Persistent exposure often leads to further adverse effects on health (Chan-Yeung 1977, Chan-Yeung *et al.* 1982, 1987, Cote *et al.* 1990).

In addition, a case of asthma which developed after only a few weeks exposure to *Thuja occidentalis* has been described. After provocation for 4 hours with wood dust from *Thuja occidentalis* or 2 hours with *Thuja plicata*, a steady decrease in the forced expiratory volume in the first second (FEV_1) was observed (39 % with *Thuja occidentalis* after 7 hours and 25 % with *Thuja plicata* after 4 hours). Provocation for 2 hours with dust from *Thuja occidentalis* caused only a 12 % decrease in FEV_1. Provocation for 7 minutes with plicatic acid (5 mg/ml, 1.25 ml) caused a decrease in FEV_1 after 8 to 9 hours delay (38 % after 14 hours). In addition, with the radioallergosorbent test (RAST), IgE specific for a conjugate of plicatic acid and human serum albumin was detected (Cartier *et al.* 1986).

An employee of a factory processing *Thuja occidentalis* reacted to provocation with plicatic acid (5 mg/ml, maximum volume 1 ml) with an immediate 43 % decrease in FEV_1; a second employee reacted to a 2-hour provocation with *Thuja plicata* dust with a delayed 22 % decrease. After provocation, an increased unspecific bronchial reactivity to methacholine was seen in another patient. Provocation with dust from *Thuja occidentalis* was not carried out with these patients. Details of any previous exposure to *Thuja plicata* are not given (Malo *et al.* 1994).

Positive results obtained in a Prausnitz-Küstner test, a skin test and after provocation with a protein-containing fraction of a dialysed aqueous extract of *Thuja plicata* (Mue *et al.* 1975) have not yet been confirmed by other authors. Evidence against the involvement of a high molecular weight antigen in these reactions was provided by the finding that provocation with a denatured extract induced a reaction but provocation with a denatured and dialysed extract did not (Chan-Yeung *et al.* 1973). Removal of the volatile components from the raw extract did not prevent the reaction after provocation in 2 patients (Chan-Yeung *et al.* 1973).

Although the positive results obtained in provocation tests make it seem very likely that plicatic acid plays an important role in the induction of the asthmatic disorders described above, the possibility that other factors are involved in the development of *Thuja plicata* asthma cannot be excluded. The fact that the specific reaction to plicatic acid is seen in only a small proportion of exposed persons, the generally long latency periods, and the fact that even small amounts of plicatic acid can induce the generally delayed or dual asthmatic reactions suggest that an immunological reaction is involved, but the mechanism is not yet understood.

Table 2. Case reports of the effects of *Thuja plicata* on the airways

Patients with reactions	Test, test material (concentration) [control]	Observations	References
2	provocation, aqueous extract (n.s.), wood dust [abachi, beech, pine, lactose, coca solution]	first symptoms after about 9 and 21 months; provocation with fine wood dust, but not coarse wood dust, caused in both patients a marked decrease in PEF and FEV_1 (> –30 %) after 3–5 h delay; one patient also reacted strongly to the extract; one patient atopic, both patients hyperreactive to histamine	Blainey *et al.* 1981
3	prick test, aqueous extract (25 mg/ml); provocation, denatured aqueous extract (2.5 mg/ml, 0.2–1.8 ml)	positive results for 2/3 patients, no reaction in 10 control persons; provocation caused immediate reaction in one patient, > –30 % FEV_1 which normalized after 30 min; in 2 patients after a 3–4 h delay a persistent (24 h and 48 h) reaction (> –60 % FEV_1); no reactions in 2 healthy persons or 2 asthma patients who were not exposed	Chan-Yeung *et al.* 1971
15	provocation, extract (2.5–10 mg/ml, 0.2–1.6 ml) and plicatic acid (2.5 or 5 mg/ml, 0.56–1.5 ml); prick test, extract and plicatic acid (25 mg/ml); intracutaneous, extract and plicatic acid (2.5 mg/ml) [Douglas fir and other woods]	after provocation mostly dual or delayed reactions, > –30 % FEV_1, results with extract and plicatic acid not always congruent; no significant reaction to Douglas fir extract; no reactions in 10 control persons (including 4 with bronchitis and 4 with asthma); in the prick test 3 persons with weak immediate reactions to *Thuja plicata* extract, no reactions in control persons; in all patients and control persons immediate reactions	Chan-Yeung 1973, Chan-Yeung *et al.* 1973
5	provocation; plicatic acid (5 mg/ml)	2 dual reactions, 2 delayed and 1 persistent reaction (FEV_1 > –20 %); 10 min after provocation, increase in histamine concentration and release of leukotriene E_4 in the bronchioalveolar lavage fluid in all patients; no increase in histamine release in the methacholine test in 6 control persons with non-occupational asthma or 2 healthy persons after provocation with plicatic acid	Chan-Yeung *et al.* 1989
4	provocation; plicatic acid (2.5–10 mg/ml)	FEV_1 > –15 %; after diagnosis or manifestation of the disorder, bronchial hyperreactivity detected in all 4 patients	Chan-Yeung and Desjardins 1992
1	provocation, wood dust [fir wood dust (*Picea* sp.)]	reaction delayed by 3–4 h, 40 % to 50 % decrease in specific airway conductance, for up to 30 h afterwards unspecific bronchial hyperreactivity to histamine, no reaction to fir	Cockcroft *et al.* 1979
2	provocation, wood dust [fir wood dust (*Picea* sp.)]	1 dual reaction (–40 % and –48 % FEV_1); 1 patient with delayed reaction, –16 % FEV_1); in both patients unspecific bronchial hyperreactivity after provocation in the methacholine test	Cockcroft *et al.* 1984

Table 2. continued

Patients with reactions	Test, test material (concentration) [control]	Observations	References
1	provocation, wood dust	delayed reaction with dyspnoea, functional residual capacity and airway resistance determined (no data); no reactions in 2 control persons	Colas *et al.* 1985
48	provocation; plicatic acid (5 mg/ml)	in the follow-up examination (> –15 % FEV_1) in 18 of 48 persons still exposed to *Thuja plicata* further decrease in average FEV_1 and FVC observed; in 11/18 patients and 13 of the other 30 patients IgE specific for a conjugate of plicatic acid and human serum albumin detected	Cote *et al.* 1990
9	provocation; plicatic acid (5 mg/ml)	at least –20 % FEV_1, in 4/9 patients also positive results in skin tests with grass pollen; eosinophil count in the alveolar fluid increased in 9 patients and 6 persons with atopic asthma above that in 6 healthy persons (3 % and 2.5 % compared with 0.5 %); number of T cells in biopsy material increased (CD3+: 9.8× and 2.6×)	Frew *et al.* 1995
4	intradermal, wood extract; provocation, extract (undiluted and 1:10)	1× immediate reaction (30 min) and 2× weak dual reaction (24 h) to undiluted extract in the skin test; 6–8 h after provocation > 50 % decrease in FEV_1 in all cases; in addition 4 patients reported with rhinitis (3 weak immediate reactions to undiluted extract in the skin test)	Gandevia and Milne 1970
1	skin test, raw extract (n.s.); provocation, wood dust	skin tests yielded negative results; provocation caused reaction after 1 h delay, 20–25 % decrease in FEV_1; no reaction in healthy control persons; symptoms reported after first contact	Gozalo Reques 1982
1	provocation, wood dust [sand]	weak immediate reaction (–17 % FEV_1) and marked delayed (2–4 h) reaction (–50 % FEV_1, salbutamol administered after 5 h); for up to 3 weeks afterwards unspecific bronchial hyperreactivity in the histamine test	Hamilton *et al.* 1979
206	provocation; plicatic acid (5 mg/ml); prick test, conjugate of plicatic acid with human serum albumin	retrospective study: at least –20 % FEV_1, 18× immediate reaction, 88× delayed reaction, 100× dual reaction; prick tests yielded negative results in patients and controls (no other details); increased levels of IgE specific for a conjugate of plicatic acid and human serum albumin detected in 7 patients with delayed reactions and in 6 with dual or immediate reactions among 44 tested patients; increased levels of IgE specific for *Thuja plicata* extract in only 11/44; IgE concentration not related to severity of provoked reaction	Lam *et al.* 1983
44	provocation; plicatic acid (5 mg/ml)	at least 20 % decrease in FEV_1, 21× delayed reaction, 23× dual reactions; no reactions in 2 control persons	Lam *et al.* 1987

Table 2. continued

Patients with reactions	Test, test material (concentration) [control]	Observations	References
1	prick test and provocation, wood extract (no other details)	2× negative result in prick test, immediate reaction after intradermal injection of 0.02 ml extract; immediate reaction after provocation (–53 % FEV_1); repeat test in a symptom-free period caused –23 % FEV_1 after a delay of 5 h	Mitchell 1970
1	intracutaneous and provocation, wood extract (no other details)	no reactions in intracutaneous test, positive results after provocation (no other details)	Oertmann and Bergmann 1993
1	scratch test and intradermal, aqueous wood extract (no other details)	positive results in both tests; in addition reaction to muninga (*Pterocarpus angolensis*) and to "Congo hardwood", no other tests carried out	Ordman 1949
332	provocation; wood extract and plicatic acid (5 mg/ml, at intervals, maximum 1.25 ml)	31× immediate reaction, 144× delayed reaction and 157× dual reaction (> –20 % FEV_1); IgE (at least class 2) specific for a conjugate of plicatic acid with human serum albumin detected in 32/138 patients (22.2 %), bronchial hyperreactivity (methacholine test) more marked in patients with dual reactions, study included 206 patients from Lam *et al.* (1983)	Paggiaro and Chan-Yeung 1987
1	provocation, wood dust	reaction delayed by 6 h, rhinitis and decrease in lung performance by 35 %, bronchospasm persisted for another 4 h	Pauli *et al.* 1983
1	prick test, aqueous wood extract (1 and 10 mg/ml); provocation, wood dust	2 prick tests yielded negative results; reaction after provocation delayed by 2–3 h (–33 % FEV_1 after 5 h), no reaction to iroko (*Chlorophora* sp.) dust, no reactions in healthy control persons or in 2 non-exposed persons with asthma	Pickering *et al.* 1972
18	provocation; plicatic acid (2.5 or 5 mg/ml)	in 28 patients with suspected *Thuja plicata* asthma 2× immediate reaction, 7× dual reaction and 9× delayed reaction (no other details); in 8/18 patients IgE specific for a conjugate of plicatic acid with human serum albumin was determined (RAST values on average 2.5–3.5 compared with 1.1–1.2 in 16 control persons and in the patients who did not react after provocation)	Tse *et al.* 1982

n.s. not specified

Table 3. Epidemiological studies of the effects of *Thuja plicata* on the airways

Prevalence	Test, test material (concentration) [control]	Observations	References
10/73 (13.5 %)	anamnesis and lung function tests at the workplace	high prevalence of *Thuja plicata* asthma postulated on the basis of anamnesis and > 10 % $-FEV_1$ at the workplace; asthma also diagnosed in 3/8 employees who were exposed to other dusts (Douglas fir, hemlock, alder, etc); although levels of inhalable dust were low, prevalence correlated with (total) dust exposure	Brooks *et al.* 1981
2/405 (0.5 %)	provocation, extract (n.s.)	5/405 patients with asthma symptoms in the anamnesis, 2 confirmed by provocation; asthma also diagnosed in 4/82 (4.9 %) previously exposed control persons (no other details); prevalence of dyspnoea and coughing about twice that in non-exposed persons	Chan-Yeung *et al.* 1978
22/1320 (1.7 %)	intracutaneous, dialysed extract [extract of fir wood (*Picea* sp.), meranti (*Shorea* sp.)]	positive results in 22/26 patients with suspected *Thuja plicata* asthma, positive results with the extract also in 3/20 non-exposed asthma patients; no reactions to control extracts; high prevalence of positive results with ubiquitous allergens; lung function tests and provocation tests not mentioned, number of exposed persons not given	Ishizaki *et al.* 1973
19/619 (3.1 %)	anamnesis, questioning	prevalence of development of asthmatic symptoms after starting work in the job about 4 times that in the control collective (9/1127, 0.8 %)	Siracusa *et al.* 1995
43/625 (6.7 %)	IgE specific for a plicatic acid/human serum albumin conjugate (RAST)	no direct relationship with respiratory symptoms; bronchial hyperreactivity in the methacholine test detected in only 16 of the 35/43 persons examined	Vedal *et al.* 1986a, Vedal *et al.* 1986b
8 % of 652	anamnesis, questioning	8 % of 652 employees described symptoms of occupational asthma, 8 % ocular symptoms and 14 % nasal symptoms	
5 % of 652	medical diagnosis (no other details)	occupational asthma diagnosed in 5 % of 652 (no other details); prevalence correlated with the exposure duration and dust concentration; FEV_1 lower in employees exposed to higher levels	

In spite of the fact that IgE specific for plicatic acid can be detected in about 7 % to 44 % of the patients who yield positive results in the provocation test (Lam *et al.* 1983, Paggiaro and Chan-Yeung 1987, Tse *et al.* 1982, Vedal *et al.* 1986a), it does not appear to be correlated with the clinical symptoms (Vedal *et al.* 1986a) nor with hyperreactivity to plicatic acid (Lam *et al.* 1983). *In vitro* (serum from previously not exposed persons), plicatic acid induced direct, immunoglobulin-independent activation of the complement system (Chan-Yeung *et al.* 1980). Ten minutes after provocation with plicatic acid, direct release of histamine and leukotriene E_4 was detected in the alveolar fluid of patients with *Thuja plicata* asthma (Table 1) (Chan-Yeung *et al.* 1989). *In vitro* studies with biopsy material—cells from the bronchioalveolar lavage fluid and peripheral basophils from 11 patients with *Thuja plicata* asthma, 7 patients with atopic asthma and 7 healthy persons (Frew *et al.* 1993b)—confirmed the results of the *in vivo* studies. Plicatic acid in a concentration of 100 μg/ml caused release of histamine in the cell material obtained from the patients and also in the cell material from the control persons without asthma, but much less in the cell material from the patients with atopic asthma. The results obtained with lower concentrations (plicatic acid 10 μg/ml and plicatic acid/human serum albumin conjugate 100 μg/ml (corresponding to 13 μg/ml plicatic acid)) suggest that the effects could be specific but, because of the small number of samples, could not prove this. Likewise, in peripheral mononuclear cells from 9 patients with *Thuja plicata* asthma, neither plicatic acid (100 μg/ml) nor plicatic acid/human serum albumin conjugate (100 μg/ml) increased the levels of histamine-releasing factors significantly more than in cultures of cells from 5 patients with atopic asthma or from 11 patients without asthma including 5 atopic persons (Frew *et al.* 1993a).

1.2 Results of animal studies

The only available information as to the effects of *Thuja plicata* on the skin is from an insufficiently documented study in which provocation revealed a high level of sensitization (no other details) after epicutaneous induction (3× weekly, 2 weeks) of 6 guinea pigs with an ethyl acetate extract of *Thuja plicata* (no other details) (Suskind 1967).

In an animal model, injection of a mixture of plicatic acid/ovalbumin conjugate and $Al(OH)_3$ (i.p. injection of 50 ng doses of plicatic acid/ovalbumin conjugate on days 0, 7, 14, 21, 35, 49 and 63 and s.c. injection of 200 ng of the same conjugate on day 77) induced specific anti-plicatic acid IgE and IgG in New Zealand White rabbits (no other details). 14 days after the last injection, small amounts of specific IgE were detected but only in 5/19 animals (titre found by passive cutaneous anaphylaxis: 3× 1:20, 1× 1:10, 1× 1:1); IgG, on the other hand, was detected in the radioimmunoassay in all animals. In the serum of two animals immunized with plicatic acid/human serum albumin conjugate and Freund's adjuvant, precipitating antibodies were also detected. Bronchial provocation of 4 animals (titre at least 1:10) with plicatic acid/human serum albumin conjugate (provocation with plicatic acid was apparently not carried out) was not successful, whereas intravenous administration of the conjugate caused after 1 minute a transient increase in respiration rate and after 5 minutes an increase in pulmonary resistance (Chan *et al.* 1987). After intraperitoneal injection of 10 μg doses of plicatic acid/ovalbumin conjugate (adsorbed on 1 mg $Al(OH)_3$) monthly for 6 months, IgG_1 and IgG_2 antibodies to plicatic acid and ovalbumin were detected by enzyme-linked immunosorbent assay (ELISA) in male Cam-Hartley guinea pigs (number not specified). On isolated tracheal tissue from the treated animals, ovalbumin in a concentration of only 1 μmol/l induced a contraction which was about 5 times as intensive as that induced by plicatic acid in concentrations of more than 1 mmol/l. Likewise, the plicatic acid-induced release of histamine and leukotriene D_4 from lung fragments and of histamine and prostaglandin D_2 from leukocytes of the treated animals was only 30 % to 50 % of that caused by ovalbumin although the concentration of plicatic acid used was more than 1000 times higher (lung fragments and leukocytes of untreated animals were used as the negative control) (Salari *et al.* 1994).

2 Manifesto (sensitization)

Dust of *Thuja plicata* wood has been described as the cause of asthmatic disorders in a large number of individual case reports. In addition, several epidemiological studies have revealed an increased prevalence of asthma among persons employed in the processing of *Thuja plicata.* Although there are still no prospective studies available, the causal factors have not yet all been identified and an immunological mechanism of action of the wood component, plicatic acid, has not been proved, the clinical and immunological data available at present suggest that the wood of *Thuja plicata* is very probably a respiratory allergen.

The evidence for contact allergy in man caused by *Thuja plicata* wood is provided by only a few reports of effects in man and one animal study which does not meet present-day requirements. Although the data for allergenic effects of the terpenoid substances contained in the wood are inconsistent, it must be expected that the wood is a contact allergen. Therefore, *Thuja plicata* wood is designated with "Sah".

There are only a few case reports of contact allergy and respiratory allergy caused by the wood of *Thuja occidentalis* and, also because the presence of plicatic acid in this wood has not yet been proved, they do not justify designation of the wood as sensitizing.

3 References

Ashley MJ, Corey P, Chan-Yeung M, MacLean L, Maledy H, Grzybowski S (1978) A respiratory survey of cedar mill workers – II. Influence of work-related and host factors on the prevalence of symptoms and pulmonary function abnormalities. *J Occup Med 20*: 328–332

Blainey AD, Graham VA, Phillips MJ, Davies RJ (1981) Respiratory tract reactions to western red cedar. *Hum Toxicol 1*: 41–51

Bleumink E, Mitchell JC, Nater JP (1973) Allergic contact dermatitis from cedar wood (*Thuja plicata*). *Br J Dermatol 88*: 499–504

Bleumink E, Nater JP (1974) Allergic reactions to (tropical) wood. *Contact Dermatitis Newslett 16*: 436–437

Brooks SM, Edwards JJ, Apol A, Edwards FH (1981) An epidemiological study of workers exposed to western red cedar and other wood dusts. *Chest 80 (Suppl)*: 30S–32S

Burry JN, Kirk J, Reid JG, Turner T (1973) Environmental dermatitis: patch tests in 1,000 cases of allergic contact dermatitis. *Med J Aust II*: 681–685

Calnan CD (1972) Dermatitis from cedar wood pencils. *Trans St. John's Hosp Dermatol Soc 58*: 43–47

Cartier A, Chan H, Malo J-L, Pineau L, Tse KS, Chan-Yeung M (1986) Occupational asthma caused by eastern white cedar (*Thuja occidentalis*) with demonstration that plicatic acid is present in this wood dust and is the causal agent. *J Allergy Clin Immunol 77*: 639–645

Chan H, Tse KS, van Oostdam J, Moreno R, Pare PD, Chan-Yeung M (1987) A rabbit model of hypersensitivity of plicatic acid, the agent responsible for red cedar asthma. *J Allergy Clin Immunol 79*: 762–768

Chan-Yeung M (1973) Maximal expiratory flow and airway resistance during induced bronchoconstriction in patients with asthma due to western red cedar (*Thuja plicata*). *Am Rev Respir Dis 108*: 1103–1110

Chan-Yeung M (1977) Fate of occupational asthma – A follow-up study of patients with occupational asthma due to western red cedar (*Thuja plicata*). *Am Rev Resp Dis 116*: 1023–1029

Chan-Yeung M, Desjardins A (1992) Bronchial hyperresponsiveness and level of exposure in occupational asthma due to western red cedar (*Thuja plicata*). Serial observations before and after development of symptoms. *Am Rev Respir Dis 146*: 1606–1609

Chan-Yeung M, Barton GM, McLean L, Grzybowski S (1971) Bronchial reactions to western red cedar (*Thuja plicata*). *Can Med Assoc J 105*: 56–61

Chan-Yeung M, Barton GM, MacLean L, Grzybowski S (1973) Occupational asthma and rhinitis due to western red cedar (*Thuja plicata*). *Am Rev Respir Dis 108*: 1094–1102

Chan-Yeung M, Ashley MJ, Corey P, Willson G, Dorken E, Grzybowski S (1978) A respiratory survey of cedar mill workers – I. Prevalence of symptoms and pulmonary function abnormalities. *J Occup Med 20*: 323–327

Chan-Yeung M, Giclas PC, Henson PM (1980) Activation of complement by plicatic acid, the chemical compound responsible for asthma due to western red cedar (*Thuja plicata*). *J Allergy Clin Immunol 65*: 333–337

Chan-Yeung M, Lam S, Koener S (1982) Clinical features and natural history of occupational asthma due to western red cedar (*Thuja plicata*). *Am J Med 72*: 411–415

Chan-Yeung M, MacLean L, Paggiaro PL (1987) Follow-up study of 232 patients with occupational asthma caused by western red cedar (*Thuja plicata*). *J Allergy Clin Immunol 79*: 792–796

Chan-Yeung M, Chan H, Tse KS, Salari H, Lam S (1989) Histamine and leukotrienes release in bronchoalveolar fluid during plicatic acid-induced bronchoconstriction. *J Allergy Clin Immunol 84*: 762–768

Cockcroft DW, Cotton DJ, Mink JT (1979) Nonspecific bronchial hyperreactivity after exposure to western red cedar. *Am Rev Respir Dis 119*: 505–510

Cockcroft DW, Hoeppner VH, Werner GD (1984) Recurrent nocturnal asthma after bronchoprovocation with western red cedar sawdust: association with acute increase in non-allergic bronchial responsiveness. *Clin Allergy 14*: 61–68

Colas M, Grosclaude M, Balland S, Perrin-Fayolle M (1985) Intérêt des tests réalistes d'exposition dans le diagnostic des asthmes professionnels aux bois exotiques (The value of realistic exposure tests in the diagnosis of occupational asthma due to exotic woods) (French). *Rev Pneumol Clin 41*: 39–46

Cote J, Kennedy S, Chan-Yeung M (1990) Outcome of patients with cedar asthma with continuous exposure. *Am Rev Respir Dis 141*: 373–376

Frew AJ, Chan H, Chan-Yeung M (1993a) Lack of role for mononuclear cell-derived histamine releasing factors in occupational asthma due to western red cedar. *Clin Exp Allergy 23*: 861–867

Frew A, Chan H, Dryden P, Salari H, Lam S, Chan-Yeung M (1993b) Immunologic studies of the mechanisms of occupational asthma caused by western red cedar. *J Allergy Clin Immunol 92*: 466–478

Frew AJ, Chan H, Lam S, Chan-Yeung M (1995) Bronchial inflammation in occupational asthma due to western red cedar. *Am J Respir Crit Care Med 151*: 340–344

Fujita M, Aoki T (1983) Allergic contact dermatitis to pyridoxine ester and hinokitiol. *Contact Dermatitis 9*: 61–65

Gandevia B, Milne J (1970) Occupational asthma and rhinitis due to western red cedar (*Thuja plicata*), with special reference to bronchial reactivity. *Br J Ind Med 27*: 235–244

Gardner JAF, Barton GM, MacLean H (1957) Occurrence of 2,7-dihydroxy-4-isopropyl-2,4,6-cycloheptatrien-1-one (7-hydroxy-4-isopropyltropolone) in western red cedar (*Thuja plicata* Donn.). *Can J Chem 35*: 1039–1048

Gardner JAF, Barton GM (1958) Occurrence of ß-Dolabrin (4-isopropenyltropolone) in western red cedar (*Thuja plicata*). *Can J Chem 36*: 1612–1615

Gardner JAF, Barton GM, MacLean H (1959) The polyoxyphenols of western red cedar (*Thuja plicata* Donn.). I. Isolation and preliminary characterization of plicatic acid. *Can J Chem 37*: 1703–1709

Gozalo-Reques F (1982) Asma por cedro rojo (*Thuja plicata*): dos casos (Asthma from red cedar: two cases) (Spanish). *Allergol Immunopathol Madr 10*: 441–447

Grimm I (1991) Kontaktallergie auf einen *Thuja occidentalis*-Extrakt (Allergic contact dermatitis from a *Thuja occidentalis* extract) (German). *Allergologie 14*: 272–274

Hamilton RD, Crockett AJ, Ruffin RE, Alpers JH (1979) Bronchial reactivity in western red cedar induced asthma. *Aust N Z J Med 9*: 417–419

Hausen BM (1981) *Woods injurious to human health – A manual*, Walter de Gruyter & Co, Berlin

Ishizaki T, Shida T, Miyamoto T, Matsumara Y, Mizuno K, Tomaru M (1973) Occupational asthma from western red cedar dust (*Thuja plicata*) in furniture factory workers. *J Occup Med 15*: 580–585

Lam S, Tan F, Chan H, Chan-Yeung M (1983) Relationship between types of asthmatic reaction, nonspecific bronchial reactivity, and specific IgE antibodies in patients with red cedar asthma. *J Allergy Clin Immunol 72*: 134–139

Lam S, LeRiche J, Phillips D, Chan-Yeung M (1987) Cellular and protein changes in bronchial lavage fluid after late asthmatic reaction in patients with red cedar asthma. *J Allergy Clin Immunol 80*: 44–50

Lovell CR (1993) *Plants and the skin*, Blackwell Scientific Publications, London, 245, 252

MacLean H, Murakami K (1966) Lignans of western red cedar (*Thuja plicata* Donn). IV. Thujaplicatin and thujaplicatin methyl ether. *Can J Chem 44*: 1541–1545

MacLean H, MacDonald BF (1967) Lignans of western red cedar (*Thuja plicata* Donn). VII. Dihydroxythujaplicatin. *Can J Chem 45*: 739–740

MacLean H, MacDonald BF (1969a) Lignans of western red cedar (*Thuja plicata* Donn). VIII. Plicatinaphthol. *Can J Chem 47*: 457–460

MacLean H, MacDonald BF (1969b) Lignans of western red cedar (*Thuja plicata* Donn). IX. Plicatinaphthalene. *Can J Chem 47*: 4495–4498

Malo JL, Cartier A, L'Archeveque J, Trudeau C, Courteau JP, Bherer L (1994) Prevalence of occupational asthma among workers exposed to eastern white cedar. *Am J Respir Crit Care Med 150*: 1697–1701

Milne J, Gandevia B (1969) Occupational asthma and rhinitis due to western (Canadian) red cedar (*Thuja plicata*). *Med J Aust 2*: 741–744

Mitchell C (1970) Occupational asthma due to western or Canadian red cedar (*Thuja plicata*). *Med J Aust 2*: 233–235

Mitchell JC, Chan-Yeung M (1974) Contact allergy from *Frullania* and respiratory allergy from *Thuja*. *Can Med Assoc J 110*: 653–657

Mue S, Ise T, Ono Y, Akasaka K (1975) A study of western red cedar sensitivity: workers' allergy reactions and symptoms. *Ann Allergy 35*: 148–152

Nault J (1987) A capillary gas chromatographic method for thujaplicins in western red cedar extractives. *Wood Sci Technol 21*: 311–316

Nault J (1988) Radial distribution of thujaplicins in old growth and second growth western red cedar (*Thuja plicata* Donn). *Wood Sci Technol 22*: 73–80

Oertmann CH, Bergmann K-C (1993) Atemwegserkrankungen bei Arbeitern im Holzgewerbe (Airway diseases in woodworkers) (German). *Allergologie 16*: 334–340

Ordman D (1949) Bronchial asthma caused by the inhalation of wood dust. *Ann Allergy 7*: 492–496/505

Orsler RJ (1969) The effects of irritant timbers. *Woodworking Ind 26*: 28–29

Paggiaro PL, Chan-Yeung M (1987) Pattern of specific airway response in asthma due to western red cedar (*Thuja plicata*): relationship with length of exposure and lung function measurements. *Clin Allergy 17*: 333–339

Pauli G, Abbas F, Bessot JC, Pujol JL (1983) Asthma induced by red cedar wood. Report of an endemic case. *Rev Fr Allergol Immunol Clin 23*: 205–206

Pickering CAC, Batten JC, Pepys J (1972) Asthma due to inhaled wood dusts – western red cedar and iroko. *Clin Allergy 2*: 213–218

Puig L, Alomar A, Randazzo L, Cuatrecasas M (1994) Erythema multiformlike reaction caused by topical application of *Thuja* essential oil. *Am J Contact Dermatitis 5*: 94–97

Salari H, Howard S, Chan H, Dryden P, Chan-Yeung M (1994) Involvement of immunologic mechanisms in a guinea pig model of western red cedar asthma. *J Allergy Clin Immunol 93*: 877–884

Siracusa A, Kennedy SM, DyBuncio A, Lin FJ, Marabini A, Chan-Yeung M (1995) Prevalence and predictors of asthma in working groups in British Columbia. *Am J Ind Med 28*: 411–423

Suskind RR (1967) Dermatitis in the forest product industries. *Arch Environ Health 15*: 322–326

Tse KS, Chan H, Chan-Yeung M (1982) Specific IgE antibodies in workers with occupational asthma due to western red cedar. *Clin Allergy 12*: 249–258

Vedal S, Chan-Yeung M, Enarson DA, Chan H, Dorken E, Tse KS (1986a) Plicatic acid-specific IgE and nonspecific bronchial hyperresponsiveness in western red-cedar workers. *J Allergy Clin Immunol 78*: 1103–1109

Vedal S, Chan-Yeung M, Enarson D, Fera T, MacLean L, Tse KS (1986b) Symptoms and pulmonary function in western red cedar workers related to duration of employment and dust exposure. *Arch Environ Health 41*: 179–183

completed 05.05.1997

Woods: *Tieghemella africana* A.Chev., *Tieghemella heckelii* Pierre

Sensitization		–
Family		Sapotaceae
Synonyms	*Tieghemella africana*	douka *Dumoria africana* A.Chev. *Mimusops africana* Lecomte
	Tieghemella heckelii[1]	*Dumoria heckelii* A.Chev. macore makoré *Mimusops heckelii* Hutch & Dalz

The makoré and douka wood which comes from central West Africa has many uses because it is strong, resistant to fungal attack, and comes in large sizes. In the form of veneer it is used, for example, for furniture, panelling and partitions, as marine plywood in boat building and as coated plywood for boxing panels. The solid wood is used for outdoor construction purposes for windows, doors, garden furniture and in boat building, and indoors for stairs, furniture and parquet flooring.

1 Allergenic Effects

From makoré wood a crude saponin mixture was isolated in 3 % to 5 % yield; its main component was a glycoside with the aglycone bassic acid (King *et al.* 1955, Sandermann and Barghoorn 1955). The haemolytic crude saponin, the crude ethanolic extract of the wood, and makoré wood dust did not cause skin irritation when applied to two test persons for 3 to 5 days in an Eucerin[2]-water emulsion (no other details) (Sandermann and Barghoorn 1955). Makoré wood also contains small amounts of 2,6-dimethoxy-1,4-benzoquinone (Hausen 1978).

[1] In the literature these species are also called *Tieghemella africana* Pierre and *Tieghemella heckelii* Pierre ex A.Chev. Other names such as African cherry and cherry mahogany are sometimes used, but these are suggestive of inappropriate botanical relationships. Because the two *Tieghemella* species yield very similar woods, both are sold under the name makoré.

[2] Eucerin is a lanolin-alcohol cream used for pharmaceutical and cosmetic purposes.

1.1 Effects in man

In the literature there are only very few mostly inadequately documented case reports of the effects of makoré and douka wood dust on the skin (Table 1) and a few cases tabulated without any other details: 3 cases (Anonymous 1975), 3 cases of skin disorders registered with the Swiss Accident Insurance Association (SUVA) (Hartmann and Schlegel 1980) and one case recognized as an occupational disease (Wolf 1991). From these reports, however, a significant potential to cause contact dermatitis cannot be deduced.

Makoré and douka are listed in several reviews as woods which cause irritation of the mucosa and airways (e.g. Anonymous 1955, Orsler 1969, Zafiropoulo *et al.* 1968). The single cases described in the literature without details of any tests (Dantin-Gallego *et al.* 1952, Hartmann and Schlegel 1980, 2 cases; Preißler 1959, 4 cases; Wolf 1991, 3 tabulated cases with a recognized occupational disease) and the reported positive test results (Table 2) do not permit the conclusion that the wood is a respiratory allergen. In summary, the findings are rather suggestive of a marked irritant effect of the dust.

1.2 Results of animal studies

After induction treatment with 0.01 M 2,6-dimethoxy-1,4-benzoquinone in Freund's complete adjuvant (FCA), provocation with a 0.02 M solution did not produce any reactions in 10 Pirbright White guinea pigs (Hausen 1970, Schulz *et al.* 1979).

Table 1. Case reports of dermatitis caused by contact with makoré

Number of patients	Test, test material (concentration)	Observations	References
2	patch test, damp wood dust from steamed veneers	weak positive reaction after 72 h in 2/38 wood workers	Krille 1980
1	–	person with recurrent facial eczema after working with makoré	Preißler 1959
1	patch test, wood dust (no other details)	case of dermatitis especially on contact surfaces and damp skin areas; positive result in the patch test (no other details) and change of job	Sandermann and Barghoorn 1956
1	patch test, damp wood dust	positive result for a cabinet maker (no other details) and for this patient also positive result in the provocation test (see Table 2)	Schleicher 1974
2	patch test and intracutaneous test, wood dust (no other details)	2 employees who developed blepharitis, oedema of the eyelids and generalized pruritis after only a few hours contact with makoré; no reaction in the tests with wood dust	Turc *et al.* 1950

Table 2. Results of intracutaneous and provocation tests with makoré extracts

Patients with reactions	Test, test material (concentration)	Observations	References
19	intracutaneous test, wood extract (0.1 ml, 1/1000)	positive result in intracutaneous test for 19 of a total of 100 tested persons, in one case suggestive anamnesis	Girard *et al.* 1980, Surber *et al.* 1978
30 / 14	intracutaneous and provocation tests, wood extract (no other details)	in 157 medical reports intracutaneous test: 30× positive, provocation test: 14× positive	Kersten and von Wahl 1994
22 / 4	intracutaneous and provocation tests, wood extract (no other details)	positive results in skin test for 22/101 patients with suspected respiratory disease; provocation test carried out on 26 patients yielded positive results for 4; these 4 cases were recognized as having an occupational disease	Kirsten *et al.* 1985
1	intracutaneous test, aqueous wood extract (no other details)	positive results in intracutaneous test for 1/2 exposed persons among 113 examined employees	Nava 1974
3	intracutaneous test, wood extract (no other details)	3× positive results in intracutaneous tests for persons with suggestive anamnesis; in addition for 11/220 persons positive results in tests with makoré without current clinical relevance	Oehling 1963
15 / 4	intracutaneous and provocation tests, wood extract (no other details)	15/55 tested patients produced a reaction in the intracutaneous test, 4 with current clinical relevance; 4/5 provocation tests yielded positive results	Oertmann and Bergmann 1993
8	intracutaneous test, wood extract (0.1 ml, 1/1000)	immediate reaction seen in 8/19 employees; reading after 4–6 h revealed reactions in 3/12 patients, after 48 h in 0/13 patients	Pigearias *et al.* 1989
1	provocation, wood dust	positive result in provocation test for 1 female patient with rhinitis on contact with makoré dust (no other details); patch test also yielded positive result (see Table 1)	Schleicher 1974

In inadequately documented studies, epicutaneous sensitization of guinea pigs was achieved with a crude ethanolic extract of *T. heckelii* wood, but reactions could only be provoked with the fraction of the extract containing 2,6-dimethoxy-1,4-benzoquinone (no other details, Hausen 1978).

In a modified maximization test (induction: intracutaneous 0.005 M 2,6-dimethoxy-1,4-benzoquinone with FCA, epicutaneous 0.2 M), non-occlusive epicutaneous provocation with 0.1 M and 0.01 M solutions of 2,6-dimethoxy-1,4-benzoquinone produced reactions after 48 hours in all animals (6× strong positive, 3× positive, 1× weak positive) and in 5/10 animals (weak to clearly positive), respectively (Buske 1974). Because in this maximization test positive results were also obtained with substances which have been shown in other studies not to be allergenic, the results described above are not sufficient to document a sensitizing effect of the wood dust of *T. heckelii* or of the wood constituent 2,6-dimethoxy-1,4-benzoquinone in animals.

Intraperitoneal injection of an extract of douka (saponin content 1.5 mg/ml, no other details) and of the concentrated saponin (3 mg/ml) caused release of histamine in rats (Subiza Martin 1971).

2 Manifesto (sensitization)

The inadequate data available for the skin reactions to the wood of the two *Tieghemella* species which have been described occasionally in the literature do not demonstrate an allergic genesis of the reactions. Likewise, the studies of sensitization in animals suggest at most a weak sensitization potential; allergenic components have not yet been demonstrated unequivocally. Although positive results have been obtained in several provocation and intracutaneous tests, the frequently described respiratory symptoms are very probably the result of marked irritant effects of the saponin-containing wood. Although the possibility that the substance has significant allergenic potential in the airways cannot be deduced from the available data, it can also not be excluded.

Therefore for the present the two *Tieghemella* species have not been designated with an "S".

3 References

Anonymous (1955) Douka. *Rev Bois For Trop 42*: 37–40

Anonymous (1975) Fiches d'allergologie et dermatologie professionnelles: bois exotiques (Occupational allergology and dermatology datasheets: exotic woods) (French). *Dec. 1975, 2. Doc Med Trav*

Buske HH (1974) *Untersuchungen über die Sensibilisierungsfähigkeit natürlich vorkommender Chinone* (Sensitization potential of natural quinones) (German). Doctoral thesis, Hautklinik der Universität Hamburg

Dantin-Gallego J, Armayor AF, Riesco J (1952) Some new toxic woods: some new manifestations of toxicity. *Ind Med Surg 21*: 41–46

Girard JP, Surber R, Guberan E (1980) Allergic manifestations due to wood dusts. in: Frazier CA (Ed.) *Occupational asthma*, Van Nostrand Reinhold Company, New York, 91–101

Hartmann A, Schlegel H (1980) Durch Holz verursachte Gesundheitsschäden in der Schweiz (Adverse effects on health caused by wood in Switzerland) (German). *Schweiz Med Wochenschr 110*: 278–281

Hausen BM (1970) *Untersuchungen über gesundheitsschädigende Hölzer* (Studies of woods injurious to health) (German). Doctoral thesis, Fachbereich Biologie der Universität Hamburg

Hausen BM (1978) Sensitizing capacity of naturally occurring quinones. V. 2,6-Dimethoxy-*p*-benzoquinone: occurrence and significance as a contact allergen. *Contact Dermatitis 4*: 204–213

Kersten W, von Wahl PG (1994) Allergische Atemwegserkrankungen in der holzverarbeitenden Industrie (Occupational disease in the wood-processing industry) (German). *Allergologie 17*: 55–60

King FE, Baker JA, King TJ (1955) The chemistry of extractives from hardwoods. Part XXIV. A saponin constituent of Makoré (*Mimusops heckelii*). *J Chem Soc*: 1338–1342

Kirsten D, Liebetrau G, Meister W (1985) Holzstaub als inhalative Noxe (Wood dust as inhalative noxious agent) (German). *Z Erkr Atmungsorgane 165*: 235–241

Krille M (1980) *Zur arbeitsdermatologischen Bedeutung exotischer Hölzer unter besonderer Berücksichtigung der Holzindustrie der DDR* (The dermatological significance of exotic woods at the workplace, in particular in the wood industry of the GDR) (German). Doctoral thesis, Zentralinstitut für Arbeitsmedizin der DDR

Nava C (1974) Patologia da polveri di legno (Pathology caused by wood dust) (Italian). *Med Lav 65*: 1–7

Oehling A (1963) Berufsallergie im Holzgewerbe (Occupational allergy in the wood industry) (German). *Allerg Asthma 9*: 312–322

Oertmann C, Bergmann KC (1993) Atemwegserkrankungen bei Arbeitern im Holzgewerbe (Airway diseases in woodworkers) (German). *Allergologie 16*: 334–340

Orsler RJ (1969) The effects of irritant timbers. *Woodworking Ind 26*: 28–29

Pigearias B, Dembele K, Bonny JS, Bogui P, Lonsdorfer J (1989) Pneumallergènes domestiques et professionnels des travailleurs de bois tropicaux à Abidjan (Côte-d'Ivoire) (Domestic and occupational allergens affecting workers with tropical wood in Abidjan) (French). *Allerg Immunol 21*: 338–340

Preißler R (1959) *Reihenuntersuchungen in Exotenhölzer verarbeitenden Betrieben infolge Verdacht auf eine schädigende Einwirkung von Gurjun- und Makoré-Staub* (Serial tests in factories working with exotic woods following suspected adverse effects of gurjun and makoré dust) (German). Doctoral thesis, Medizinische Akademie Carl Gustav Carus, Dresden

Sandermann W, Barghoorn AW (1955) Über die Inhaltsstoffe von Makoré und Peroba-Holz sowie ihre gesundheitsschädigende Wirkung (The components of makoré and peroba wood and their effects on health) (German). *Holzforschung 9*: 112–117

Sandermann W, Barghoorn AW (1956) Gesundheitsschädliche Hölzer (Woods injurious for health) (German). *Holz Roh- Werkst 14*: 87–94

Schleicher H (1974) Über phytogene allergische Kontaktekzeme (Phytogenic allergic contact dermatitis) (German). *Dermatol Monatsschr 160*: 433–444

Schulz KH, Garbe I, Hausen BM, Simatupang MH (1979) The sensitizing capacity of naturally occurring quinones. Experimental studies in guinea pigs. II. Benzoquinones. *Arch Dermatol Res 264*: 275–286

Subiza Martin E (1971) Études expérimentales avec le bois d'Oukola (*Dumoria africana*) (Experimental studies with the wood of *Dumoria africana*) (French). *Bronches 21*: 517–525

Surber R, Guberan M, Girard JP (1978) Allergies respiratoires aux poussières de bois – Cas cliniques et études épidémiologiques (Respiratory allergies due to wood dust – Clinical cases and epidemiologic study) (French). *Rev Fr Allergol 17*: 193–198

Turc H, Brunel R, Tolot F (1950) Accidents allergiques au bois de Makoré (Allergy to makoré wood) (French). *Arch Mal Prof Med Trav Secur Soc 11*: 490–492

Wolf J (1991) Gesundheitsgefahren vermeiden (Avoiding health hazards) (German). *HBG-Mitteilungen 65*: 3–22

Zafiropoulo A, Audibert A, Charpin J (1968) A propos des accidents dus a la manipulation des bois exotiques (Accidents during manipulation of exotic wood) (French). *Rev Fr Allergol 8*: 155–170

completed 02.03.1999

Woods: *Triplochiton scleroxylon* K. Schum.

Sensitization (1996)	**Sah**
Family	Sterculiaceae
Synonyms	abachi African maple African whitewood ayous obeche samba wawa

Triplochiton scleroxylon from West Africa yields a yellowish-white to golden yellow light wood which is used for veneers, as a non-loadbearing construction wood, and for the manufacture of musical instruments and such.

1 Allergenic Effects

To date no components of *T. scleroxylon* which could be the cause of the allergic reactions to this wood have been isolated. It is conceivable that a protein with a molecular weight between 10 and 30 kDalton is responsible for the pulmonary reaction (Tuppurainen *et al.* 1995). In recent studies, immunoblotting with sera from patients with asthmatic reactions to *T. scleroxylon* wood dust has led to the detection of possibly allergenic protein fractions with molecular weights of about 17, 28 and approximately 60 kDalton (Quirce *et al.* 2000).

1.1 Effects in man

There are only very few inadequately documented reports of allergic contact dermatitis caused by *T. scleroxylon* wood (Woods and Calnan 1976) but no data suitable for an assessment. By 1978, nine cases of skin disorders after contact with obeche wood (no other details) had been reported to the Swiss Accident Insurance Association (SUVA) (Hartmann 1978, Hartmann and Schlegel 1980) and the trade association of the German wood-working industry recognized one case of an occupational skin disorder during the years between 1976 and 1988 (Wolf 1991). In serial tests of 200 employees, not very useful weak positive results were obtained for 12 persons in patch tests with wood dust from steamed veneers (Krille 1980). The progression of these reactions was not uniform and they were probably either contact urticaria or unspecific irritative reactions. Urticarial reactions, which are always seen when there is a marked reaction in the intracutaneous test (Gronemeyer and Debelic 1967), were demonstrated as early as 1963 by rubbing skin with the wood of *T. scleroxylon* (Oehling 1963). In the more recent literature too, contact urticaria caused by this wood has been reported repeatedly (Table 1).

Table 1. Case reports of urticaria caused by contact with *T. scleroxylon* in the rubbing test

Patients with reactions	Test material	Observations	References
1	wood	marked urticarial reaction after a few minutes	Böckers and Bork 1990
21	wood	21 tabulated cases (including the 11 patients from Oehling 1963)	Gronemeyer and Debelic 1967
8	wood dust	reaction with urtica, erythema and pruritis in all patients; no reactions in the 10 exposed persons without symptoms or in the 10 non-exposed control persons exposure: 1–25 years, symptoms began after 3 months to 23 years, severe urticaria especially on the arms and neck, in 2 cases severe generalized urticaria	Hinojosa *et al.* 1990
11	wood	positive results for 11/19 employees who produced reactions with abachi extract in the intracutaneous test	Oehling 1963
1	wood	positive result for 1/11 patients who reacted in the intracutaneous test	Oertmann and Bergmann 1993
1	wood	positive result for 1/6 patients with current abachi allergy	Schwarting 1984
1	wood	severe reaction with erythema	Weber and Häußinger 1988

In numerous cases, the allergenic effects of *T. scleroxylon* wood dust on the airways could be demonstrated in the intracutaneous or bronchial provocation tests (Table 2). The immunological genesis of the reaction has also been confirmed by the detection of specific IgE with both validated methods (RAST, RAST-inhibition) and with methods which have not yet been validated (reverse enzyme immunoassay: REIA) (Hinojosa *et al.* 1984, 1986, 1990) as well as by the positive results obtained at an early date with the Prausnitz-Küstner test (Oehling 1963). The available data (Hinojosa *et al.* 1995, Oehling 1963) do not, however, indicate a possibility of desensitization by medium-term or long-term administration of abachi extracts.

Table 2. Case reports of the effects of *Triplochiton scleroxylon* wood dust on the airways

Patients with reactions	Test, test material (concentration)	Observations	References
1	prick test and nasal provocation test, wood extract (no data)	carpenter with perennial rhinitis and asthma; prick test and provocation test yielded positive results (no other details); symptoms regressed after change of job	Böckers and Bork 1990
7	prick test, wood extract (no data)	positive results in the prick test for 7/90 exposed persons, only in one case was specific IgE detected in the RAST (class 2); negative results in the prick test for 53 non-exposed patients	Carosso *et al.* 1987
5	intracutaneous test, wood extract (no data)	results for 55 tested employees included 4× strong and 1× weak reaction to obeche; RAST carried out for one patient also yielded positive results	Fasani *et al.* 1982
2	intracutaneous test, wood extract (0.1 ml, 1/1000)	positive results in intracutaneous test for 12/96 employees, two of whom had suggestive anamnesis; precipitating antibodies detected in 6 persons	Girard *et al.* 1980
–	intracutaneous and provocation tests (no data)	13 tabulated cases of registered respiratory disease; of these 5× immediate reaction in the skin test and 3× positive result in provocation test	Hartmann 1978, Hartmann and Schlegel 1980
2	prick test and provocation test, wood extract	total IgE: 1000 and 1400 IU/ml; positive results in the prick test with 0.1 % extract for both patients (heavy smokers); negative results in the prick test with 10 % extract for 10 control persons (including 5 atopic); provocation test (2 min inhalation of the extract) carried out with one patient: 15 min after provocation with 0.01 % extract decrease in FEV_1 (32 %) and FVC (20 %), initial values regained after 3 h; precipitating antibodies not detectable	Hinojosa *et al.* 1984
4	prick test and provocation test, wood extract	marked immediate reactions in the prick test (2 patients with 0.1 % extract, 2 with 1 %), no reaction to 10 % in 30 control persons; provocation test yielded positive results with 1 % extract for 2 persons (decrease in FEV_1 after 30 min 30 % and 46 %), with 0.1 % for one person (–54 % FEV_1 after 10–15 min) and an unclear result for one person (maximum decrease in FEV_1 21 %, initial value regained after 1 h); no reactions to 10 % extract in 6 control persons; precipitating antibodies not detectable; 2 patients were smokers	Hinojosa *et al.* 1986
8	provocation (no data); prick test, wood extract (1/100, 1/1000)	immediate asthmatic reaction after provocation in all patients (no other details); prick test yielded positive results for all patients, in 6 cases with pseudopods; no reactions to the extract (1/10) in the control persons; total IgE: 45–1000 IU/ml; specific IgE (REIA) in all patients	Hinojosa *et al.* 1990

Table 2. continued

Patients with reactions	Test, test material (concentration)	Observations	References
1	prick test, scratch test (no data)	organ maker with rhinitis, conjunctivitis and dyspnoea; weak positive result in the prick test, positive result in the scratch test; total IgE: 343 IU/ml; patch tests with aqueous and acetone extracts yielded negative results	Hunzelmann 1994
1	provocation, wood dust	35 % decrease in FEV_1 after 10 min, after 1 h returned to initial value; also decrease in FVC (23 %) and PEF (50 %); control test with lactose yielded negative result	Innocenti and Angotzi 1980
20	skin test and provocation test, wood extract (no data)	in reports on 157 persons: positive results in the skin test for 31 and in the provocation test for 20; no data for clinical relevance	Kersten and von Wahl 1994
6	prick test and provocation test, wood extract (no data)	anamnesis suggestive for 8/10 patients with respiratory disease thought to be caused by wood dust and not suggestive for 2/10; positive results in the skin test for 6/8 and 0/2 persons, positive results in the provocation test for 5/8 and 1/2, for 2/8 and 1/2 in the test for histamine release; specific IgE detectable in 4/8 and 0/2	Merget *et al.* 1987
6	intracutaneous test, aqueous wood extract (no data)	of the 6 exposed persons yielding positive results in the intracutaneous test 4× asthma and 2× rhinitis	Nava 1974
19	intracutaneous test, wood extract (no data)	positive results obtained in the intracutaneous test for 19/220 asthmatics from woodworking factories; in 14 other persons weak positive test results without suggestive anamnesis	Oehling 1963
6	intracutaneous and provocation tests, wood extract (no data)	6/8 provocation tests produced acute airway obstruction; positive results obtained in the intracutaneous test or rubbing test for 5 patients with current sensitization and 6 patients with clinically silent sensitization	Oertmann and Bergmann 1993
2	prick test, damp wood dust and provocation test, wood dust	immediate reaction in the prick test in both patients (atopic and smokers), no reactions in 10 control persons; 20 min after provocation dyspnoea and decrease in FEV_1 (35 % and 22 %) and PEF (45 % and 35 %); total IgE: 2560 and 7 IU/ml, specific IgE: 17.5 and 0.3 IU/ml	Reijula *et al.* 1994
1	skin test, nasal, conjunctival and inhalative provocation test (no data)	positive result in the skin test for 1/6 patients with current abachi allergy; positive result in provocation test; after inhalative provocation obstructive reaction and 30 % decrease in FEV_1	Schwarting 1984

Table 2. continued

Patients with reactions	Test, test material (concentration)	Observations	References
18	prick test and provocation test, wood dust	positive results in the prick test for 15/18 patients (including 6 atopic persons and 7 smokers), immediate asthmatic reaction after provocation (no other details) in all 16 tested persons; total IgE: 13–3500 IU/l; specific IgE detected (RAST) in all 16 patients but not in 9 control persons; in 3 patients RAST inhibition of 89–95 %	Tuppurainen *et al.* 1995
1	intracutaneous, wood dust; inhalative provocation without a mask, wood dust	marked positive reaction in the intracutaneous test; dyspnoea 30 sec after start of provocation, increase in bronchial airway resistance from 0.43 to 0.94 kPa/l/s, after administration of a beta-adrenergic agent decrease to initial value and then another increase to 0.75 kPa/l/s after 6 h; total IgE (RIA) 1000 kU/l, RAST: class 1; no precipitating antibodies	Weber and Häußinger 1988
–	no data	15 cases of obstructive respiratory disease recognized by the trade association (no other details)	Wolf 1991
1	bronchial provocation, wood dust	in 1/21 employees tested in an exposure chamber immediate pulmonary reaction and decrease in FEV_1 of at least 15 %	deZotti *et al.* 1995

RAST radioimmunosorbent test
REIA reverse enzyme immunoassay
RIA radioimmunoassay
FEV_1 forced expiratory volume in the first second
FVC forced vital capacity
PEF peak expiratory flow

2 Manifesto (sensitization)

The wood of *Triplochiton scleroxylon* has marked sensitizing potential. The immunological genesis of the skin and airway reactions has been demonstrated in *in vitro* and *in vivo* studies. Therefore the wood is designated with "Sah".

3 References

Böckers M, Bork K (1990) Soforttypallergie auf Abachi-Holz (Immediate type allergy against obeche wood) (German). *Allergologie 13*: 71

Carosso A, Ruffino C, Bugiani M (1987) Respiratory diseases in wood workers. *Br J Ind Med 44*: 53–56

Fasani F, Pisati A, La-Manna A (1982) Fattori di rischio allergico nella lavorazione del legno (Factors of allergic risk in wood working) (Italian). *G Ital Med Lav 4*: 211–214

Girard JP, Surber R, Guberan E (1980) Allergic manifestations due to wood dusts. in: CA Frazier (Ed.) *Occupational asthma*, Van Nostrand Reinhold, New York, 91–101

Gronemeyer W, Debelic M (1967) Der sogenannte 'Reibtest', seine Anwendung und klinische Bedeutung (The so-called 'rubbing test', its use and clinical significance) (German). *Dermatologica 134*: 208–218

Hartmann AL (1978) *Das Holz als gesundheitliche Noxe – Eine Untersuchung anhand des Krankengutes der SUVA* (Wood as a health hazard – a study of SUVA cases) (German). Doctoral thesis, Medizinische Fakultät der Universität Zürich

Hartmann A, Schlegel H (1980) Durch Holz verursachte Gesundheitsschäden in der Schweiz (Adverse effects on health caused by wood in Switzerland) (German). *Schweiz Med Wochenschr 110*: 278–281

Hinojosa M, Losada E, Moneo I, Dominguez J, Carrillo T, Sanchez-Cano M (1986) Occupational asthma caused by African maple (obeche) and ramin: evidence of cross reactivity between these two woods. *Clin Allergy 16*: 145–153

Hinojosa M, Moneao I, Dominguez J, Delgado E, Losada E, Alcover R (1984) Asthma caused by African maple (*Triplochiton scleroxylon*). *J All Clin Immunol 74*: 782–786

Hinojosa M, Subiza J, Moneo I, Puyana J, Diez ML, Fernandez-Rivas M (1990) Contact urticaria caused by obeche wood (*Triplochiton scleroxylon*). Report of eight patients. *Ann Allergy 64*: 476–479

Hinojosa M, Subiza JL, Losada E (1995) Immunotherapy has proved to be effective in the treatment of African maple wood occupational asthma (AMWOA). *J Allergy Clin Immunol 95*: 317

Hunzelmann N (1994) Typ-I-Sensibilisierung gegen Abachi-Holz bei einem Orgelbauer (Type 1 sensitization to obeche wood in an organ maker) (German). *Z Haut Geschlechtskr 69*: 640–641

Innocenti A, Angotzi G (1980) Asma professionale da ensibilizzazione a *Triplochiton scleroxylon* (samba, obeche) (Occupational asthma due to *Triplochiton scleroxylon* (samba, obeche) (Italian). *Med Lav 71*: 251–254

Kersten W, von Wahl P-G (1994) Allergische Atemwegserkrankungen in der holzverarbeitenden Industrie (Occupational disease in the wood-processing industry) (German). *Allergologie 17*: 55–60

Krille M (1980) *Zur arbeitsdermatologischen Bedeutung exotischer Hölzer unter besonderer Berücksichtigung der Holzindustrie der DDR* (The dermatological significance of exotic woods at the workplace, in particular in the wood industry of the GDR) (German). Doctoral thesis, Zentralinstitut für Arbeitsmedizin der DDR

Merget R, Knapp U, Rexin S, Zachgo W, Maasch HJ, Wahl R, Schultze-Werninghaus G (1987) Beitrag zur Immunologie der beruflichen Abachiholzallergie (Immunology of occupational allergy to obeche wood) (German). *Allergologie 10*: 417–418

Nava C (1974) Patologia da polveri di legno (Pathology caused by wood dust) (Italian). *Med Lav 65*: 1–7

Oehling A (1963) Berufsallergie im Holzgewerbe (Occupational allergy in the wood industry) (German). *Allerg Asthma 9*: 312–322

Oertmann C, Bergmann K-C (1993) Atemwegserkrankungen bei Arbeitern im Holzgewerbe (Airway diseases in woodworkers) (German). *Allergologie 16*: 334–340

Quirce S, Hinojosa M, Maranon F, Ferrer A, Fernandez-Caldas E, Sastre J (2000) Identification of obeche wood (*Triplochiton scleroxylon*) allergens associated with occupational asthma. *J All Clin Immunol 106*: 400–401

Reijula K, Kujala V, Latvala J (1994) Sauna builder's asthma caused by obeche (*Triplochiton scleroxylon*) dust. *Thorax 49*: 622–623

Schwarting HH (1984) Wawa-Holzallergie (Allergy to wawa wood) (German). *Allergologie 7*: 356

Tuppurainen M, Keskinen H, Tupasela O, Nordman H (1995) Occupational asthma caused by Obeche wood dust. *Allergy 50 (Suppl)*: 38–39

Weber N, Häußinger K (1988) Asthma bronchiale bei Allergie gegen Abachiholz (Asthma bronchiale due to allergy against African maple) (German). *Prax Klin Pneumol 42*: 759–761

Wolf J (1991) Gesundheitsgefahren vermeiden (Avoiding health hazards) (German). *HGB-Mitteilungen 65*: 3–22

Woods B, Calnan CD (1976) Toxic woods. *Br J Dermatol 95, Suppl 13*: 1–97

deZotti R, Gubian F, Peresson M, Molinari S (1995) Asthma and rhinitis in woodworkers. *Allergy 50*: 38

completed 22.04.2002

Zinc chloride and zinc chloride fume

MAK value (2000)	**not yet established, see Section IIb of the *List of MAK and BAT Values***
Peak limitation	**–**
Absorption through the skin	**–**
Sensitization	**–**
Carcinogenicity	**–**
Prenatal toxicity	**–**
Germ cell mutagenicity	**–**
Synonyms	zinc dichloride
Chemical name (CAS)	zinc chloride
CAS number	7646-85-7
Structural formula	$ZnCl_2$
Molecular weight	136.29
Melting point	283°C
Boiling point	732°C

The present document is based mainly on the documentation of the American Conference of Governmental Industrial Hygienists (ACGIH 1992) and on a draft by the Committee on Updating OELs of the Health Council of the Netherlands which has not been published because of plans to assess this substance in the EU programme on existing chemicals.

1 Toxic Effects and Mode of Action

Zinc chloride can be taken up by man and animals by inhalation of zinc chloride fume and also after ingestion and, to a lesser extent, by skin absorption of zinc chloride solutions. Although zinc is an essential trace element, in high concentrations it can be toxic. Inhaled zinc chloride fume can cause airway irritation in man and, at high concentrations, metal fume fever and pulmonary oedema and fibrosis. The oral LD_{50} for the mouse and rat is between 350 and 1200 mg/kg body weight. In animals too the most important target organ for zinc chloride is the lung. Zinc chloride is highly irritating for the skin and mucous membranes. Zinc chloride has been tested in various *in vitro* test systems for genotoxicity but positive results were obtained only in a *Salmonella* mutagenicity test with the strain TA97 and after preincubation in water or HEPES buffer. *In vivo*, a clastogenic potential in the micronucleus test in the mouse has been reported, but only at toxic doses.

2 Mechanism of Action

Zinc is an essential trace element. The daily human intake with the diet is about 8 to 15 mg zinc. Zinc chloride is readily soluble in water and body fluids and acts in the organism mainly as ionic zinc. Zinc is required for the function of more than 70 enzymes such as dehydrogenases, aldolases and peptidases. In addition, by reducing levels of lipid peroxidation, zinc has a stabilizing effect on cell membranes. Zinc can inhibit histamine release from mast cells and macrophage migration (Arlette 1983).

3 Toxicokinetics and Metabolism

Zinc chloride can be taken up orally, dermally and by inhalation. After ingestion of the substance, the zinc ion is absorbed in the small intestine. The level of absorption is determined decisively by the nutritional status of the individual with respect to zinc, calcium and phosphorus and by the phytic acid content of the diet. Normally 8 % to 10 % of the zinc ingested in the diet is absorbed in the rat, cat and dog and the remainder is excreted with the faeces (FDA 1982). Of the zinc circulating in the blood plasma, 60 % is bound only weakly to albumins and amino acids; the remaining 40 % is bound strongly to α-globulins and practically not bioavailable. A person weighing 70 kg contains 2 to 3 g zinc, of which 50 % is in the bones. Although there may be more zinc in the bones and skin than in some other tissues, the element can be detected in every cell in the organism. The average biological half-life of the element in an adult person is 154 days. Zinc is excreted mainly in the faeces (Arlette 1983, FDA 1982).

The uptake of $ZnCl_2$ through human skin has been measured *in vitro* in the presence of $CuCl_2$. From 5 % solutions (w/w) of zinc chloride in petrolatum and in hydrogel (hydroxypropyl methyl cellulose), less than 10 % of the applied substance was absorbed in 72 hours and only 3 % in 24 hours. The authors do not discuss the possibility of interaction between the metal ions (Pirot *et al.* 1996).

$^{65}ZnCl_2$ penetrated the skin of rats very rapidly, but only in small quantities. Within 10 minutes after dermal application of $ZnCl_2$ (either 0.5–1.1 mCi/ml, > 100 mCi/mg Zn, pH 1 or adjusted to pH 4 with NaOH), radioactivity was detected in coagulated blood, in serum, liver and heart tissue. The values reached a maximum 4 hours after application.

During the 24 hours after application, the radioactivity in these tissues decreased slightly. 10 minutes after application, the tissue radioactivity was lowest in the testes where it increased slightly during the next 24 hours. Within 2 hours of dermal application, 4.1 % and 1.6 %, respectively, of the applied radioactivity was absorbed. Of that, 25 % to 29 % was detected in the liver, 21 % to 22 % in the gastrointestinal tract and 50 % to 53 % in the remaining carcass (Hallmans and Lidén 1979).

In guinea pigs too, the level of skin penetration is low. Guinea pigs treated dermally with $^{65}ZnCl_2$ concentrations of 5 to 4870 mM (dissolved in water and HCl, pH 1.8–6.1) generally absorbed 1 % to 2 % of the applied substance within 5 hours (Wahlberg 1965, Skog and Wahlberg 1964).

After administration of 15 µCi $^{65}ZnCl_2$ to male Wistar rats by intravenous injection, the plasma radioactivity decreased rapidly to 50 % of its initial value after 1 minute and 5 % after about 30 minutes. This level then remained constant in all assays carried out during the subsequent 5 hours. The ^{65}Zn levels detected in the tissues decreased in the following order: liver > kidney > spleen > stomach > gall bladder > lung > heart > bones > testes > blood cells > muscles > brain (Pullen *et al.* 1990).

Zinc can cross the placental barrier in the rat. When single doses of 10 µCi $^{65}ZnCl_2$ were injected intravenously into pregnant rats, the amount of zinc taken up by the whole foetus during the 24 hours after the injection increased as a function of application time from day 15 to day 21 of gestation by a factor of more than 25. The ^{65}Zn concentration in foetal tissue, on the other hand, was only doubled and had its maximum after application on day 19 of gestation. The level of ^{65}Zn retention in the whole placenta was also shown to increase at later application times. In comparison with the increase in the whole foetus, however, this effect was very small. In pregnant rats, the uptake of ^{65}Zn by the liver, spleen, kidney and thigh bone was lower than in the control group of not pregnant rats. With longer duration of gestation, the tissue concentrations in the dams 24 hours after dosing tended to decrease as a result of the increasing transfer of zinc into the foetus and placenta (Żylicz *et al.* 1975).

4 Effects in Man

4.1 Single exposures

Brief exposure to zinc chloride fume, for example, 5 hours at 600 mg/m^3, can cause metal fume fever (Doig and Challen 1964, Nemery 1990, Walsh *et al.* 1994). A sweet metallic taste and irritation in the throat are noticed 4 to 12 hours after the exposure. The influenza-like symptoms characteristic of metal fume fever can then develop.

Persons exposed to a zinc chloride fume concentration of 40 mg/m^3 experience a metallic taste. Experimental exposure to 80 mg/m^3 for 2 minutes caused slight nausea and coughing in most test persons. In persons exposed to 120 mg/m^3 for 2 minutes, irritation of the nose and airways developed (Schenker *et al.* 1981).

The results of studies of toxic effects in persons exposed briefly to zinc chloride fume (produced in the reaction of zinc or zinc oxide with hexachloroethane) are summarized below. With the exception of one study in which a concentration of 4075 mg/m^3 was reported, quantitative exposure data are not given. The symptoms seen after inhalative exposure are breathlessness, acute respiratory distress, sore throat, pains in the chest, tightness in the chest, acute airway irritation, cyanosis, painful coughing with sputum, bronchopneumonia, pulmonary oedema, pulmonary fibrosis, nausea and vomiting, and headaches. In a few cases, exposure to such high concentrations was lethal within 24 to 32 days. In some cases, the symptoms disappeared about two hours after the exposure but then became more severe later (after a few hours, 10 days or 2 weeks). In persons exposed to lower concentrations the effects were reversible and most symptoms disappeared within a few days, although some persisted for 1 to 2 weeks or even for 12 months (ACGIH 1992).

One patient who was splashed around his eyes, nose and throat with a concentrated solution of zinc chloride (see Section 4.3) permanently lost his sense of smell in spite of medical treatment (Houle and Grant 1973). Accidental ingestion of zinc chloride solutions (e.g. soldering fluid) by two persons caused local corrosion in the mouth, throat, oesophagus, stomach and trachea and also nausea, vomiting, abdominal pains and lethargy (Chobanian 1981, McKinney *et al.* 1994, 1995).

4.2 Local effects on skin and mucous membranes

A patch test with 2% zinc chloride solution resulted in the formation of pustules and vesicles on the skin of 5 of 15 dentists. However, the study did not include a control group. In addition, the reactions could not be reproduced in a second patch test and were therefore interpreted as skin irritation (Shigeto *et al.* 1989).

In an accident, a concentrated zinc chloride solution was splashed into both eyes of one person and into one eye of a second person. In the affected eyes, corneal oedema developed and some persistent corneal damage. It took 6 to 28 weeks before maximum visual acuity returned (Houle and Grant 1973).

4.3 Allergenic effects

Two cases of occupational asthma caused by soldering fluid have been described. However, both persons were exposed to ammonium chloride as well as zinc chloride (Weir *et al.* 1989).

5 Animal Experiments and *in vitro* Studies

5.1 Acute toxicity

Exposure of rats for 10 minutes to zinc chloride aerosols (droplets with a mass median aerodynamic diameter (MMAD) of 2.3–2.6 μm) at $ZnCl_2$ concentrations of 1980 mg/m^3 or more was lethal for the animals within 3 days (10-minute LC_{50} ($ZnCl_2$): > 1200 mg/m^3 and < 2000 mg/m^3). No symptoms of airway irritation were observed during the exposure, but shortness of breath developed gradually. Autopsy revealed atelectasis (collapsed lung areas), hyperaemia, haemorrhage and oedema in the lungs. The symptoms and macroscopically visible changes were like those seen in animals exposed to a pyrotechnic mixture (zinc and hexachloroethane) (Karlsson *et al.* 1986).

Groups of 10 New Zealand White rabbits and Porton Wistar rats were exposed for 30 minutes to zinc chloride fume generated from pyrotechnic mixtures of zinc oxide and hexachloroethane. The two exposure concentrations were given as 2750 and 4000 mg/m^3 total solids and 600 and 810 mg/m^3 zinc, equivalent to about 1250 and 1690 mg/m^3 $ZnCl_2$. In the lower concentration group, 7 rabbits and 5 rats died during the first 24 hours; the remaining animals survived until the end of the 14-day observation period. In the higher concentration group, 5 rabbits and 5 rats died during the first 24 hours and the remaining 5 rabbits during the 7 days after the exposure. The histological examination of the animals which died revealed changes only in the respiratory passages. Acute inflammation and necrosis were detected in the larynx and trachea, and alveolitis, petechial haemorrhages, oedema and bronchopneumonia in the lungs. In the surviving animals the effects regressed practically completely (Marrs *et al.* 1983).

The oral LD_{50} ($ZnCl_2$) is 1100 mg/kg body weight for the rat and 1260 mg/kg body weight for the mouse. Intraperitoneal injection of $ZnCl_2$ revealed an LD_{50} of 58 mg/kg body weight for the rat and 91 mg/kg body weight for the mouse (Domingo *et al.* 1988).

5.2 Subacute, subchronic and chronic toxicity

Groups of 100 female Porton mice, 50 female Porton Wistar rats and 50 female Dunkin-Hartley guinea pigs were exposed to zinc concentrations of 0, 1.3, 12.8 or 119 mg/m^3 (equivalent to 2.71, 26.7 and 254 mg/m^3 $ZnCl_2$), 1 hour daily, 5 days weekly for 29 weeks (in all 100 single exposures, but guinea pigs in the high concentration group only 15 single exposures) (Marrs *et al.* 1988). The animals were killed 13 months later. The zinc chloride fume (MMAD about 2 μm) was generated from a pyrotechnic mixture (hexachloroethane 44 %, potassium nitrate 2 %, zinc oxide 32–47 %, calcium silicide 7–22 %); zinc comprised only 20 % w/w of the fumes and the animals were exposed simultaneously to other substances. The results of this study are shown in Table 1.

Table 1. Results of a long-term study in which animals inhaled zinc chloride fume generated from a pyrotechnic mixture (Marrs *et al.* 1988)

Species:	Porton mice, 100 ♀ per group Porton Wistar rats, 50 ♀ per group Dunkin-Hartley guinea pigs, 50 ♀ per group
Exposure:	inhalation of zinc chloride fume generated from a pyrotechnic mixture of zinc and hexachloroethane mouse, rat: 0, 1.3 ± 0.4, 12.8 ± 2.6, 121.7 ± 17.8 mg/m^3 $ZnCl_2$ guinea pig: 0, 1.3 ± 0.4, 12.8 ± 2.6, 119.3 ± 31.7 mg/m^3 $ZnCl_2$
Duration:	1 h/day, 5 days/week, 29 weeks – a total of 100 single exposures, recovery period 13 months, except for the guinea pigs of the 119.3 mg/m^3 group: 1 h/day, 5 days/week, 3 weeks – a total of 15 single exposures, recovery period 26 weeks + 13 months
Toxicity:	mouse: no effects on body weight gains; 121 mg/m^3: 50 % premature deaths (less than 4 months after start of exposure) rat: no effects on body weight gains or mortality; no tumours guinea pigs exposed to 1.3 and 12.8 mg/m^3: no effects on body weight gains or mortality; no tumours guinea pigs exposed to 119.3 mg/m^3: very high mortality at the start of exposure and therefore exposure limited to a total of 15 sessions

Results of the histopathological examination:			control	1.3 mg/m^3	12.8 mg/m^3	121.7 mg/m^3
mouse	lung	carcinoma originating in the alveoli	6/78	7/74	8/76	15/50**
		focal lymphocytic infiltration	0/78	4/74	0/76	4/50*
	larynx	submucosal lymphocytic infiltration	34/78	17/74	19/76	7/50*
	kidney	chronic interstitial nephritis	26/78	36/74	20/76	20/50
	liver	fatty degeneration	11/78	20/74	28/76*	16/50*
		lymphocytic infiltration	6/78	8/76	15/76	12/50*
rat	lung	lymphocytic infiltration (peribronchial)	26/43	32/44	10/48**	7/44**
	kidney	nephropathy	28/43	35/44	31/48	30/44
guinea pig	lung	lymphocytic infiltration	44/45	38/39	44/46	39/39
	kidney	chronic interstitial nephritis	28/45	28/39	37/46	26/39

* $p < 0.05$
** $p < 0.01$

The target organ for zinc chloride fume was the respiratory tract. The incidence of lung carcinoma (7/74, 8/74, 15/50, control: 6/78) was increased in the mice in the highest concentration group; mortality was markedly higher in this group. In rats and guinea pigs, however, no lung tumours were found. Other target organs in the mice were the liver and kidneys. The study is of limited usefulness in the present assessment for the reasons given below. An increased incidence of lung tumours was found only in mice and only in the highest concentration group in which mortality was high, 50 % during the first four months. The incidence of lung tumours in historical controls in this strain of mouse is very high, up to 18 %. The histological examination was carried out 13 months after the end of the 5-month exposure period. This study period of about 18 months is long enough as a latency period for the development of tumours in mice, but not in rats or guinea pigs. Only female animals of all species were tested.

5.3 Local effects on skin and mucous membranes

Zinc chloride causes severe irritation of the skin. The application of a 10 % solution of zinc chloride without occlusion to the skin of AG2 mice, New Zealand White rabbits and Dunkin-Hartley guinea pigs resulted in dermal and epidermal irritation, ulcers, follicular acanthosis, parakeratosis and hyperkeratosis (Lansdown 1991).

There are no studies available of the allergenic effects of zinc chloride.

5.4 Reproductive and developmental toxicity

28 ICR/SIM mice were given $ZnCl_2$ doses of 150 mg/kg body weight and day by gavage on days 8 to 12 of gestation. The average increase in body weights of the dams was significantly lower than that of the control animals (n = 28). Zinc chloride caused a slight but significant reduction in litter sizes but only when administered in maternally toxic doses (Seidenberg *et al.* 1986).

10 pregnant Syrian golden hamsters were given a single $ZnCl_2$ dose of 2 mg/kg body weight by intravenous injection on day 8 of gestation. Neither changes in the dams nor embryotoxic or foetotoxic effects were observed (Hartsfield *et al.* 1992).

In an intraperitoneal teratogenicity test, pregnant CF-1 albino mice were given $ZnCl_2$ doses of 12.5, 20.5 or 25 mg/kg body weight on day 8, 9, 10 or 11 of gestation. Doses of 20.5 mg/kg body weight and more caused maternal toxicity and an increase in the incidence of skeletal anomalies. The anomalies detected were supernumerary and ripple ribs and delayed ossification of sternebrae and toe bones (Chang *et al.* 1977).

5.5 Genotoxicity

5.6.1 *In vitro*

The results of the genotoxicity studies are summarized in Table 2.

Negative results were obtained in a plate incorporation test with various *Salmonella typhimurium* strains in the presence and absence of S9 mix (Karlsson *et al.* 1991, Wong 1988). Only after preincubation with deionized water or HEPES buffer was weak mutagenicity of $ZnCl_2$ detected in the absence of S9 mix in the *Salmonella typhimurium* strain TA97 (revertants/plate: with water and 250.5 μM $ZnCl_2$ 219, control 92; with HEPES buffer and 62.5 μM $ZnCl_2$ 217, control 137) (Pagano and Zeiger 1992). In a mouse lymphoma test, $ZnCl_2$ in concentrations of 1.21–12.13 μg/ml induced a slight increase in the number of mutants but the effect was not statistically significant and not dose-dependent (Amacher and Paillet 1980). Other genotoxicity tests including the rec assay, induction of DNA strand breaks, DNA repair or chromosomal aberrations all yielded negative results.

5.6.2 *In vivo*

Table 2 also shows the results of the *in vivo* genotoxicity studies.

Administration of zinc chloride to mice for one month at a concentration of 0.5 % in the diet (about 750 mg/kg body weight and day) induced chromosomal aberrations in bone marrow cells only in the animals given a calcium-deficient diet, not in those given a standard diet. However, because the body weights of the animals were markedly reduced and some of them died, these results cannot be interpreted (DeKnudt 1979).

Zinc chloride administered to Swiss albino mice (5 per group) as single intraperitoneal doses of 7.5, 10 and 15 mg/kg body weight, corresponding to 25 %, 33 % and 50 % of the LD_{50} (30 mg/kg body weight), induced a significant dose-dependent increase in the incidence of chromosomal aberrations in bone marrow cells. The clastogenic effect was also seen after repeated intraperitoneal injection of zinc chloride solutions. Doses of 2 mg/kg and 3 mg/kg body weight (7 % and 10 % of the LD_{50}) were injected into groups of 5 mice every second day and the animals were killed after the injection on day 8, 16 or 24. The incidence of chromosomal aberrations in the bone marrow cells of the animals was increased significantly as a function of dose and duration of treatment. After repeated intraperitoneal injection of $ZnCl_2$ doses of 7.5, 10 or 15 mg/kg body weight and day (25 %, 33 % and 50 % of the LD_{50}) for 5 days, the incidence of sperm head anomalies was increased significantly in a dose-dependent manner (Gupta *et al.* 1991). In this publication, toxic effects in the treated animals are not documented. It must be assumed that the doses used were mainly in the toxic range because they were high relative to the LD_{50}. Therefore the study must be interpreted with caution.

In an inadequately documented micronucleus test in which condensed fumes from a pyrotechnic mixture of zinc and hexachloroethane were administered to mice by intraperitoneal injection, no evidence of clastogenic effects was found (Karlsson *et al.* 1991).

Table 2. Genotoxicity studies with zinc chloride

Test system ***in vitro***	Test organism or animal species, administration route	Concentration or dose	Results −S9/+S9	Comments	References
SMT (plate incorporation test)	*S. typhimurium* TA98, TA100, TA1535, TA1537	no data	−/−	condensate of fumes from a pyrotechnic Zn/hexachloroethane mixture	Karlsson *et al.* 1991
SMT (no other details)	*S. typhimurium* TA92, TA94, TA98, TA100, TA1535, TA1537, TA1538	no data	−	inadequately documented	Sokolowska and Jongen 1984
SMT (plate incorporation test)	*S. typhimurium* TA98, TA102, TA1535, TA1537	120 ppm	−/−	cytotoxic: 50 % growth inhibition	Wong 1988
SMT (no other details)	*Salmonella typhimurium* no other details	no data	+	inadequately documented, results not usable	Kalinina and Polukhina 1977
SMT (preincubation test)	*S. typhimurium* TA97	15–1000 µM	(+)/n.t.	reproducible increase in mutants only after preincubation in double distilled water or HEPES buffer	Pagano and Zeiger 1992
Vitotox test for genotoxicity	*S. typhimurium* TA98 and TA104 modified with genes whose induction is dependent on SOS repair	no data	−/n.t.		van der Lelie *et al.* 1997
induction of λ prophages	*E. coli* $WP2_s(\lambda)$	3200 µM	−/n.t.		Rossman *et al.* 1984
rec assay (differential killing assay)	*Bacillus subtilis* H_{17}, M_{45}	2.5 µmol/assay	−/n.t.		Nishioka 1975
gene mutation (TK)	mouse L5178Y/$TK^{+/-}$ lymphoma cells	1.21–12.13 µg/ml	−/n.t.	at 12.13 µg/ml survival > 20 %	Amacher and Paillet 1980
DNA strand breaks	CHO cells	10, 100 µM	−/n.t.		Robison *et al.* 1982, 1984
DNA repair	CHO cells	1000 µM	−/n.t.		Robison *et al.* 1984
chromosomal aberrations	human lymphocytes	3×10^{-5} M	−/n.t.	3×10^{-3} M cytotoxic	Deknudt and Deminatti 1978

Table 2. continued

Test system *in vivo*	Test organism or animal species, administration route	Concentration or dose	Results – S9/+S9	Comments	References
CA in the bone marrow	mouse, C57Bl oral 25 ♂, chromosomal aberrations determined in 10 surviving ♂ from each group	0.5 % in standard diet (1.1 % Ca), about 750 mg/kg body weight and day for 1 month	–	marked toxicity: significantly reduced body weights (–40 %)	Deknudt and Gerber 1979
		0.5 % in Ca reduced diet (0.03% Ca), about 750 mg/kg body weight and day for 1 month	+	marked toxicity: significantly reduced body weights (–45%), dose was in the range LD_{30}–LD_{50} results not usable	
CA in the bone marrow	Swiss Albino mice 5 per dose	7.5, 10, 15 mg/kg body weight, once i.p.	+	doses 25 %, 33 % and 50 % of the LD_{50}, toxic effects not documented CA: significantly increased, dose-response relationship;	Gupta *et al.* 1991
		2 and 3 mg/kg body weight and day, i.p., every second day during a period of 8, 16 or 24 days	+	doses 7 % and 10% of the LD_{50}, toxic effects not documented CA: significantly increased, dose-response relationship between effects increased with duration of exposure	
sperm head anomalies	Swiss Albino mice 5 per dose	7.5, 10, 15 mg/kg body weight and day for 5 days i.p.	+	doses 25 %, 33 % and 50 % of the LD_{50}, toxic effects not documented sperm head anomalies: significantly increased, dose-response relationship	Gupta *et al.* 1991
micronucleus test in bone marrow cells	CBA mouse 5 ♂ or 5 ♀ per group	no data i.p.	–	0.1 ml of a condensate of fumes from a pyrotechnic mixture of zinc and hexachloroethane: MTD results not usable	Karlsson *et al.* 1991

(+), weakly positive result; CA chromosomal aberrations; MTD, maximum tolerable dose; n.t., not tested; TK thymidine kinase

5.7 Carcinogenicity

5.7.1 Long-term studies

Female Porton mice, female Porton Wistar rats and female Dunkin-Hartley guinea pigs were exposed to zinc chloride fume by inhalation (see Table 1 and Section 5.2); lung tumours developed only in the mice of the highest concentration group (Marrs *et al.* 1988). For a number of reasons, this study is only of limited value (Section 5.2) and cannot be taken as evidence of a carcinogenic effect of zinc chloride.

After injection of 0.05 ml aliquots of a zinc chloride solution (40 mg/ml; dose 4 mg per animal) into both testes of each of 49 Syrian hamsters (2 months old, injection during the annual phase of rapid gonad growth, animals killed about 10 weeks later), the histopathological examination revealed focal necrosis (about 25 % of each testis, in 6 animals the testes were completely destroyed) and in two animals an embryonal carcinoma in the testis adjacent to necrotic areas (Guthrie and Guthrie 1974). Because of the massive local effects and the choice of an application route which is irrelevant for man, the results of this study do not permit conclusions as to a potential carcinogenic effect of zinc chloride.

5.7.2 Cell transformation studies

Zinc chloride (tested in concentrations of 1–30 µg/ml) yielded positive results in a cell transformation test after exposure of SHE cells for 7 days to concentrations of 20 and 30 µg/ml (LeBoeuf *et al.* 1996). In accordance with the above results, in another cell transformation test with SHE cells incubated for 7 days with zinc chloride in a single concentration of 3.0 µg/ml, no increase in cell transformation rate was detected (Rivedal and Sanner 1981).

6 Manifesto (MAK value/classification)

Zinc chloride and zinc chloride fume cause irritation of the airways, skin and eyes in man and experimental animals. Data for an irritation threshold from studies of single or multiple exposures of persons or experimental animals which would be suitable for the derivation of a MAK value are not available. Zinc chloride and zinc chloride fume are therefore included in Section IIb of the *List of MAK and BAT Values*.

There are no valid carcinogenicity studies available for zinc chloride. Zinc chloride can induce cell transformation *in vitro*. Zinc chloride is mutagenic in the *Salmonella* mutagenicity test only in the strain TA97 and only under special test conditions (preincubation in deionized water or HEPES buffer). No genotoxic effects have been seen in mammalian cells *in vitro*. *In vivo* studies have yielded occasional indications of a genotoxic potential of zinc chloride, but mainly after doses in the toxic range. Further study of the genotoxic and carcinogenic effects of zinc chloride are therefore necessary for a conclusive evaluation of the carcinogenic potential.

On the basis of the available data, it cannot be decided whether designation with "S" or "H" is necessary.

7 References

ACGIH (American Conference of Governmental Industrial Hygienists) (1992) Zinc chloride fume. in *Documentation of the threshold limit values and biological exposure indices.* ACGIH, Cincinnati, Ohio, USA

Amacher DE, Paillet SC (1980) Induction of trifluorothymidine-resistant mutants by metal ions in L5178Y/TK$^{+/-}$ cells. *Mutat Res 78*: 279–288

Arlette JP (1983) Zinc and the skin. *Pediatr Clin North America 30*: 583–596

Chang CH, Mann DE jr, Gautieri RF (1977) Teratogenicity of zinc chloride, 1,10-phenanthroline, and a zinc-1,10-phenanthroline complex in mice. *J Pharmaceut Sci 66*: 1755–1758

Chobanian SJ (1981) Accidental ingestion of liquid zinc chloride: local and systemic effects. *Ann Emerg Med 10*: 91–93

Deknudt und Deminatti (1978) Chromosome studies in human lymphocytes after *in vitro* exposure to metal salts. *Toxicology 10*: 67–75

DeKnudt G, Gerber GB (1979) Chromosomal aberrations in bone-marrow cells of mice given a normal or a calcium-deficient diet supplemented with various heavy metals. *Mutat Res 68*: 163–168

Doig AT, Challen PJR (1964) Respiratory hazards in welding. *Ann Occup Hyg 7*: 223–231

Domingo JL, Llobet JM, Paternain JL, Corbella J (1988) Acute zinc intoxication: comparison of the antidotal efficacy of several chelating agents. *Vet Hum Toxicol 30*: 224–228

FDA (Food and Drug Administration) (1982) Zinc salts: Proposed affirmation of GRAS status. *Fed Reg 457*: 47441–47445

Gupta T, Talukder G, Sharma A (1991) Cytotoxicity of zinc chloride in mice *in vivo. Biol Trace Elem Res 30*: 95–101

Guthrie J, Guthrie OA (1974) Embryonal carcinomas in Syrian hamsters after intratesticular inoculation of zinc chloride during seasonal testicular growth. *Cancer Res 34*: 2612–2614

Hallmans G, Lidén S (1979) Penetration of ^{65}Zn through the skin of rats. *Acta Derm Venereol (Stockh) 59*: 105–112

Hartsfield JK Jr, Lee M, Morel JG, Hilbelink DR (1992) Statistical analysis of the effect of cadmium and zinc on hamster teratogenesis. *Biochem Med Metabol Biol 48*: 159–173

Houle RE, Grant WM (1973) Zinc chloride keratopathy and cataracts. *Am J Ophthalmol 75*: 992–996

Kalinina LM, Polukhina GH (1977) Mutagenic effect of heavy metal salts on *Salmonella* in activation systems *in vivo* and *in vitro*. (Abstract). *Mutat Res 46*: 223–224

Karlsson N, Cassel G, Fangmark I (1986) A comparative study of the acute inhalation toxicity of smoke from TiO_2-hexachloroethane and Zn-hexachloroethane pyrotechnic mixture. *Arch Toxicol 59*: 160–166

Karlsson N, Fängmark I, Häggqvist I, Karlsson B, Rittfeldt L, Marchner H (1991) Mutagenicity testing of condensates of smoke from titanium dioxide/hexachloroethane and zinc/hexachloroethane pyrotechnic mixtures. *Mutat Res 260*: 39–46

Lansdown ABG (1991) Interspecies variations in response to topical application of selected zinc compounds. *Food Chem Toxicol 29*: 57–64

LeBoeuf RA, Kerckaert GA, Aardema MJ, Gibson DP, Brauninger R, Isfort RJ (1996) The pH 6.7 Syrian hamster embryo cell transformation assay for assessing the carcinogenic potential of chemicals. *Mutat Res 356*: 85–127

Marrs TC, Colgrave HF, Edgington JAG, Brown RFR, Cross NL (1988) The repeated dose toxicity of a zinc oxide/hexachloroethane smoke. *Arch Toxicol 62*: 123–132

Matarese SL, Matthews JI (1986) Zinc chloride (smoke bomb) inhalational lung injury. *Chest 89*: 308–309

McKinney PE, Brent J, Kulig K (1994) Acute zinc chloride ingestion in a child: local and systemic effects. *Ann Emerg Med 23*: 1383–1387

McKinney PE, Brent J, Kulig K (1995) Zinc chloride ingestion in a child: exocrine pancreatic insufficiency. *Ann Emerg Med 25*: 562

Nemery B (1990) Metal toxicity and the respiratory tract. *Eur Respir J 3*: 202–219

Nishioka H (1975) Mutagenic activities of metal compounds in bacteria. *Mutat Res 31*: 185–189

Pagano DA, Zeiger E (1992) Conditions for detecting the mutagenicity of divalent metals in *Salmonella typhimurium*. *Environ Mol Mutagen 19*: 139–146

Pirot F, Panisset F, Agache P, Humbert P (1996) Simultaneous absorption of copper and zinc through human skin *in vitro*. Influence of counter-ion and vehicle. *Skin Pharmacol 9*: 43–52

Pullen RGL, Franklin PA, Hall GH (1990) ^{65}Zinc uptake from blood into brain and other tissues in the rat. *Neurochem Res 15*: 1003–1008

Rivedal E, Sanner T (1981) Metal salts as promoters of *in vitro* morphological transformation of hamster embryo cells initiated by benzo(*a*)pyrene. *Cancer Res 41*: 2950–2953

Robison SH, Cantoni O, Costa M (1982) Strand breakage and decreased molecular weight of DNA induced by specific metal compounds. *Carcinogenesis 3*: 657–662

Robison SH, Cantoni O, Costa M (1984) Analysis of metal-induced DNA lesions and DNA-repair replication in mammalian cells. *Mutat Res 131*: 173–181

Rossman TG, Molina M, Meyer LW (1984) The genetic toxicology of metal compounds. I. Induction of lambda prophage in E. coli WP2s(λ). *Environ Mutagen 6*: 59–69

Schenker MB, Speizer FE, Taylor JO (1981) Acute upper respiratory symptoms resulting from exposure to zinc chloride aerosol. *Environ Res 25*: 317–324

Seidenberg JM, Anderson DG, Becker RA (1986) Validation of an *in vivo* developmental toxicity screen in the mouse. *Teratogen Carcinogen Mutagen 6*: 361–374

Shigeto N, Numata T, Abe M, Murata H, Hamada T, Kioka M (1989) Metal hypersensitivity in dentists: a patch test study. *Hiroshima J Med Sci 38*: 187–189

Skog E, Wahlberg JE (1964) A comparative investigation of the percutaneous absorption of metal compounds in the guinea pig by means of the radioactive isotopes: ^{51}Cr, ^{56}Co, ^{65}Zn, ^{110m}Ag, ^{115m}Cd, ^{200}Hg. *J Invest Dermatol 43*: 187–192

Sokolowska DM, Jongen WM (1984) Heavy metals and *Salmonella typhimurium*: mutagenicity and interaction with model compounds (Abstract). *Mutat Res 130*: 168

van der Lelie D, Regniers L, Borremans B, Provoost A, Verschaeve L (1997) The VITOTOX test, an SOS bioluminescence *Salmonella typhimurium* test to measure genotoxicity kinetics. *Mutat Res 389*: 279–290

Wahlberg JE (1965) Percutaneous toxicitiy of metal compounds. *Arch Environ Health 11*: 201–204

Walsh CT, Sandstead HH, Prasad AS, Newberne PM, Fraker PJ (1994) Zinc: health effects and research priorities for the 1990s. *Environ Health Perspect 102, Suppl 2*: 5–46

Weir DC, Robertson AS, Jones S, Sherwood Burge P (1989) Occupational asthma due to soft corrosive soldering fluxes containing zinc chloride and ammonium chloride. *Thorax 44*: 220–223

Wong PK (1988) Mutagenicity of heavy metals. *Bull Environ Contam Toxicol 40*: 597–603

Żylicz E, Zablotna R, Szot Z (1975) Zinc-65 transfer across the placenta at late pregnancy in rat. *Nukleonika 20*: 887–896

completed 14.01.2000

Zinc oxide fume

MAK value (2000)	**1 mg/m^3 R (respirable dust fraction)**
Peak limitation	–
Absorption through the skin	–
Sensitization	–
Carcinogenicity	–
Prenatal toxicity	–
Germ cell mutagenicity	–
Synonyms	C.I. 77947 C.I. pigment 4 flowers of zinc philosopher's wool red oxide of zinc spartalite sterlingite zinc white zincite
Chemical name (CAS)	zinc oxide
CAS number	1314-13-2
Structural formula	ZnO
Molecular weight	81.37
Melting point	1975°C

1 Characterization of the Substance, Production, Use

1.1 Physiochemical characterization of the substance

Zinc is a divalent, bluish-white, soft heavy metal, which is a poor conductor of electricity; it is abundant in the earth's crust (about 0.02 %).

Burning of zinc yields zinc oxide. When zinc is heated to temperatures above 500°C in the flame during oxygen cutting or in the arc during welding, zinc oxide particles are released into the air as zinc oxide fume and can be inhaled (Mueller and Seger 1985).

Zinc oxide occurs naturally as the mineral zincite in coarsely particulate masses. The density of zinc oxide is 5.61 g/cm^3. The weakly alkaline zinc oxide powder is practically insoluble in water and alcohol but dissolves in dilute acetic acid and dilute mineral acids.

1.2 Production

Zinc has been produced commercially since the 18th century. In 1986 more than 6.8 million tons of zinc were produced. The main producers were Canada, Mexico, Honduras and Peru. First, the zinc ore is milled. With flotation techniques, a fraction with a metal content of 50 % to 60 % is isolated. This concentrate is roasted to remove sulfur. Then it is processed further by smelting or by electrolytic purification processes. The zinc smelt contains other metals such as lead, copper and cadmium as impurities. The electrolytic purification process yields very pure zinc (99.99 % Zn) which can be used for alloys and injection moulding (Fisher 1992). Zinc oxide is produced by oxidation of metallic zinc vapour or in wet chemical processes from various zinc salts. The diameter of the particles in the usual ZnO pigments is about 250 nm, in transparent micronized ZnO for cosmetics between 10 and 70 nm (Tichy 1993).

1.3 Uses

The main use of zinc is in galvanization of iron and steel to prevent corrosion and oxidation. Metallic zinc is also used in injection moulding for parts for cars, electrical appliances, tools, metal objects, jewellery and toys. Since ancient times zinc has been used in alloys with copper and tin. Zinc alloys can contain lead, cadmium, iron, tin, copper, aluminium, magnesium and titanium.

Zinc oxide is the most used zinc compound and most zinc oxide is used in the rubber industry (as an activator in the vulcanization process). Zinc oxide is also used as a grey colouring (zinc grey) or a white pigment (zinc white) in the production of paints, primers and fillers. It is used in ceramics and in photocopying.

At the height of the Egyptian civilization and in other Mediterranean cultures, zinc was probably used as a component of ointments. Today too, zinc oxide is used in cosmetics (UV absorber, powder, make-up), in zinc pastes, ointments, mixtures to be shaken, as a component of sticking plaster and an additive in dental preparations (Fisher 1992).

2 Exposure

It has been estimated that in the USA about 700000 workers run the risk of developing metal fume fever, and world-wide about 3 million workers. The NIOSH (National Institute for Occupational Safety and Health) has estimated that about 50000 workers are exposed to zinc fume in foundries, during welding of galvanized materials, during soldering or oxygen cutting of brass or galvanized metals, or in galvanizing processes. Most authors express the opinion that metal fume fever can only be induced by freshly formed zinc fume. Some researchers, on the other hand, suggest that high level exposures to aged zinc fume can also cause zinc fume fever (Perry 1994). Examination of one third of the cases of metal fume fever reported in Switzerland between 1987 and 1992, selected at random, revealed that about 60 % of patients were persons who had used oxygen cutting on galvanized boilers and hot water systems in small, poorly ventilated cellar rooms; most of the workers affected were from the heating and plumbing industries. In about 30 % of cases, the metal fume fever could be ascribed to welding of galvanized metals (Rüegger 1995).

During sanding and welding of galvanized automobile bodywork, the average ZnO concentrations in the air were 1.08 and 2.22 mg/m^3, respectively; during work with sheet zinc and steel, maximum zinc oxide concentrations averaged 0.64 and 0.03 mg/m^3. Physiological urine and serum zinc concentrations vary widely; thus these parameters were found to be not useful for monitoring persons exposed to concentrations in the range of the then valid maximum workplace concentration of 5 mg/m^3 (Knecht *et al.* 1983).

3 Toxic Effects and Mode of Action

Zinc is an essential element for all living creatures. The amount required by man (10–15 mg/day) is generally present in the diet.

Whereas metallic zinc is practically non-toxic, inhalation of zinc oxide during smelting or welding can cause acute intoxication (metal fume fever). This disorder develops only after inhalation of zinc oxide, not after oral, parenteral or dermal application of the substance. The symptoms of metal fume fever first appear between four and twelve hours after exposure as a sweetish metallic taste and a feeling of irritation in the throat. Unspecific influenza-like symptoms (general feeling of unwellness, coughing, weakness, etc.) can follow. Some of the affected persons develop none of these signs but sudden chills and fever, a feeling of tightness in the chest and mostly mild respiratory symptoms (dyspnoea). During the further course of the disease, various organ systems can be affected. In addition to rapidly increasing temperature, the symptoms may include frontal headaches, visual disorders, pain in the abdomen, muscles and joints, nausea, vomiting and dysuria. Some exposed persons develop asthmatiform reactions. Sometimes provocation tests yield evidence of a hyper-reactive bronchial system.

The inhalation of zinc oxide fume causes inflammatory changes in the lungs of man and animals; ZnO concentrations of 2.5 mg/m^3 and more cause changes in inflammation markers in the bronchoalveolar lavage fluid (BAL), in lung function and histological parameters.

4 Mechanism of Action

In early studies of the pathogenesis of metal fume fever, direct toxic effects on the bronchi and lung parenchyma were sometimes assumed. Mechanisms such as denaturation of protein and transient adrenal insufficiency were also thought to be responsible for the metal fume fever (Mueller and Seger 1985).

As exposure to zinc oxide fume does not result in a clear increase in the blood zinc level, it must be considered unlikely that inhaled zinc is absorbed and distributed in quantities sufficient to cause peripheral effects directly (Fine *et al.* 1997). It has been suggested that the pathogenesis of metal fume fever involves the increased formation of oxygen radicals in human neutrophils (detected with a chemoluminescence immunoassay) in the presence of Zn^{2+} and ZnO (Lindahl *et al.* 1998).

The results of animal studies and the known effects in man suggest that the systemic effects of zinc oxide fume are indirect effects of the massive inflammatory changes in the lungs with release of mediators and cytokines (Blanc *et al.* 1991, 1993, Fine *et al.* 1997). There is no agreement as to which mediators play the main role in the pathogenesis of metal fume fever. Suggestions include the tumour necrosis factor (TNF) (Blanc *et al.* 1993, Kuschner *et al.* 1997) and interleukin-6 (IL-6) (Fine *et al.* 1997).

In test persons exposed experimentally to welding fumes (cumulative zinc exposure level in the range between 1.8 and 2.6 g × min/m^3, exposure for 15–30 min), an association was demonstrated between the recovery period (3, 8 or 22 hours) and the levels of inflammation markers in the BAL (see Table 1; Blanc *et al.* 1993). Whereas the counts of polymorphonuclear leukocytes, macrophages and leukocytes, and the level of IL-6 increased with increasing time after the exposure (in some cases significantly), the levels of IL-1 and TNF in the BAL decreased during the same period. This is seen as confirmation of the hypothesis that a network of mediators and cytokines is causally involved in the pathogenesis of metal fume fever. Probably TNF is the key mediator in the pathophysiology of metal fume fever because its concentration is significantly increased as little as 3 hours after the exposure to zinc oxide and decreases again soon after (Blanc *et al.* 1993, Kuschner *et al.* 1997) at a time at which the typical symptoms of metal fume fever begin to appear. In contrast, the maximum concentrations of IL-6 were found after 22 hours, IL-8 after 8 hours and IL-1 after 3 hours and none of these was increased significantly. In addition, 8 and 22 hours after exposure, the level of TNF was shown to correlate with that of IL-8 and of IL-6 respectively; this suggests that the release of these two cytokines was stimulated by TNF (Blanc *et al.* 1991, 1993, Kuschner *et al.* 1997). *In vitro* studies with mononuclear cells (human U937 cells) incubated with solutions of zinc oxide (0.1–1.0 mM) revealed a concentration-dependent increase in TNF and, after a time-lag, a significant increase in the level of IL-8 in the culture medium at 1 mM ZnO. This increase in IL-8 correlated with the TNF increase (Kuschner *et al.* 1998).

Another research group found that 3 and 6 hours after experimental exposures to 2.5 and 5 mg/m^3 zinc oxide fume, the plasma levels of IL-6 increased with the time since exposure (significantly after 6 hours) but without significant changes in the TNF levels. Therefore these authors suggested that the fever-inducing cytokine IL-6 which was detected in the plasma and whose production in the lungs is stimulated by IL-1 and TNF could also have been released from the liver or from cells of the circulatory system. *In vitro* studies with human monocytes incubated with blood and trace elements including zinc revealed a significant increase in the endotoxin-mediated release of IL-6. The authors concluded that not only zinc but also endotoxins could play a role in the induction of the fever symptoms (Fine *et al.* 1997).

It has long been known that it is possible for persons affected with metal fume fever to develop tolerance or adaptation after repeated exposures. The symptoms only become apparent upon exposures following several exposure-free days (Nemery 1990). An association between adaptive reactions of the lung tissue after exposure to zinc oxide and increased metallothionein protein synthesis, which is stimulated by cytokines, is conceivable (Cosma *et al.* 1992, Beckett *et al.* 1996).

In several case reports, an immunological origin of metal fume fever and similar disorders has been postulated. In the BAL of a patient with metal fume fever after exposure to zinc oxide, the cell count was increased to 1300 cells/mm^3 with 28 % neutrophils and 25 % eosinophils (Castet and Bouillard 1992). After welding galvanized metal, a 34-year old worker manifested reproducible symptoms of an immediate allergic reaction and a delayed reaction (urticaria, Quincke's oedema) which were accompanied by a slight increase in the IgE level and associated with symptoms of metal fume fever (Farell 1987). In a smelter worker who reacted to zinc oxide fumes with repeated severe episodes of coughing, dyspnoea and fever, increased lymphocyte counts were detected in the BAL (mainly CD-8 T lymphocytes) and an increased level of zinc in the alveolar macrophages. The authors suggested that the clinical course of the disorder and the results of the BAL analysis make this the first published case of exogenous allergic alveolitis after zinc fume exposure (Ameille *et al.* 1992).

During the burning of fossil fuels and during smelting of metal ores, sulfuric acid is released and can be absorbed onto the surface of ultrafine particles containing zinc. This situation was simulated in animal studies with guinea pigs exposed to zinc oxide fume and sulfur dioxide. Exposure for 3 hours or 1 hour to ultrafine zinc oxide particles with adsorbed sulfuric acid resulted in significant changes in lung function parameters at an H_2SO_4 concentration of 60 µg/m^3, and at 20 to 30 mg/m^3 in unspecific bronchial hyperreactivity (detected by provocation with acetylcholine; effect like that produced by 200 µg/m^3 free sulfuric acid, no significant effect with ZnO alone) (Amdur *et al.* 1988, Amdur and Chen 1989, Chen *et al.* 1992). Repeated exposures (generally 3–5) resulted in a significant increase in neutrophil count, protein level and enzyme activities in the BAL above that found in animals exposed to zinc oxide alone (Conner *et al.* 1989). These results suggest that the observed unspecific bronchial hyperreactivity is not induced by the ultrafine zinc oxide particles themselves but rather by the sulfuric acid adsorbed on their surfaces (Chen *et al.* 1992).

The above review of the studies of the pathomechanism of metal fume fever makes it clear that the pathogenetic processes are still not fully understood even after 100 years of research. However, many authors are of the opinion that zinc fume fever is caused by stimulation of an inflammatory reaction in the lung. In toxicity studies, therefore, not only clinical parameters and lung function changes are of interest but also the levels of inflammation markers are determined.

5 Toxicokinetics

The data for the uptake of Zn through the lungs are inconsistent. For example, in the urine of workers exposed to zinc oxide fume in concentrations of 3 to 5 mg/m^3, zinc levels of 0.6 to 0.7 mg/l urine were found (ACGIH 1992). On the other hand, in persons exposed to zinc oxide fume, no clear increase in the blood zinc levels were detected (Fine *et al.* 1997, Knecht *et al.* 1983).

Zinc is distributed in the whole organism. High concentrations are found in the muscles, the highest in the prostate gland (Nordiska Expertgruppen 1981).

After exposure (head and nose only) of rats (3 hours, 4.3 mg/m^3), guinea pigs (3 hours 11.3 mg/m^3) and rabbits (6 hours, 6.0 mg/m^3) to zinc oxide fume (ultrafine particles: median diameter 0.06 µm, mass median diameter 0.17 µm) marked differences were found between the species in the proportion of the zinc dose retained in the lungs. In the guinea pigs, 19.8 % of the inhaled zinc dose was found in the lung tissue, in the rats 11.5 % and in the rabbits only 4.7 % (Gordon *et al.* 1992). In rats exposed for 24 hours by inhalation of zinc oxide fume (zinc concentration about 7.5 mg/m^3; mass median aerodynamic diameter 0.71–0.94 µm), significant increases in zinc concentration were found only in the lungs and plasma (not in liver, kidney, spleen or muscle) and only until 4 hours after the end of exposure, not after 24 or 48 hours. The half-time for lung clearance was given as 4.8 hours (rate of clearance of zinc oxide particles: 7×10^{-5} g/cm^2 surface area and day) (Hollinger *et al.* 1979).

Elimination took place mainly in the faeces (about 1–2 mg/day) as well as in the urine and sweat (each about 0.5 mg/day) (Nordiska Expertgruppen 1981).

6 Effects in Man

6.1 Single exposures

The results obtained after experimental exposure of test persons to welding fumes or pure zinc oxide fume are shown in Table 1.

Metal fume fever – symptoms

After exposure for 2 hours to zinc oxide fume in a concentration of 5 mg/m^3, almost all test persons (4/4: Gordon *et al.* 1992; 11/12: Fine *et al.* 1997; all persons without previous exposure to zinc oxide fume) developed the classical symptoms of metal fume fever. The symptoms first appeared 4 to 8 hours after the exposure, were most severe after 9 hours and no longer detectable after 24 hours (Fine *et al.* 1997, Gordon *et al.* 1992). Even in persons exposed to the lowest ZnO concentration (2.5 mg/m^3), body temperature was raised significantly by 1.23 ± 0.32°F after 6 to 12 hours (Fine *et al.* 1997). Airway resistance was not significantly increased in any of the 4 test persons (Gordon *et al.* 1992).

Table 1. Effects of experimental exposures of test persons to inhaled zinc oxide fume

Collective	Exposure	Results	References
14 test persons: welders with 1–20 years in the job, 8 currently working as welders and 6 with previous symptoms of metal fume fever	exposure to welding fumes (mixture of substances): 15–30 minutes welding in closed exposure chamber average cumulative ZnO exposure level 2.3 ± 1.7 g × min/m^3 individual ZnO exposure levels: 0.6–5.1 g × min/m^3 BAL: 8 h (n = 5) and 22 h (n = 9) after exposure	serum zinc levels in the normal range **results of the clinical examination**: at ZnO levels > 3.5 g × min/m^3: myalgia in 4/4 at ZnO levels > 5.0 g × min/m^3: fever in 2/2 **lung function parameters**: 1 h after exposure: slight but significant decrease in FEV_1 and FEF_{25-75} but TLC, DL_{CO} and SR_{AW} unchanged 6 and 20 h after exposure: all lung function parameters normal **BAL**: 8 h: dose-dependent increase in PMN 22 h: dose-dependent increase in PMN, macrophage count, lymphocyte count and total leukocyte count Lymphocyte typing after 8 h: dose-dependent increase in natural killer cells; after 22 h: dose-dependent increase in various T cell subtypes IL-1 and TNF unchanged	Blanc *et al.* 1991
23 test persons [a]: 14 welders with 1–22 years in the job, 11 with previous symptoms of metal fume fever	exposure to welding fumes (mixture of substances): 15–30 minutes welding in a closed exposure chamber exposure levels (as Zn) **group 1** (n = 6): 1.8 ± 0.2 g × min/m^3 (1.4–2.5), BAL **3 h** after exposure; **group 2** (n = 11): 2.0 ± 0.4 g × min/m^3 (< 0.1–5.0), BAL **8 h** after exposure; **group 3** (n = 9): 2.6 ± 0.6 g × min/m^3 (0.6–5.1), BAL **22 h** after exposure	**BAL**: TNF: 3 h after exposure (group 1) significantly higher than at 8 or 22 h after exposure (groups 2 and 3) IL-1, IL-6 and IL-8 not significantly changed: max. conc. of IL-1 after 3 h, of Il-6 after 22 h and of IL-8 after 8 h; significant correlation between TNF and IL-6 after 22 h and between TNF and IL-8 after 8 h albumin, PMN and macrophage count increased significantly with the time since exposure (group 1 < 2 < 3) **blood**: PMN increased significantly with the time since exposure (group 1 < 2 < 3)	Blanc *et al.* 1993

Table 1. continued

Collective	Exposure	Results	References
14 test persons without previous exposure to zinc oxide fume	exposure to pure zinc oxide fume via a respirator Zn concentration: 0 or 16.4 ± 12.5 mg/m^3 (mean, range 2.76–37.0) for 45 ± 28 min (mean, range 15–120) negative control: test persons before exposure to ZnO cumulative Zn dose: 0.537 ± 0.232 g × min/m^3 (mean, range: 0.166–1.110) **low level exposure** (n = 7) Zn: < 0.458 g × min/m^3; **high level exposure** (n = 7) Zn: > 0.458 g × min/m^3; lung function parameters 18 h after exposure; BAL 20 h after exposure	**results of clinical examination**: no temperature increase or symptoms of metal fume fever **lung function parameters**: only FEV_1 minimally decreased (p = 0.03) **BAL**: PMN significantly increased for all persons (n = 14) and for high exposure level; lymphocyte count significantly increased only for all test persons (n = 14); macrophage count and lymphocyte typing: no significant effects significant dose-response relationship for TNF and IL-8 (but not for IL-1 or IL-6) with a threshold level (Zn) at about 0.500 g × min/m^3, no effects of smoking habits **blood**: PMN unchanged	Kuschner *et al.* 1995
15 test persons without previous exposure to zinc oxide fume	exposure to pure zinc oxide fume via a respirator Zn concentration: 0 or 33 mg/m^3 (median) for 10, 15 or 30 min (n = 2, 5 and 8) cumulative Zn dose: 0.769 ± 0.566 g × min/m^3 (mean) BAL 3 h after exposure negative control: test persons before exposure to ZnO positive control: collective from Kuschner *et al.* (1995) with BAL 20 h after exposure	**BAL**: PMN, macrophage count, lymphocyte count unchanged TNF, IL-1, IL-6 and IL-8 significantly increased TNF at 3 h after exposure significantly higher than at 22 h after exposure (positive control)	Kuschner *et al.* 1997

Table 1. continued

Collective	Exposure	Results	References
12 test persons without previous exposure to zinc oxide fume	zinc oxide fume via a respirator ZnO concentrations: 0, 2.5, 5 mg/m^3 2 hours exposure on various days (blind study) cumulative zinc dose: 0.3, 0.6 $g \times min/m^3$	**results of clinical examination**: significant increase in temperature at 2.5 mg/m^3 3–10 h after exposure in 9/12 persons, at 5 mg/m^3 10–12 h after exposure in 7/12 symptoms: tiredness, muscle pains, coughing significantly more frequent only after 5 mg/m^3 **plasma**: IL-6 significantly increased 6 h after exposure to ZnO concentrations of 2.5 mg/m^3 or more, TNF not changed significantly no correlation with IL-6, temperature increases or symptoms	Fine *et al.* 1997

[a] includes the collective of Blanc *et al.* 1991

BAL bronchoalveolar lavage fluid
DL_{CO} diffusion capacity for carbon monoxide
FEF_{25-75} forced expiratory flow at between 25 % and 75 % of the vital capacity
FEV_1 forced expiratory volume in the first second
IL interleukin
PMN polymorphonuclear leukocytes
SR_{AW} specific airway resistance
TLC total lung capacity
TNF tumour necrosis factor

The effects found in peripheral blood of persons exposed to zinc oxide fume are unspecific. Those generally seen during acute metal fume fever include leukocytosis, an increase in the polymorphonuclear neutrophil (PMN) count and in the erythrocyte sedimentation rate (Blanc *et al.* 1993, Gordon and Fine 1993, Nemery 1990, Rüegger 1995). The usual clinical course of the disorder suggests that metal fume fever does not have long-term effects. However, this aspect has not yet been studied in detail (Rüegger 1995).

Inflammation markers

In persons exposed experimentally to ZnO fume in concentrations between 2.5 and 33 mg/m^3, changes were found in inflammation markers and mediators in the BAL such as PMN count, and concentrations of TNF, IL-1, IL-6 and IL-8 (see Table 1 and Section 4: Blanc *et al.* 1991, 1993, Fine *et al.* 1997, Kuschner *et al.* 1995, 1997).

Lung function parameters

In persons exposed experimentally to welding fumes (cumulative ZnO exposure level up to 5.1 g × min/m^3) or to pure zinc oxide fume (Zn concentration up to 37 mg/m^3) for periods averaging up to 45 minutes, the only significant changes in lung parameters were slight alterations in forced expiratory volume in one minute (FEV_1) (Blanc *et al.* 1991, Kuschner *et al.* 1995).

The unspecific bronchial hyperreactivity determined in welders with the metacholine provocation test was twice as high before the shift and three times as high after the shift as that found for control persons who were not exposed to zinc oxide. The most severe reactions were seen in the sub-collective who welded metals coated with a pigmented zinc dust epoxide primer (highest determined concentration of ZnO in air: 2.22 mg/m^3) (Emmerling *et al.* 1987).

In welders exposed to total fume concentrations averaging 4.39 ± 6.91 and ZnO fume concentrations averaging 0.032 ± 0.072 mg/m^3 while working in a shipyard, the arithmetic average values of the lung function parameters forced vital capacity (FVC), FEV_1 and the transfer factor for CO (TL) decreased significantly but not in a dose-dependent manner in the course of the daily shift. Without showing the results of his calculations, the author states that when individual dust exposure levels are taken into account in a simple regression analysis, the individual regression coefficients for Cu, ZnO, Mn and NO_2 show that these substances have no more pronounced effects on the reduction of the lung function parameters than do other air contaminants found at these workplaces. Multiple regression analysis to examine the importance of simultaneous exposure to several atmospheric pollutants suggested that Cu, ZnO, Mn and NO_2 had more effects than the other pollutants but the coefficients were not significantly different from zero (Akbar-Khanzadeh 1993). Small but statistically significant changes in lung function parameters (vital capacity (VC), FEV_1) were detected in persons exposed during a night shift to zinc oxide fume (ZnO: 5.1 ± 4.2 mg/m^3, maximum total dust: 35 ± 24 mg/m^3 after static sampling and 8 ± 7 mg/m^3 after personal sampling). This was interpreted as a subclinical reaction to the inhaled zinc oxide, after several cases of metal fume fever had been observed (Pasker *et al.* 1997).

In a case report describing a patient with marked peak flow reduction, it was stated that a pulmonary involvement in zinc fume fever was not uncommon (Nemery and Demedts 1991). In a welder during the febrile phase of metal fume fever, the diffusion capacity for CO (DL_{CO}) was reduced by more than 40 % and the partial pressure of oxygen reduced by 9 mm Hg. Typically the radiological examination revealed practically no changes (Vogelmeier *et al.* 1986). In a shipyard worker who developed metal fume fever after work involving spraying zinc, changes were detected in clinical and laboratory medical parameters and transient multiple nodular opacities (Brown 1988). Other authors described leukocytosis, fever, reductions in FEV_1 and VC and transient diffuse nodular infiltrations in the thorax radiograph of a patient with zinc fume fever (Malo *et al.* 1990).

Authors who carried out a "follow-up BAL", in each case with one patient suffering from metal fume fever, found no signs of persistent inflammatory changes (Blanc *et al.* 1991, Castet and Bouillard 1992, Vogelmeier *et al.* 1986). On the other hand, a case has been described of a 48-year-old welder who suffered long-term effects after acute zinc poisoning caused by working for four days with an oxygen cutter on galvanized pipes at an insufficiently ventilated workplace (Andrzejak *et al.* 1992). One year after the intoxication, the patient still showed signs of impairment of brain function.

In summary, it should be pointed out that the changes in lung function parameters during the shift (Pasker *et al.* 1997) and the evidence of increased bronchial hyperreactivity (Emmerling *et al.* 1987) were detected in studies of welders who are exposed to other substances as well as ZnO. Therefore the zinc oxide concentrations described in these publications cannot be used in the derivation of a threshold.

6.2 Repeated exposures

In a Finnish study of 234 workers who had been exposed to zinc oxide for 5.5 years on average (total dust: 2.5–4.6 mg/m^3, zinc content 90 %, mainly zinc oxide), no impairment of lung function and no asthmatic symptoms were reported. Only for the symptom "chronic phlegm production" was the incidence significantly higher in the exposed persons than in the unexposed control group (Roto 1980).

Shortly after introduction of a new production process in a car factory, health monitoring was carried out on a group of 35 bodywork welders who were exposed to inhaled zinc oxide fume in concentrations slightly below the MAK value valid in 1987. No evidence of symptoms or lung function disorders "typical of zinc" were found. The only observed effect was an increase in the reports of noticing a metallic taste. The welding fumes contained 30 % to 70 % zinc oxide. The general threshold limit value for dust was repeatedly exceeded for brief periods at individual workplaces and therefore supplementary protective measures were introduced for the workers (Hilla and Tiller 1987).

Of 7 employees of an American firm who worked in the galvanization process, only one reported having previously suffered symptoms of zinc fume fever. The others had not experienced such episodes. The metal fume concentrations (ZnO: 0.27–2.8 mg/m^3, shift average values, 14 determinations after personal sampling) were clearly below the threshold value applying at the time (Miller *et al.* 1992).

In an earlier environmental medical study, 30000 persons insured by the one health insurance company were questioned. Among the inhabitants of a suburb of Mülheim who were exposed to annual average zinc oxide concentrations of 67 to 120 mg/m^3 and day (emissions of a galvanizing works and a foundry), the number of persons aged 60 years or more who suffered from bronchitis was twice that in suburbs without such emissions. However, parameters such as social status, population density, smoking habits, were not analysed (Langmann and Kettner 1968).

In summary, the reports of chronic inhalative exposure to zinc oxide describe mainly cohorts exposed to mixtures of substances (welding fumes) and contain insufficient documentation of the zinc oxide concentrations involved; thus these publications provide no clear information about effective concentrations.

6.3 Reproductive and developmental toxicity

Analysis of samples collected with personal dust samplers worn by twelve persons who handled zinc alkyl dithiophosphate (an oil additive) at work revealed maximum ZnO concentrations of 0.3 mg/m^3 (clearly below the TLV). Under these exposure conditions, no effects on fertility or spermatogenesis were detected (Hollett and Aw 1982).

6.4 Carcinogenicity

A prospective mortality study of 4802 persons employed in 9 American copper and zinc smelters revealed no evidence of a significant increase in cancer mortality among the 978 zinc smelting workers (Logue *et al.* 1982).

6.5 Other effects

A prospective mortality study of 4802 persons employed in 9 American copper and zinc smelters (see Section 6.4) revealed an increased standardized mortality ratio (SMR) for cerebrovascular disorders. The results were consistent in all 9 works and, in the opinion of the authors, deserve further study. The authors also pointed out that similar results have been obtained in Swedish copper smelters (Logue *et al.* 1982).

In several case reports (e.g. Ameille *et al.* 1992, Castet and Bouillard 1992, Farell 1987; see Section 4), effects on the immune system are suggested to be the cause of zinc fume fever.

Zinc fume fever can be associated with tiredness (Fisher 1992), headaches and visus disorders (Gordon and Fine 1993). A case has been described of a 48-year-old welder who suffered long-term effects after acute zinc poisoning caused by working for four days with an oxygen cutter on galvanized pipes at an insufficiently ventilated workplace (Andrzejak *et al.* 1992). One year after the intoxication, the patient still showed signs of chronic encephalopathy.

7 Animal Experiments and *in vitro* Studies

In the animal studies which have been carried out to date, mainly rodents (guinea pigs, rats and rabbits) have been exposed by inhalation of zinc oxide fume in exposure chambers and then dose-response relationships, mostly for inflammation markers, lung function parameters or histological findings, have been investigated.

There are no data available for the effects of zinc oxide fume on skin and mucous membranes, and there are no studies available of the allergenicity, reproductive toxicity or genotoxicity of this substance.

7.1 Acute toxicity

In inhalation studies, guinea pigs and rats were exposed for 3 hours and rabbits for 2 hours by head and nose exposure to zinc oxide fume (ultrafine particles: median diameter 0.06 µm, mass median diameter 0.17 µm) in zinc oxide concentrations of 2.5 and 5 mg/m^3 (Gordon *et al.* 1992). There were significant differences between the species in their reactions to the zinc oxide. Guinea pigs and rats were more sensitive than rabbits. 24 hours after the end of exposure, in the BAL of guinea pigs and rats exposed to 2.5 mg/m^3 there were significant increases in total protein, LDH and beta-glucuronidase; in the animals exposed to 5 mg/m^3 the total cell count was also increased. Significant increases in LDH and beta-glucuronidase were also detected in rats 4 hours after the 5 mg/m^3 exposure (assays carried out at this time only in rats). In rabbits, on the other hand, no evidence of inflammatory changes in the lungs was found. These species differences seem to be the result of differences in zinc retention in the lungs. Guinea pigs retained about 20 % of the total inhaled dose of zinc oxide, rats about 12 % and rabbits only 5 % (see Section 5; Gordon *et al.* 1992).

Single exposures of rats for 3 hours by inhalation of zinc oxide fume (median diameter 0.06 µm, geometric standard deviation 1.8) in zinc oxide concentrations of 1.0, 2.5 or 5.0 mg/m^3, was shown to induce the expression of two metal-responsive genes, metallothionein (MT; expression concentration-dependent) and haem oxygenase (HO; expression independent of concentration) (Cosma *et al.* 1992).

These animal studies revealed dose-response relationships between the dose taken up during brief exposures to zinc oxide fume in concentrations of 2.5 mg/m^3 or more and the appearance of inflammation markers in the BAL. The changes were greater 24 hours after the exposure than immediately after or 4 hours after. A NOEL cannot be derived from the results of these studies.

In guinea pigs exposed for one hour (head and nose exposure) by inhalation of zinc oxide in a concentration of about 1 mg/m^3 (ultrafine particles, average diameter 0.05 µm, geometric standard deviation 2.0), the pulmonary compliance was reduced slightly but significantly and continued to decrease for up to 2 hours after the end of exposure. No changes in airway resistance, respiration rate or respiratory minute volume were detected in animals exposed to zinc oxide fume, unlike in those exposed to an irritating zinc sulfate aerosol (Amdur *et al.* 1982). The authors suggested that the progressive change in compliance could be a result of the fact that the zinc oxide particles deposited in the peripheral regions of the lung persist longer than do the particles of more readily soluble zinc salts. Symptoms of metal fume fever are not induced by such low level exposures.

7.2 Subacute, subchronic and chronic toxicity

Guinea pigs were exposed (head and nose only) on 1, 2 or 3 consecutive days, each time for 3 hours, to ZnO in concentrations of 2.3, 5.9 or 12.1 mg/m^3 (ultrafine particles, diameter 0.05 µm, geometric standard deviation 2.0). The second and third exposures to 5.9 and 12.1 mg/m^3 resulted in significant increases in protein, neutrophil count and the activities of angiotensin-converting enzyme (ACE) and beta-glucuronidase in the BAL; the activities of alkaline and acid phosphatase and LDH were increased significantly also after exposure to 2.3 mg/m^3. Histological examination for signs of centroacinar inflammation (number of foci of inflammation in the lungs) first revealed significant effects after 2 exposures to 12.1 mg/m^3; 2.3 mg/m^3 was the no observed effect level (NOEL). The severity of inflammation was closely correlated with the protein concentration and with the ACE activity (Conner *et al.* 1986, 1988).

In subsequent studies, guinea pigs inhaled zinc oxide fume on 6 days (5.5 mg/m^3, 3 hours daily; ultrafine particles, "projected area diameter" 0.05 µm, geometric standard deviation 2.0) and lung function parameters were determined in 8 to 9 animals per time interval until up to 72 hours after the last exposure. Airway resistance was increased transiently (until 24 hours after the end of exposure), compliance (until 48 hours) and total lung capacity (TLC) (until 1 hour) reduced. After 72 hours the VC, functional residual capacity and DL_{CO} were still reduced. Microscopic examination of the lungs revealed inflammatory changes in the *ductuli alveolares* and adjacent alveoli, some of which had regressed after 48 hours. The interstitium was thickened as a result of infiltration of macrophages, lymphocytes, neutrophils and pulmonary fibroblasts and the effects were correlated with the relative lung weight which was markedly increased during the whole observation period. In the bronchiolar epithelium, but not in the bronchi, significantly increased cell proliferation rates were measured in the period up to 48 hours after exposure. This effect was no longer detectable after 72 hours (Lam *et al.* 1985, 1988). In inhalation studies with guinea pigs exposed to ZnO concentrations of 7 mg/m^3 on average (ultrafine particles, diameter 0.05 µm, geometric standard deviation 2.0, head and nose exposure) for 3 hours daily on each of 5 days, the TLC and VC decreased progressively with increasing exposure duration. The DL_{CO} was initially unaffected but from day 4 was reduced to a value 30 % less than the control value. After brief peak exposures (25 and 34 mg/m^3) the lung volumes were reduced abruptly and to a greater extent than in the experiments without concentration peaks. At the same time the airway resistance was increased and the compliance reduced (Lam *et al.* 1988).

7.3 Carcinogenicity

There are no studies of the carcinogenicity of zinc or zinc oxide.

In their review of synergistic and antagonistic effects of metals on the carcinogenic activity of other metals, Rodriguez and Kasprzak (1989) describe an inhibitory effect of zinc oxide and zinc acetate when injected intramuscularly or subcutaneously together with nickel sulfide (Ni_3S_2). The latency period of the tumours induced by nickel sulfide was 4 to 6 weeks longer in the animals treated simultaneously with the zinc compounds but the tumour incidence was unchanged. The findings after simultaneous injection of cadmium and zinc are not consistent; both an increase and a decrease in the carcinogenic activity of cadmium have been reported. The mechanism could possibly involve an increase in metallothionein synthesis.

In long-term inhalation studies (18 months) of the carcinogenicity of cadmium compounds in rats, simultaneous inhalation of zinc oxide reduced the incidence of tumours induced by cadmium oxide (Heinrich 1992).

7.4 Other effects

In a study with rats it was investigated whether orally administered zinc and inhaled zinc oxide fume have pharmacological effects on the toxicity of paraquat. Exposure to zinc oxide fume aerosols increased the toxicity of paraquat (Hollinger *et al.* 1979).

8 Manifesto (MAK value/classification)

The available data demonstrate that in persons exposed briefly to zinc oxide fume there are acute changes in parameters relevant for health such as clinical symptoms, inflammation markers in the BAL and in serum, and lung function parameters. For example, in test persons exposed once for 2 hours to zinc oxide fume in a concentration of 2.5 mg/m^3 (LOEL), body temperature and the plasma IL-6 level were increased significantly; in persons exposed to concentrations of 5 mg/m^3 or more, symptoms of the classical metal fume fever developed. In the BAL, TNF and IL-8, which are at the beginning of the cascade of inflammatory changes and are the most sensitive inflammation markers, reached their maximum levels as little as 3 hours after the end of the exposure (33 mg/m^3 for 30 minutes). These exposure conditions correspond to cumulative exposure in the ZnO concentration range from about 300 to 700 mg $\times$ min/m^3.

At present the data for effects of long-term occupational exposure to zinc oxide fume are still inadequate. Apparently persons exposed repeatedly can develop tolerance; then symptoms appear only as a result of exposures taking place after several exposure-free days. Studies of the employees of a Finnish zinc factory have shown that occupational exposure to zinc oxide at levels not exceeding a threshold of 5 mg/m^3 for several years causes neither chronic bronchitis nor impairment of lung function. In persons exposed to welding fumes, lower zinc oxide concentrations (e.g. < 2.2 mg/m^3) were associated with biological effects. These results for exposures to mixtures of substances, however, are not suitable for the derivation of a MAK value for zinc oxide.

The data from animal studies demonstrate that inhalation of zinc oxide fume (studies with ultrafine particles) causes physiological, biochemical and cellular changes in the pulmonary tract of rodents. Reduction in pulmonary compliance was detected in guinea pigs exposed to zinc oxide in a concentration of 1 mg/m^3 after as little as 1 hour. However, this result is not suitable for the derivation of a threshold value for man because compliance determination in guinea pigs is considered to be subject to error because of the anatomy of these animals. With rats and guinea pigs but not with rabbits exposed for 3 hours to zinc oxide fume, dose-response relationships were demonstrated for the appearance of acute inflammation markers in the BAL. Significant changes in the protein concentration and the activities of LDH and beta-glucuronidase were found after a single exposure to ZnO in a concentration of 2.5 mg/m^3 (LOEL). In the BAL from guinea pigs exposed to zinc oxide fume (ZnO: 2.3–12.1 mg/m^3, 3 hours daily, 2–3 days) significant changes in the activities of LDH, acid phosphatase and alkaline phosphatase were seen in animals exposed to 2.3 mg/m^3 or more, and from 5.9 mg/m^3 significant changes in the beta-glucuronidase activity, neutrophil count, protein concentration and ACE activity. The last two parameters were closely correlated with histological signs of centroacinar inflammation in the lungs. The NOEL for these changes was found to be 2.3 mg/m^3 (exposure 3 times 3 hours, cumulative exposure to ZnO: 1240 mg $\times$ min/m^3) and a significant effect was seen after two 3-hour exposures to 12.1 mg/m^3 (cumulative exposure to ZnO: 4350 mg $\times$ min/m^3) or more. After exposure of guinea pigs for 6 days (3 hours daily) to ZnO in a concentration of 5.5 mg/m^3, lung function changes (some of which were reversible) and inflammatory changes in lung histology were observed.

A NOEL has not been documented by either human medical investigations or experimental animal studies. In test persons exposed once for two hours, a LOEL for inflammation markers of 2.5 mg/m^3 and a LOEL for symptoms of metal fume fever of 5 mg/m^3 were determined for zinc oxide fume. The effects of the development of tolerance on the LOEL and the effects of zinc oxide fume during long-term occupational exposure have not yet been adequately studied. In animal studies, the NOEL for histologically detectable inflammatory changes in the lung was 2.3 mg/m^3 (ZnO concentration, exposure 3 times for 3 hours each) with a LOEL of 12.1 mg/m^3 in the guinea pig, the most sensitive species in which the zinc retention by the lung is highest. On the other hand, the LOEL for an increase in the concentration of inflammation markers was 2.3 mg/m^3. It must be taken into account that there are enormous differences between species in the systemic dose resulting from the one exposure concentration. Calculated from the respiratory minute volume and the lung weight, the dose in the target tissue lung of the guinea pig is 8 to 9 times that in man (Conner *et al.* 1988).

On the basis of these data, a provisional MAK value for zinc oxide fume (respirable fraction) of 1 mg/m^3 is established without an excursion factor. Further studies are urgently necessary, especially studies of the effects of repeated exposure to concentrations below 2.5 mg/m^3 under workplace conditions and of the effects of the development of tolerance. In these studies, changes in the inflammation markers in the BAL, especially in TNF and interleukins, should be recorded at the times of maximum sensitivity.

There are no studies relevant for a classification in one of the categories for germ cell mutagens. To date no information is available as to the ability of zinc oxide to enter the human organism through the intact skin. Therefore it cannot be decided whether a designation with "H" is necessary. There are also no data which would make it possible to decide whether designation with an "S" is necessary.

10 References

ACGIH (American Conference Governmental Industrial Hygienists) (1992) Zinc Oxide. in: *Documentation of the threshold limit values and biological exposure indices.* ACGIH, Cincinnati, Ohio, USA

Akbar-Khanzadeh F (1993) Short-term respiratory function changes in relation to workshift welding fume exposures. *Int Arch Occup Environ Health 64*: 393–397

Amdur MO, McCarthy JF, Gill MW (1982) Respiratory response of guinea pigs to zinc oxide fume. *Am Ind Hyg Assoc J 43*: 887–889

Amdur MO, Chen LC, Guty J, Lam HF, Miller PD (1988) Speciation and pulmonary effects of acidic SO_x formed on the surface of ultrafine zinc oxide aerosols. *Atmosph Environm 22*: 557–560

Amdur MO, Chen LC (1989) Furnace-generated acid aerosols: speciation and pulmonary effects. *Environm Health Perspect 79*: 147–150

Ameille J, Brechot JM, Brochard P, Capron F, Dore MF (1992) Occupational hypersensitivity pneumonitis in a smelter exposed to zinc fumes. *Chest 101*: 862–863

Andrzejak R, Antonowicz J, Andreasik Z (1992) An atypical case of acute zinc poisoning (Polish). *Med Pracy 43*: 329–333

Beckett WS, Chen LC, Cosma G, Fine J, Garte S, Gordon T, Kinney P, Sparer J (1996) *Metal fume fever.* US Department of Commerce, Yale University, New Haven, 26/7/96. NTIS PB 97-206379

Blanc P, Wong H, Bernstein MS, Boushey HA (1991) An experimental human model of metal fume fever. *Ann Int Med 114*: 930–936

Blanc PD, Boushey HA, Wong H, Wintermeyer SF, Bernstein MS (1993) Cytokines in metal fume fever. *Am Rev Respir Dis 147*: 134–138

Brown JJL (1988) Zinc fume fever. *Br J Radiol 61*: 327–329

Castet D, Bouillard J (1992) Pneumopathie aigue au cours d'une exposition à l'oxyde de zinc (Acute pneumopathy caused by exposure to zinc oxide) (French). *Rev Mal Resp 9*: 632–633

Chen LC, Miller MO, Amdur MO, Gordon T (1992) Airway hyperresponsiveness in guinea pigs exposed to acid-coated ultrafine particles. *J Tox Environ Health 35*: 165–174

Conner MW, Flood WH, Rogers AE, Amdur MO (1986) Pulmonary damage in guinea pigs caused by ultrafine zinc oxide: evaluation by light and electron microscopy and analysis of pulmonary lavage fluid. in: Romig AD jr, Chambers WF (Eds) *Microbeam Analysis–1986*: 589–590

Conner MW, Flood WH, Rogers AE, Amdur MO (1988) Lung injury in guinea pigs caused by multiple exposures to ultrafine zinc oxide: changes in pulmonary lavage fluid. *J Tox Environ Health 25*: 57–69

Conner MW, Flood WH, Rogers AE (1989) Changes in pulmonary lavage fluid of guinea pigs exposed to ultrafine zinc oxide with absorbed sulfuric acid. *J Tox Environ Health 24*: 223–234

Cosma G, Fulton H, DeFeo T, Gordon T (1992) Rat lung metallothionein and heme oxygenase gene expression following ozone and zinc oxide exposure. *Toxicol Appl Pharmacol 117*: 75–80

Emmerling G, Zschische W, Schaller K-H, Weltle D, Valentin H, Zober A (1987) *Arbeitsmedizinische Untersuchungen zur Problematik bronchialer Überempfindlichkeit bei Lichtbogenschweißarbeiten an oberflächenbeschichteten Stählen* (Occupational medical studies of bronchial hypersensitivity during arc welding of coated steel) (German). DVS-Verlag, Düsseldorf: 1–219

Farell FJ (1987) Angioedema and urticaria as acute and late phase reactions to zinc fume exposure, with associated metal fume fever-like symptoms. *Am J Ind Med 12*: 331–337

Fine JM, Gordon T, Chen LC, Kinney P, Falcone G, Beckett WS (1997) Metal fume fever: characterization of clinical and plasma IL-6 responses in controlled human exposures to zinc oxide fume at and below the TLV. *J Occup Environ Med 39*: 722–726

Fisher D (1992) Zinc. in: Sullivan JB, Krieger GR (Eds) *Hazardous Materials Toxicology.* Williams & Wilkins, Baltimore: 865–868

Gordon T, Chen LC, Fine JM, Schlesinger RB, Su WY, Kimmel TA, Amdur MO (1992) Pulmonary effects of inhaled zinc oxide in human subjects, guinea pigs, rats, and rabbits. *Am Ind Hyg Assoc J 53*: 503–509

Gordon T, Fine JM (1993) Metal fume fever. *Occup Med 8*: 505–517

Heinrich U (1992) Pulmonary carcinogenicity of cadmium by inhalation in animals. in: Nordberg GF, Herber RFM, Alessio L (Eds) *Cadmium in the human environment: toxicity and carcinogenicity. IARC Scientific Publications No. 118*, Lyon: 405–413

Hilla W, Tiller RE (1987) Gesundheitsmonitoring einer Gruppe zinkrauchexponierter Schweißer in einem Automobilwerk (Health monitoring of a group of welders exposed to zinc fume in a car factory) (German). in: Norpoth K (Ed.) *Bericht über die 27. Jahrestagung der Deutschen Gesellschaft für Arbeitsmedizin e.V.* unter Mitwirkung des Verbandes Deutscher Betriebs- und Werksärzte e.V. - Berufsverband Deutscher Arbeitsmediziner - in Verbindung mit dem Hauptverband der gewerblichen Berufsgenossenschaften e.V. Gentner Verlag, Stuttgart: 315–318

Hollett BA, Aw TC (1982) *Health hazard evaluation report No. HETA-80-228-1241*, Edwin Cooper Company, Sauget, Illinois. Hazard Evaluations and Technical Assistance Branch, NIOSH, U.S. Department of Health and Human Services, Cincinnati, Ohio, NTIS PB 84-172568

Hollinger MA, Raabe OG, Giri SN, Freywald M, Teague SV, Tarkington B (1979) Effect of the inhalation of zinc and dietary zinc on paraquat toxicity in the rat. *Toxicol Appl Pharmacol 49*: 53–61

Knecht U, Borris D, Tiller R, Header M, Woitowitz HJ (1983) Schweiß- und Schleifarbeiten an verzinkten Karosserieblechen: Inhalative Belastung und Bio-Monitoring (Welding and sanding of galvanized bodywork: inhalation exposure and biomonitoring) (German). in: *Bericht über die 23. Jahrestagung der Deutschen Gesellschaft für Arbeitsmedizin e.V.*, gemeinsam mit dem Verband Deutscher Betriebs- und Werksärzte e.V., Berufsverband Deutscher Arbeitsmediziner, dem Hauptverband der gewerblichen Berufsgenossenschaften e.V., Göttingen, May 4–7 1983, Gentner Verlag, Stuttgart: 497–504

Kuschner WG, D'Alessandro A, Wintermeyer SF, Wong H, Boushey HA, Blanc PD (1995) Pulmonary responses to purified zinc oxide fume. *J Invest Med 43*: 371–378

Kuschner WG, D'Alessandro A, Wong H, Blanc PD (1997) Early pulmonary cytokine responses to zinc oxide fume inhalation. *Environ Res 75*: 7–11

Kuschner WG, D'Alessandro A, Hambleton J, Blanc PD (1998) Tumor necrosis factor-alpha and interleukin-8 release from U 937 human mononuclear cells exposed to zinc oxide *in vitro. J Occup Environ Med 40*: 454–459

Lam HF, Conner MW, Rogers AE, Fitzgerald S, Amdur O (1985) Functional and morphologic changes in the lungs of guinea pigs exposed to freshly generated ultrafine zinc oxide. *Toxicol Appl Pharmacol 78*: 29–38

Lam HF, Chen LC, Ainsworth D, Peoples S, Amdur MO (1988) Pulmonary function of guniea pigs exposed to freshly generated ultrafine zinc oxide with and without spike concentrations. *Am Ind Hyg Assoc J 49*: 333–341

Langmann R, Kettner H (1968) Gefahren und Bekämpfung des Zinkoxids in der Luft. *Öffentl Gesundheitswesen 30*: 195–198

Lindahl M, Leanderson P, Tagesson C (1998) Novel aspect on metal fume fever: zinc stimulates oxygen radical formation in human neutrophils. *Hum Exp Toxicol 17*: 105–110

Logue JN, Koontz MD, Hattwick MAW (1982) A historical prospective mortality study of workers in copper and zinc refineries. *J Occup Med 24*: 398–408

Malo JL, Malo J, Cartier A, Dolovich J (1990) Acute lung reaction due to zinc inhalation. *Eur Respir J 3*: 111–114

Miller A, Echt A, Jewell G (1992) *Health hazard evaluation Report HETA 92-140-2264*, T-L Irrigation Company, Hastings, Nebraska. National Institute for Occupational Health and Safety, Cincinnati, OH, NTIS PB 93-214104

Mueller EJ, Seger DL (1985) Metal fume fever - a review. *J Emerg Med 2*: 271–274

Nemery B (1990) Metal toxicity and the respiratory tract. *Eur Respir J 3*: 202–219

Nemery B, Demedts M (1991) Respiratory involvement in metal fume fever. *Eur Respir J 4*: 764–765

Nordiska Expertgruppen (1981) 22. Zinc. *Arbete och Hälsa 13*: 1–18

Pasker HG, Peeters M, Genet P, Clément J, Nemery B, van de Woestijne KP (1997) Short-term ventilatory effects in workers exposed to fumes containing zinc oxide: comparison of forced oscillation technique with spirometry. *Eur Respir J 10*: 1523–1529

Perry GF (1994) Occupational medicine forum. *J Occup Med 36*: 1061–1063

Rodriguez RE, Kasprzak KS (1989) Antagonists to metal carcinogens. *J Am Coll Toxicol 8*: 1265–1269

Roto P (1980) Asthma, symptoms of chronic bronchitis and ventilatory capacity among cobalt and zinc production workers. *Scand J Work Environ Health 6, Suppl. 1*: 1–49

Rüegger M (1995) Lungenschäden durch Metalle (Lung disorders due to metals) (German). *Schweiz Med Wochenschr 125*: 467–474

Tichy S (1993) Transparentes ZnO zum Haut- und Sonnenschutz. *SÖFW-J 119*: 482–490

Vogelmeier C, König G, Bencze K (1986) Pulmonale Beteiligung bei Metallrauch-Fieber (Pulmonary effects in metal fume fever) (German). in: Szadkowski D (Ed.) *Bericht über die 26. Jahrestagung der Deutschen Gesellschaft für Arbeitsmedizin e.V.* unter Mitwirkung des Verbandes Deutscher Betriebs- und Werksärzte e.V., Berufsverband Deutscher Arbeitsmediziner, in Verbindung mit dem Hauptverband der gewerblichen Berufsgenossenschaften e.V., Gentner Verlag, Stuttgart, Vol 1: 559–563

completed 14.01.2000

Authors of Documents in Volume 18

Professor Dr.med. F. Bahmer, Zentralkrankenhaus St.-Jürgen-Straße, 28205 Bremen

Dr. R. Bartsch, Senatskommission der DFG zur Prüfung gesundheitsschädlicher Arbeitsstoffe, Technische Universität München, D-85350 Freising-Weihenstephan

Frau Dr. B. Brinkmann, Senatskommission der DFG zur Prüfung gesundheitsschädlicher Arbeitsstoffe, Technische Universität München, D-85350 Freising-Weihenstephan

Professor Dr. Dr. H.-P. Gelbke, BASF Aktiengesellschaft, Abteilung Toxikologie, 67063 Ludwigshafen

Professor Dr. Ernst Hallier, Abteilung Arbeits- und Sozialmedizin, Zentrum Umwelt- und Arbeitsmedizin, Georg-August-Universität Göttingen, Waldweg 37, 37073 Göttingen

Professor Dr. U. Heinrich, Fraunhofer-Institut für Toxikologie und Aerosolforschung, Nikolai-Fuchs-Straße 1, 30625 Hannover

Professor Dr.rer.nat. H. Kappus, Humbolt-Universität Berlin, Universitätsklinikum Rudolf Charité Virchow-Klinikum, Dermatologie, Augustenburger Platz 1, 13353 Berlin

Dr. H. Lessmann, Informationsverbund Dermatologischer Kliniken an der Universitätshautklinik, von-Siebold-Straße 3, 37075 Göttingen

Professor Dr. med. E. Löser, previously BAYER AG, Institut für Toxikologie, Friedrich-Ebert-Straße 217, 42117 Wuppertal

Dr. P. Morfeld, Institut für Arbeitswissenschaften der RAG Aktiengesllschaft, Wengeplatz 1, 44369 Dortmund

Professor Dr. K.H. Norpoth, Universitätsklinikum GH Essen, Institut für Hygiene und Arbeitsmedizin, Hufelandstr. 55, 4300 Essen

Frau Dr. C. Pohlenz-Michel, previously Senatskommission der DFG zur Prüfung gesundheitsschädlicher Arbeitsstoffe, Technische Universität München, D-85350 Freising-Weihenstephan

Professor Dr. D. Schneider, Bundesanstalt für Arbeitsschutz und Arbeitsmedizin, Nöldnerstr. 40–42, 10317 Berlin

Priv.-Doz. Dr. med. Axel Schnuch, IVDK an der Univ.-Hautklinik, von Siebold Str. 3, D 37075 Goettingen

Dr. N. Tenkhoff, Hüls AG, Postfach 13 20, 45743 Marl

Dr. G.A. Westphal, Abteilung Arbeits- und Sozialmedizin, Zentrum Umwelt- und Arbeitsmedizin, Georg-August-Universität Göttingen, Waldweg 37, 37073 Göttingen

Professor Dr.med. H.-J. Woitowitz, Institut und Poliklinik, für Arbeits- & Sozialmedizin der Universität Gießen, Aulweg 129/III, 35392 Gießen

Frau Dr. K. Ziegler-Skylakakis, c/- Senatskommission der DFG zur Prüfung gesundheitsschädlicher Arbeitsstoffe, Technische Universität München, D-85350 Freising-Weihenstephan

Index for Volumes 1–18